Introduction to TECHNICAL MATHEMATICS

Introduction to TECHNICAL MATHEMATICS

Gilbert M. Peter
Cuesta College

Daniel R. Peterson
West Hills College

Scott, Foresman and Company
Glenview, Illinois *Brighton, England*

TO
Kerri
Keith
Cindy
Gary
David

Library of Congress Catalog Card Number: 73-89419

ISBN: 0-673-07784-5

AMS 1970 Subject Classification: 98A99

Regional offices of Scott, Foresman and Company are located in Dallas, Texas; Glenview, Illinois; Oakland, New Jersey; Palo Alto, California; Tucker, Georgia; and Brighton, England.

Preface

This book was written for students who are preparing for further study in technological fields. Our emphasis has been on practicality and applicability rather than elegance and mathematical rigor; however, the development is sufficient to provide the mathematics necessary for successful completion of more specialized courses.

The text has been thoroughly class tested over a period of several years. The topics we have included are well-known for their applicability to technical mathematics. Numerous examples and exercises from a wide range of technical specialties, including automotive, diesel, engineering, industrial, paramedical, electrical, and chemical areas, have been included. We have particularly emphasized student competence in problem analysis and solution.

We begin with a review of basic arithmetic to provide background for students whose mathematical preparation may be limited. The first three chapters make up this review, and include material on exponents and the use of the slide rule. Also included in the review is a discussion of the metric system, along with exercises on its use.

With Chapter 4 we begin development of the basic operations of algebra. The algebra sections include an introduction to complex numbers, functions, functional notation, systems of equations, higher order equations, conic sections, exponential curves, and logarithmic curves. We have introduced matrix theory in Chapter 8 where it adds clarity to solving systems of equations and where its applicability to computers can be pointed out.

In the geometry chapter we have emphasized applicability. To that end, useful constructions are carefully detailed. Most of the traditional topics of trigonometry are developed through a right triangle rather than the circular function approach. Trigonometric identities are developed and used because of their later application in technical areas. We conclude the book with chapters on sine wave analysis and essentials of vectors, polar coordinates, and their applications.

Although the material is developed sequentially, it need not be used that way. *Introduction to Technical Mathematics* contains sufficient material for a two-semester course giving three hours of credit each semester. It can also be successful in a one-semester, four-hour course enrolling students with stronger mathematical preparation. We have used Chapters 4 through 11 in a one-semester review of high school algebra, geometry, and trigonometry. Finally, the sections on algebra, geometry, and trigonometry have been grouped to provide text materials for six-week "mini-review" courses for precalculus students.

We are grateful to Mrs. Daniel R. Peterson for typing and retyping the manuscript, to Paul N. Thompson, Larew M. Collister, W. William Nash, and Janet Forbes for their suggestions and criticisms at various stages of the manuscript, and to the staff of Scott, Foresman and Company, especially Susan Hastings, for carefully preparing the manuscript for publication.

Daniel R. Peterson
Gilbert M. Peter

Table of Contents

1 Review of Basic Operations 1

1.1 *Introduction* 1
1.2 *Sets* 1
1.3 *Finite and Infinite Sets* 2
1.4 *The Number System* 2
1.5 *Negative Numbers* 6
1.6 *Addition of Integers* 7
1.7 *Subtraction of Integers* 9
1.8 *Combined Operations* 10
1.9 *Multiplication of Integers* 12
1.10 *Division of Integers* 13
1.11 *Factors and Primes* 14
1.12 *Greatest Common Factor* 15
1.13 *Least Common Multiple* 17
1.14 *Fractions* 17
1.15 *Equivalent Fractions* 18
1.16 *Comparison of Fractions* 18
1.17 *Adding and Subtracting Fractions* 19
1.18 *Multiplication of Fractions* 20
1.19 *Division of Fractions* 21
1.20 *Complex Fractions* 22
1.21 *Decimal Fractions* 25
1.22 *Addition and Subtraction of Decimals* 25
1.23 *Multiplying Decimals* 26
1.24 *Division of Decimals* 26
1.25 *Changing the Form of Fractions* 27
1.26 *Rounding* 28
1.27 *Percent* 29
1.28 *Measurement Conversion* 33

2 Exponents and Radicals 38

2.1 *Exponents* 38
2.2 *Multiplying with Exponential Numbers* 39
2.3 *Dividing with Exponential Numbers* 39
2.4 *Combined Operations with Exponential Numbers* 41
2.5 *Negative Exponents* 43
2.6 *Scientific Notation* 45
2.7 *Square Root* 48
2.8 *Square Root Tables* 49
2.9 *Irrational Numbers* 51
2.10 *Radicals* 51
2.11 *Fractional Exponents* 55
2.12 *Imaginary Numbers* 55
2.13 *Complex Numbers* 56

3 Slide Rule 60

3.1 *Parts of the Slide Rule* 60
3.2 *Reading the Scales* 61
3.3 *Finding the Decimal Point* 66
3.4 *Multiplication* 68
3.5 *Division* 72
3.6 *Combined Operations* 73
3.7 *The CI and DI Scales* 76
3.8 *Square Root* 78
3.9 *Squaring a Number* 80
3.10 *Cube Root* 81
3.11 *Cubing a Number* 83

4 Algebra 86

4.1 *The Meaning of Algebra* 86
4.2 *Literal Numbers and Variables* 86
4.3 *Fundamental Operations* 87
4.4 *Algebraic Terminology* 87
4.5 *The Degree of a Polynomial* 89
4.6 *Operations with Algebraic Expressions* 91
4.7 *Multiplication of a Polynomial by a Monomial* 96

4.8 *Multiplication of a Polynomial by a Polynomial* 97
4.9 *Special Products* 98
4.10 *Division of a Polynomial by a Monomial* 100
4.11 *Division of a Polynomial by a Polynomial* 101
4.12 *Synthetic Division* 104
4.13 *Factoring* 107
4.14 *Factoring by Grouping* 110
4.15 *Special Types of Polynomials* 112
4.16 *Complex Numbers in Algebra* 114
4.17 *Product of Complex Numbers* 115

5 Linear Equations, Functions, and Graphs 119

5.1 *Equations* 119
5.2 *Solving Equations* 120
5.3 *Equations Solved by Addition and Subtraction* 120
5.4 *Equations Solved by Multiplication and Division* 122
5.5 *Equations Involving Multiple Operations* 124
5.6 *Ratio and Proportion* 127
5.7 *Formulas* 132
5.8 *Functions* 134
5.9 *Graphs* 136
5.10 *The Rectangular Coordinate System* 137
5.11 *Linear Equations* 139
5.12 *The Slope of a Line* 142
5.13 *The Slope Intercept Form of the Equation of a Line* 144
5.14 *The Intercept Form of the Equation of a Line* 145
5.15 *Other Forms for the Equation of a Line* 148

6 Logarithms 152

6.1 *The General Exponential Equation* 152
6.2 *The Logarithmic Equation* 153
6.3 *Common Logarithms* 155
6.4 *Negative Characteristics* 156
6.5 *Antilogarithms* 157
6.6 *Linear Interpolation* 157
6.7 *Calculations with Logarithms* 160

7 Geometry 163

7.1 *Geometry* 163
7.2 *Undefined Concepts* 163
7.3 *Rays and Line Segments* 163
7.4 *Angles* 164
7.5 *Triangles* 166
7.6 *The Pythagorean Theorem* 167
7.7 *Similarity and Congruence* 170
7.8 *Polygons* 173
7.9 *Quadrilaterals* 174
7.10 *Areas of Triangles and Quadrilaterals* 176
7.11 *Circles* 182
7.12 *Volume* 184
7.13 *Geometric Constructions* 188
7.14 *Angle Construction* 193
7.15 *Construction of Regular Polygons* 199

8 Systems of Equations and Algebraic Operations 205

8.1 *Systems of Linear Equations* 205
8.2 *Solution of Systems of Linear Equations by Graphing* 207
8.3 *Solution of Systems of Linear Equations by Addition or Subtraction* 208
8.4 *Solution of Systems of Linear Equations by Substitution* 211
8.5 *Stated Problems* 214
8.6 *Equations in More Than Two Unknowns* 219
8.7 *Solution of Systems of Equations in More Than Two Unknowns* 220
8.8 *Matrices and Determinants* 222
8.9 *Operations on Matrices* 223
8.10 *Multiplication of Matrices* 224
8.11 *Determinants* 225
8.12 *Evaluation of Determinants by Minors and Cofactors* 227
8.13 *Properties of Determinants* 228
8.14 *Systems of Equations and Cramer's Rule* 231
8.15 *The Inverse of a Matrix* 233
8.16 *Solving Equations by Use of Matrices* 236

9 Higher Order Equations 240

9.1 *Second Degree Equation* 240
9.2 *Standard Form of a Quadratic Equation* 240
9.3 *Solutions to Quadratic Equations* 241
9.4 *Solution by Factoring* 241
9.5 *Extraction of Roots* 243
9.6 *Solutions by Completing the Square* 246
9.7 *Solutions by Formula* 247
9.8 *The Discriminant of the Quadratic Formula* 250
9.9 *Roots and Coefficients: Their Relationship* 252
9.10 *The Graphs of Quadratic Equations* 254
9.11 *The Conic Sections* 255
9.12 *The Circle* 256
9.13 *The Ellipse* 258
9.14 *The Hyperbola* 262
9.15 *The Parabola* 266
9.16 *Graphs of Other Equations* 270

10 Trigonometric Operations 275

10.1 *Trigonometry* 275
10.2 *The Trigonometric Ratios* 275
10.3 *Uniqueness of the Ratio* 277
10.4 *Standard Reference Triangles* 279
10.5 *Trigonometric Tables and the Values of the Trigonometric Ratios* 282
10.6 *Use of Tables for Solving Triangles* 283
10.7 *Applications to Right Triangles* 285
10.8 *Coordinates Revisited* 291
10.9 *Standard Position of an Angle* 292
10.10 *Reference Angles* 294
10.11 *The Sine Law* 299
10.12 *An Important Identity* 305
10.13 *The Cosine Law* 306
10.14 *Trigonometric Functions of Any Angle* 311
10.15 *Trigonometric Identities* 314
10.16 *Quotient Identities* 315
10.17 *Other Pythagorean Identities* 316
10.18 *Identities for Negative Angles* 318
10.19 *Sum and Difference Identities* 319
10.20 *Double Angle and Half Angle Identities* 322
10.21 *Trigonometric Scales on the Slide Rule* 325
10.22 *Reciprocal Functions on the Slide Rule* 328

11 Radian Measure and Sine Wave Analysis 332

11.1 *Radian Measure* 332
11.2 *Conversion of Degree and Radian Measure* 334
11.3 *Arc Length and Radians* 335
11.4 *Area of a Circular Sector* 336
11.5 *Radian Measure on the Slide Rule* 337
11.6 *Linear and Angular Velocity* 339
11.7 *The Sine Wave* 343
11.8 *The Graph of the Sine Wave* 343
11.9 *The Basic Sine Wave* 345
11.10 *The Phase Angle of the Sine Wave* 348

12 Vectors and Polar Coordinates 353

12.1 *Vectors* 353
12.2 *Scalar Multiplication* 355
12.3 *Vector Addition* 356
12.4 *Vector Subtraction* 358
12.5 *Vectors in Terms of Unit Vectors* 359
12.6 *Polar Coordinates* 362
12.7 *Graphing in Polar Coordinates* 365
12.8 *Equation in Polar Form* 366
12.9 *Converting from Rectangular to Polar Form* 368
12.10 *Graphing Complex Numbers* 371
12.11 *The Trigonometric or Polar Form of Complex Numbers* 373
12.12 *The Product of Complex Numbers in Trigonometric Form* 375
12.13 *Powers of Complex Numbers* 376
12.14 *Roots of Complex Numbers* 376

Answers to Odd-Numbered Problems 380

Table I Approximate Square Roots and Cube Roots 408

Table II Common Logarithms 409

Table III Trigonometric Ratios 411

Index 412

Introduction to TECHNICAL MATHEMATICS

1
Review of Basic Operations

1.1 Introduction

The first chapter is intended as a comprehensive review of the fundamental operations of arithmetic. It is assumed that the reader is familiar with these operations but may need to review the basic properties. Throughout the chapter use is made of other systems of measurement and extensive use should be made of Table II in the Appendix to facilitate performing these problems.

1.2 Sets

In order to discuss the structure and use of our number system, we need to become familiar with some basic set notation. This notation will be used throughout the book wherever it lends itself to simplification of the material.

A set is merely a well-defined collection of objects. Each object in a set is a *member*, or *element*, of the set. To say that a set is *well-defined* means that it is possible to decide whether or not an object is a member of the set. For example, we may wish to consider the set of ignition points that are pitted. If the ignition points in your car are pitted, then they belong to the set. If not, then they are outside the set. In the same way, we can determine whether or not any given ignition points belong to the set just referred to.

A set is frequently defined by enumerating or listing its members within braces. Thus, $\{x, y\}$ is a set consisting of the members x and y, while {pliers, crescent wrench, screwdriver} defines a set of basic tools for minor repairs.

Another method commonly used to define a set is the rule. A rule is a verbal description of a set, which allows one to determine which objects belong to that set.

1.3 Finite and Infinite Sets

By now the reader has probably concluded that sets can be very small as well as very large. In fact, a set can have no members at all. Such a set is called the *empty* or *null set*. The symbol for the empty set is ∅ or { }. (Note that the symbol {∅} does not mean the empty set, but means the set having the empty set as an element.)

If there is a counting number which tells us exactly how many elements are in a set, the set is said to be *finite*. In one sense then, a set is finite if it has a last element. On the other hand, if the set does not have this property, that is if it has no last element, it is said to be *infinite*. The set of all numbers is an example of an infinite set.

1.4 The Number System

The first set of numbers used by human beings was probably the set necessary for counting. These numbers are called the set of counting numbers, or more formally the set of *natural numbers*. The set of natural numbers is an infinite set and is written symbolically as {1,2,3,4, . . .}, where the three dots indicate "and so on." Each member of the set is obtained by adding 1 to the preceding member.

Notice that the set of natural numbers does not contain the element zero. When zero is included in the set of natural numbers, the new set which results is called the set of *whole numbers*. Since zero is one of the digits of our number system, we begin our numerical operations using the set of whole numbers.

Our system of numbers uses ten digits:

0,1,2,3,4,5,6,7,8,9.

All the numbers in our system are written using combinations of these digits. The value of a digit is determined by its place within a numeral. *Place value* is the value given to a specific position in a numeral. For instance, in any whole number, the digit farthest to the right has a place value of 1 and is said to be in the units or ones place. The digit in that place indicates the number of ones. The next place, or column, to the left is the tens column, where the digit indicates the number of tens. The next column left is the hundreds column, the next the thousands column, and so on. For example, 3267 means 3 thousands + 2 hundreds + 6 tens + 7 ones.

When performing addition within the set of whole numbers, the commutative property is essential. The commutative property for addition says that two whole numbers may be added in any order without affecting the sum. That is, if a and b are whole numbers, then $a + b = b + a$.

Another essential property in the addition of whole numbers is the associative property. This property says that the manner in which three whole numbers are grouped for addition does not affect the sum. That is, if a, b and c are whole numbers, then $(a + b) + c = a + (b + c)$. Together, the commutative and associative properties allow us to rearrange numbers to form convenient number combinations in finding a sum.

When using more than one operation or operations between two or more numbers in horizontal form, symbols of grouping are used. These symbols—the parentheses ( ), brackets [ ], or braces { }—are used to indicate which operations are to be performed first.

Example 1. $[5 + (7 + 3)] + 8$ should be performed as $[5 + 10] + 8 = 15 + 8 = 23$, where the operation within the innermost symbol of grouping is performed first.

The commutative and associative properties of whole numbers also hold true for the operation of multiplication. The commutative property for multiplication of whole numbers states that the order in which two numbers are multiplied does not affect the product. That is, if a and b are whole numbers, $a \cdot b = b \cdot a$. The associative property for multiplication states that the manner in which three numbers are grouped for multiplication does not affect the product. In symbols, if a, b and c are whole numbers, $(a \cdot b) \cdot c = a \cdot (b \cdot c)$.

Various symbols are used for multiplication. $a \times b$, $a \cdot b$, $(a)(b)$ or $a(b)$ are all used at different times. One must become familiar with all these symbols of operation. In addition, when using *literal numbers* (letters used as numerals), the product of a and b can be shown as ab. This, of course, cannot be used for nonliteral numbers because of our place value system.

Notice that within the set of whole numbers the commutative and associative properties do not hold true for the operations of subtraction or division. That is, $a - b \neq b - a$ and $a \div b \neq b \div a$.

Example 2. $10 - 5 \neq 5 - 10$.

Example 3. $10 \div 5 \neq 5 \div 10.$

Since zero added to any number leaves the sum unchanged, zero is called the *additive identity*, or *the identity element for addition*. Also, multiplying any number by 1 leaves the product unchanged. Therefore, 1 is called the *multiplicative identity*, or *the identity element for multiplication*.

Example 4. $75 + 0 = 75.$

Example 5. $19 \cdot 1 = 19.$

One further property of zero should be recalled, namely, that the product of zero with any number is zero. This property is called the *zero product property*. That is, $a \cdot 0 = 0$ for any whole number a.

When the operations of multiplication and division are used in the same problem, the *distributive property* becomes necessary. This property states that the product of a number times the sum (or difference) of two or more numbers can be found by multiplying the number times each term of the sum and then adding the products. In symbols this would be written as $a(b + c) = a \cdot b + a \cdot c$.

Example 6. $5(7 + 3) = 5 \cdot 7 + 5 \cdot 3 = 35 + 15 = 50.$

When no symbols of grouping are used, and more than one operation is involved in the problem, the operations of multiplication or division are always performed before the operations of addition or subtraction.

Exercise 1

Find the sums of the following:

1) $5 + 12 + 28 + 75$
2) $12 + 7 + 11 + 5 + 8 + 2$
3) $13 + 15 + 6 + 9 + 4$
4) $27b + 18c + 12b$
5) $16a + b + 3c + 26a + 2b + 5c$

6)
$$\begin{array}{r} 2127 \\ 3124 \\ 2999 \\ \underline{9279} \end{array}$$

7)
$$\begin{array}{r} 3756 \\ 2296 \\ 4455 \\ 2935 \\ \underline{1852} \end{array}$$

8)
$$\begin{array}{r} 95{,}057 \\ 1{,}920 \\ 29{,}774 \\ 39{,}274 \\ \underline{2{,}724} \end{array}$$

9) $\begin{array}{r} 792{,}886 \\ 257 \\ 60{,}317 \\ 779{,}212 \\ \underline{212{,}111} \end{array}$

10) $\begin{array}{r} 8{,}560{,}243 \\ 264{,}212 \\ 212{,}222 \\ 9{,}993{,}124 \\ 2{,}172{,}162 \\ \underline{2{,}297{,}614} \end{array}$

Perform the following subtractions:

11) $7795 - 7721$

12) $6794 - 3729$

13) $\begin{array}{r} 277{,}921 \\ \underline{192{,}981} \end{array}$

14) $\begin{array}{r} 453{,}862 \\ \underline{127{,}695} \end{array}$

15) $\begin{array}{r} 1{,}264{,}773 \\ \underline{37{,}210} \end{array}$

16) $\begin{array}{r} 764{,}321{,}976 \\ \underline{94{,}298{,}672} \end{array}$

Perform the indicated operations, doing the operations within the symbols of grouping first

17) $(85 + 16) - 34$

18) $(92 - 21) - 17$

19) $2142 - (325 + 479) - 262$

20) $51{,}223 - \{9622 + (3140 - 217)\} + 10$

Find the following products.

21) 642×127

22) $(281)(1123)$

23) $4271 \times 65{,}132$

24) $\begin{array}{r} 54{,}321 \\ \underline{\times\ 1{,}234} \end{array}$

25) $\begin{array}{r} 21{,}782 \\ \underline{\times\quad 261} \end{array}$

26) $\begin{array}{r} 38{,}241 \\ \underline{\times\quad 57} \end{array}$

Find the following products first by applying the distributive property and then by doing the operation within the symbols of grouping first.

27) $3(5 + 7 + 12)$

28) $12(8 + 11 + 26)$

29) $28(32 + 2159 - 72 - 84)$

Find the following quotients.

30) $\dfrac{1276}{47}$

31) $\dfrac{8745}{45}$

32) $63\overline{)845}$

33) $56\overline{)3976}$

34) $47\overline{)1041}$

35) $38\overline{)1178}$

36) $298\overline{)79{,}264}$

37) $87\overline{)87{,}878}$

38) $4965\overline{)145{,}792{,}805}$

Find the solutions to the following.

39) A man has \$1275 which he wants to invest in stock costing \$75 per share. How many shares of stock can he purchase?

40) Find the perimeter of (distance around) a rectangular piece of sheet metal whose dimensions are 64 inches by 32 inches.

41) An electrician figured that he needed the following lengths of No. 12 wire to complete his job: 13 ft., 37 ft., 62 ft., 27 ft., 39 ft., and 62 ft. Allowing 10 ft. for waste and working around obstacles, how much wire should he purchase?

42) An electric meter read 72,610 kilowatt-hours on November 30. How much electricity was used during December, if the meter read 74,729 kilowatt-hours on December 31?

43) Brand A spark plugs can be purchased for $.49 each and Brand B for $1.09 each. How much money would be saved on a set of 8 spark plugs by using Brand A?

44) If a truck traveled 247 miles on 19 gallons of gasoline, how many miles per gallon did the truck get?

45) A factory, in filling an electronics order for 1,690,416 tubes, packed the tubes in cases of 6 boxes, each box holding 12 tubes. How many cases were in the shipment?

46) How much gasoline can be refined from a 4200-gallon tank of crude oil if each 42-gallon barrel of crude oil yields the following products in addition to gasoline: fuel oil, 20 gallons; kerosene, 4 gallons; lubricating oil, 2 gallons; miscellaneous, 3 gallons; waste, 2 gallons?

47) If a new picture tube for a television set costs $97.54 and the same model rebuilt costs $46.95, how much less expensive is it to purchase a rebuilt picture tube?

48) The gasoline meter at a service station reads 52,784 gallons in the morning, and at closing time the meter reads 54,641 gallons. How many gallons of gasoline were pumped during the day?

49) If during the wheat harvest farmers brought in loads of 5700 bushels, 4400 bushels, 3722 bushels, and 2641 bushels of wheat to an elevator and if during the same day the elevator sold 7250 bushels of wheat, what was the net increase in the amount of wheat in the elevator during the day?

50) The TV tube inventory of a warehouse at the beginning of the week was 5296 tubes. If 491 tubes were used Monday, 112 Tuesday, 95 Wednesday, 263 Thursday, and 387 Friday, what was the total tube inventory at the end of the week?

1.5 Negative Numbers

Sometimes we want to measure quantities which are less than zero, such as temperature, electrical current, sea level, debits, and many others. So far we have not considered numbers that would allow us to work with these situations. Let us now consider a number line containing a point representing zero, and containing other points to the right of zero representing the set of whole numbers. (See Figure 1.1.) We can also extend the number line to the left, calling the numbers to the left of zero **negative numbers.** In other words, any number less than zero is a negative number.

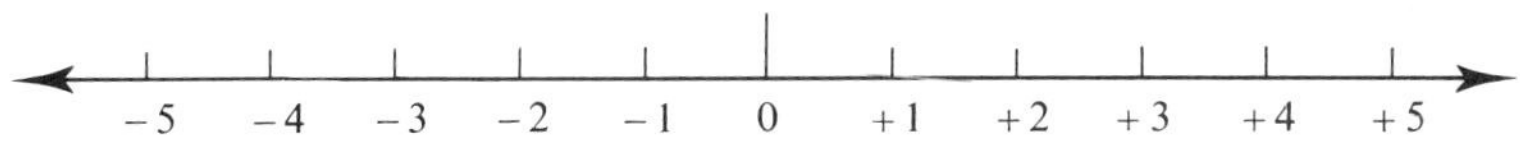

Figure 1.1

Notice that for each whole number to the right of zero there is a negative number the same distance from zero but in the opposite direction. A minus sign in front of each negative number shows it to be in a negative direction on the number line. A number greater than zero is indicated by a plus sign indicating that the number is positive. (A number written with no sign is assumed to be positive.) Using the concept of the number line, any number appearing to the left of another number on the number line is less than that number. Therefore, -7 is less than -4 (or $-7 < -4$) and $-3 < 1$. In the same manner, if one number appears to the right of another, it is greater than that number. Thus $9 > 7$ and $-1 > -5$. The positive and negative numbers, ordered around the **origin** (zero) on a number line, are called *directed numbers*. Notice that zero is neither positive nor negative.

Consider the directed numbers -3 and $+3$. They are the same distance from the origin but on different sides of it (Figure 1.1). If we wish to add these two numbers, we are in effect adding the distance from zero to $+3$ to the distance from zero to -3, and one is just the opposite of the other. This addition would place us back at zero. Negative 3 (or -3) is called the opposite of positive 3 (or $+3$) and because of the property just illustrated, these numbers are called *additive inverses* of one another. Notice that every natural number has an additive inverse or negative.

The set of natural numbers, their additive inverses, and zero is called the set of *integers*. Notice that the set of natural numbers is identical to the set of positive integers; we will see later that they have the same properties.

1.6 Addition of Integers

If we consider integers as directed numbers, operations involving them may be easily described. For instance, if walking forward is considered positive, then walking backward or in the opposite direction must be considered negative. Using this analogy, if we combine a number with its additive inverse, we could consider it as moving in one direction on the number line a certain number of units and then moving in the opposite direction the same number of units. This would put us back where we started, namely, at zero. (See Figure 1.2.) Any number then added to its additive inverse is zero, the origin on the number line. Recall that this satisfies our definition of an additive inverse of a number in Section 1.5. The additive inverse of any number is the number which when added to that number results in zero.

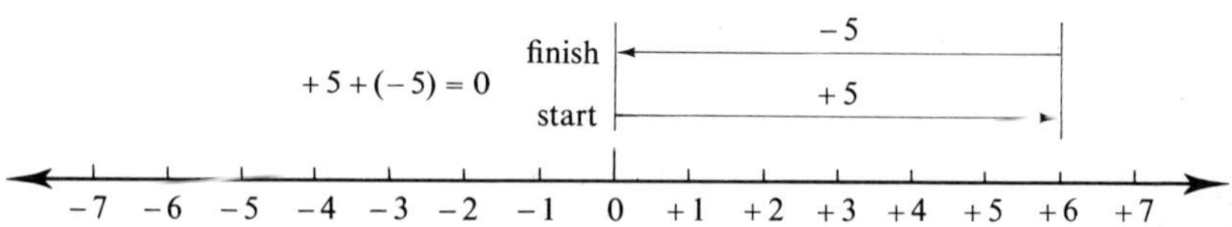

Figure 1.2

In general, to add two signed numbers having unlike signs, we take the difference in the numerical values of the numbers and give it the sign of the one having the larger numerical value. For instance, if we deposit \$25 in the bank and write a check for \$35 we will be "in the hole" by \$10. (See Figure 1.3.) Using integers, $25 + (-35) = -10$.

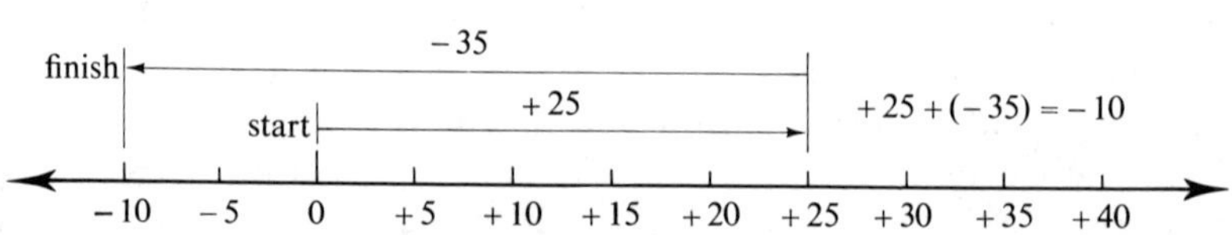

Figure 1.3

Example 1. If the temperature rises 15 degrees and then drops 18 degrees, the total change in temperature can be designated as $15 + (-18) = -3$.

Notice that the set of integers allows us to subtract more from a number than the value of the number. To move in the opposite direction is much the same as subtracting or taking away from. In fact, to add a negative number is the same as to subtract a positive number. That is, $5 - 8$ or $5 - (+8) = 5 + (-8) = -3$. In general,

$$a - b = a + (-b).$$

Using the same process on the number line, integers with the same sign can be added. In this case, since the integers have like signs, the direction indicated would be the same for each number. For instance, to add $+7$ to $+5$, we begin at the origin and move to the right 7 units. We then move 5 units further to the right, ending at $+12$. (See Figure 1.4.) Therefore, $(+7) + (+5) = +12$ or $7 + 5 = 12$.

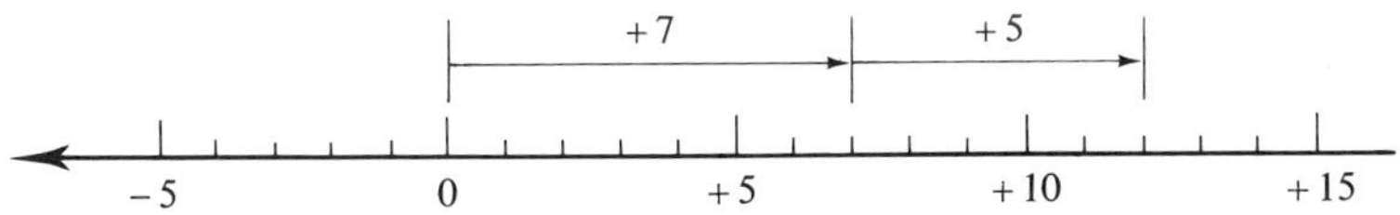

Figure 1.4

A similar situation occurs when both integers are negative, only in the opposite direction. The sum of -9 and -6 can be found by beginning at the origin and moving to the left (the negative direction) 9 units. From this point we move six more units to the left, ending at -15. (See Figure 1.5.) Hence $-9 + (-6) = -15$

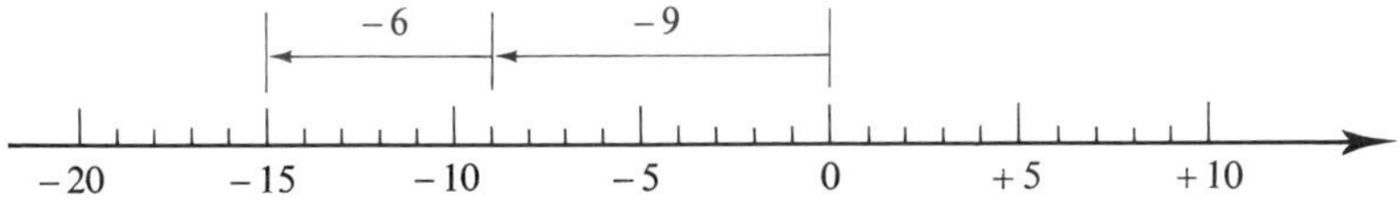

Figure 1.5

1.7 Subtraction of Integers

Subtraction is the inverse operation of addition. We have already discussed the similarity of adding a negative number and subtracting a positive number: $9 - 3 = 9 + (-3) = 9 - (+3)$. As a general rule, in fact, subtracting a number with a given sign is identical to adding the same number of opposite sign.

Example 1. $5 - (-4) = 5 + (+4) = 9$ or $5 + 4 = 9$.

Example 2. $+6 - (+3) = 6 + (-3) = 3$ or $6 - 3 = 3$.

Often the rule for the subtraction of integers is stated as: *to subtract one integer from another, change the sign of the subtrahend and follow the procedure for addition of integers.*

When writing problems involving positive and negative numbers, if there is no danger of confusion, a plus sign need not be used for positive numbers. If a number is negative, however, the minus sign always must be used. If more than one sign is used, symbols of grouping are very useful.

Example 3. $575 - (+471) = 575 - 471 = 104.$

Example 4. $5 + (-7) = 5 - 7 = -2.$

Example 5. $(-7) + 5 = -7 + 5 = -2.$

Notice from Examples 4 and 5 that the commutative law holds for addition of integers.

Example 6. $5a - 7a = a(5 - 7) = a[5 + (-7)] = a(-2) = -2a.$

Note here that the distributive law also holds for the set of integers.

Example 7. $-12 + (-7) = -12 - 7 = -19.$

Example 8. $-27 - (-11) = -27 + (+11) = -16.$

Example 9. $-[(-4) + (-3)] = -[-4 - 3] = -[-7] = +7.$

1.8 Combined Operations

Recall that when addition and subtraction of integers are involved in the sample problem, the operations are done in the order in which they appear, unless symbols of grouping are used to indicate otherwise. When many numbers are combined in the same problem, it is often easiest to combine all the positive numbers, then combine all the negative numbers, and finally combine the two quantities. This also gives us an excellent method for checking operations with signed numbers.

Example 1.
$$\begin{aligned}(+4) + (+8) + (-7) + (-3) + (+9) + (-7) \\ = (4 + 8 + 9) + [(-7) + (-3) + (-7)] \\ = +21 + (-17) = +4 \text{ or } 4.\end{aligned}$$

This is possible since the commutative and associative properties of whole numbers also hold true under the operation of addition for the set of integers.

Exercise 2

Perform the indicated operations.

1) $73 + 57$
2) $8 + 27$
3) $16 + (-48)$
4) $73 + (+75)$
5) $86 - 29$
6) $4x - (2x + 1)$
7) $5a - 3a$
8) $19y - (-3y)$
9) $43 + (-7)$
10) $3a - 5a$
11) $92 + (-26) - (-5)$
12) $7 - (-8) + 3$
13) $32 - (-16)$
14) $25 + [-(-3)]$
15) $-19 - (-19)$
16) $203 - (-203)$
17) $-45 + (-27)$
18) $-83 + (-92)$
19) $203 + (+203)$
20) $-512 - (-217)$
21) $29 - (-4) + (-6)$
22) $203 - (+203)$
23) $-87 + (-92)$
24) $-87 - (-92)$
25) $23 + (-62) - (-27)$
26) $-87 - (+92)$
27) $-89 - (-165) + (-202)$
28) $-45 - (-58) - (-26)$
29) $0 - (+175)$
30) $2a - (4 - a)$
31) $95 - (-54) + (-72) + (-72) + (-5)$
32) $27x - 3(x - 1)$
33) $82 - (+37) + (-26) - (-24) + (-9)$
34) $x - (-4x - 3) - 3$
35) $a + (-b) - (+b) + (-a)$
36) $24y - 3y - (-12y) - (-5y)$
37) $x - 2y + (-x)$
38) $75 - 4x - (86 - 3x)$
39) $2x - (-2y) + (-3x) - (-2y)$
40) $3a - (-2a) + (2b) - (-2b) + (2c) - (2c)$

Represent each of the following as a signed number.

41) A change of temperature from 95° above zero to 3° below zero.
42) A change in altitude from Death Valley at 260 feet below sea level to the top of Mount Whitney at 14,495 feet above sea level.

43) The specification for manufacture quality control on a piston to be within three-thousandths of an inch of design size.

Solve the following:

44) Longitude is expressed in degrees west or east of Greenwich, England. If 26° west longitude is written as $-26°$, express 52° east longitude, 43° west longitude, and 75° east longitude.

45) On a given day in January, the high readings in the continental United States ranged from 11° below zero at Fargo, North Dakota, to 86° above zero at Brownsville, Texas. Represent the difference in temperatures mathematically and calculate this difference.

46) Represent a bank deposit of \$475.00 and withdrawals of \$26.00, \$87.50, \$27.95, and \$96.75 using signed numbers. If this represents a day's activity in an account, calculate the net amount credited to the account during the day.

47) During 5 successive years a business lost \$50,000 the first year, made \$28,000 the second year, lost \$4,000 the third year, made \$46,000 the fourth year, and made \$21,000 the fifth year. How much profit (or loss) was made during the 5 years?

48) During 4 downs in a football game the following yardage was recorded: a loss of 2 yards, a gain of 11 yards, a loss of 1 yard, and a gain of 3 yards. Did the team make "first and ten"? Represent the yardage on a number line using the origin of the number line for the original line of scrimmage.

1.9 Multiplication of Integers

We have already referred to multiplication as a short cut method for repeated addition. This idea can be extended to encompass the set of integers. Thus $5(-2)$ can be considered as $(-2) + (-2) + (-2) + (-2) + (-2) = -10$. The commutative property of multiplication holds for integers, so $5(-2) = (-2)(5) = -10$. In general, *the product of two integers having unlike signs is a negative integer.*

Now let us consider the product of two integers having like signs. We already know that positive integers are nothing more than natural numbers, so the product of two positive integers is a positive integer. To discover a method of multiplying two negative integers we shall recall some properties discussed previously. Consider the product of -3 and -7. We know that

$$-3[(7) + (-7)] = (-3)(0) = 0,$$

by the property of the additive inverse of a number and the property of zero. Further, using the distributive property gives us

$$-3[(7) + (-7)] = (-3)(7) + (-3)(-7),$$

$$(-3)(7) + (-3)(-7) = 0,$$

$$(-21) + (-3)(-7) = 0.$$

The last equation says that $(-3)(-7)$ must be a number which when added to -21 results in zero. This number is the additive inverse of -21, namely, $+21$. Hence $(-3)(-7) = 21$, *so the product of two integers having like signs is a positive integer.*

The associative property of multiplication can be used to extend the procedure of finding the product of negative integers. For example,

$$(-3)(-2)(-5) = (-3)[(-2)(-5)]$$

$$= [(-3)(+10)] = -30.$$

In short, the product of an odd number of negative factors results in a negative integer, while the product of an even number of negative factors results in a positive integer.

Example 1. $(-5)(-6) = +30 = 30.$

Example 2. $(-18)(+12) = -216.$

Example 3. $(-1)(-1)(-1)(-1)(-1)(-1)(-1)(-1) = +1 = 1.$

1.10 Division of Integers

Division is the inverse operation of multiplication. In this respect the division of integers follows the same pattern as the multiplication of integers. Since $(5)(-2) = -10$, it follows that

$$\frac{-10}{+5} = -2 \quad \text{and} \quad \frac{-10}{-2} = +5.$$

In general we can assert that *the quotient of two integers having like signs is a positive integer and the quotient of two integers having unlike signs is a negative integer.*

Since fractions are merely the division of one number by another, the rules governing the division and multiplication of integers refer equally well to the fractions. The forms $-\frac{a}{b}$, $\frac{-a}{b}$, or $\frac{a}{-b}$ are equivalent, so that

$$-\frac{4}{2} = \frac{-4}{2} = \frac{4}{-2} = -2.$$

Exercise 3

Calculate the following:

1) $(-4)(+3)$
2) $(-7)(-2)$
3) $(-8)(-5)$
4) $(a)(-5)$
5) $-4(2+5)$
6) $3[(-2)+(4)]$
7) $(242)(-347)$
8) $(-4)(-5)(-6)$
9) $(-3)(2)(-4)(5)$
10) $(321)(-284)$
11) $(-221)(-212)$
12) $(341)(112)(-24)$
13) $(212)(-342)(267)$
14) $(-2631)(179)$
15) $(-4) \div (-2)$
16) $(+4) \div (+2)$
17) $(4) \div (-25)$
18) $(4) \div (-2)$
19) $(-217)(-21)$
20) $(-382)(-49)$
21) $(340)(21)$
22) $(27{,}603)(1321)$
23) $-262 \div -22$ to the nearest hundredth
24) $-479.26 \div 63$ to the nearest unit
25) $2763 \div 257$ to the nearest tenth
26) $[(-17)+(-9)+(31)] \div [(13)+(-2)-(-9)]$
27) $[(-1224)+(324)-(-24)] \div 2$
28) $[(-21)+(-32)-(-17)] \div -3$
29) $[32+(-41)-(26)] \div 12$
30) $[(-9)+(-15)-(-17)-(-21)] \div 3$

1.11 Factors and Primes

If two or more numbers are multiplied together, each of the numbers from which a product is obtained is called a *factor* of the product. For instance, since $10 = 5 \cdot 2$, both 5 and 2 are called factors of 10. When one number is exactly divisible by another, that is, when no remainder results from the division, the first number is said to be a *multiple* of the second. For example, $14 = 7 \cdot 2$. 14 is a multiple of 7 because 7 is contained exactly in 14. 14 is also a multiple of 2. Therefore, 7 and 2 are factors of 14.

Example 1. Find the set of all the factors of 75.
Since 75 is divisible by 5, that is $75 \div 5 = 15$, 5 and 15 are factors. However, $1 \cdot 75 = 75$, $25 \cdot 3 = 75$, and $3 \cdot 5 \cdot 5 = 75$; therefore, the set of *all* the factors of 75 is $\{1,3,5,15,25,75\}$. 75 is also said to be a multiple of each of these numbers.

When a number has no whole number factors except the number itself and 1, the number is called *prime*. An exception to this definition is the number 1 itself. Because of many special properties of 1, and since every number has a factor of 1, the number 1 is not considered to be a prime number. The smallest prime number (and the only even prime number—why?) is thus 2. The next prime is 3, then 5, 7, 11, 13, 17, 19, 23, and so on.

When a number is written as a product of factors each of which is a prime number, then the number is in *prime factored form*. To find these prime factors it is usually easiest to see which primes divide exactly into the number until all the prime factors are found.

Example 2. Find the prime factors of 80.

$$\begin{array}{r|l} 2 & 80 \\ 2 & 40 \\ 2 & 20 \\ 2 & 10 \\ & 5 \end{array}$$

Since 80 is even, it is divisible by 2, and so on until all the prime factors are found. The prime factored form of 80 is $2 \cdot 2 \cdot 2 \cdot 2 \cdot 5$. The set of prime factors of 80 is $\{2,5\}$. (Notice the new notation used in this example.)

When a number is divisible by numbers other than itself and 1, that is, when it can be written as a product of primes, it is said to be a *composite* number.

By definition, the numbers zero and one are neither prime nor composite.

1.12 Greatest Common Factor

If two numbers contain the same number as a factor, or if two numbers are a multiple of the same number, then that number is said to be a *common factor* of the two numbers. Many numbers have more than one common factor. For instance, 24 and 36 have 2 as a common factor, but they also have 3, 4, 6, 12 as common factors.

$$\begin{array}{lcl} 2 \times 12 = 24 & \text{and} & 2 \times 18 = 36 \\ 3 \times 8 = 24 & & 3 \times 12 = 36 \\ 4 \times 6 = 24 & & 4 \times 9 = 36 \\ & & 6 \times 6 = 36 \end{array}$$

Often we are interested in the greatest number contained as a factor in each of a set of numbers. This number is called the *greatest common factor* or *G.C.F.* of the set of numbers. The G.C.F. need not be a prime number. In fact, in the illustration used above, 12 is the greatest number contained as a factor in both 24 and 36; hence 12 is the G.C.F. of 24 and 36. Yet 12 is not a prime number, since in factored form it is $2 \cdot 2 \cdot 3$.

Because the terms factor and divisor are often used interchangeably, many authors refer to the G.C.F. as the *greatest common divisor.*

Example 1. Consider the numbers 15 and 25. 15 can be written as $5 \cdot 3$ and 25 can be written as $5 \cdot 5$. Inspecting the factors of each of these numbers, that is, $5 \cdot 3$ and $5 \cdot 5$, the factor 5 is the only factor common to both 15 and 25. Therefore, the G.C.F. of 15 and 25 is 5.

In this example, the G.C.F. is a prime number but again, this need not be so. In general, the G.C.F. is the greatest number common to each of the numbers. As shown by Example 1, it is usually easier to find the G.C.F. of two numbers by first factoring the numbers into a product of primes. The G.C.F. is then the product of all the primes contained as factors in each of the numbers. This process is illustrated below.

Example 2. Find the greatest common factor of 87 and 27.

$$\begin{array}{r|l} 3 & 87 \\ \hline & 29 \end{array}$$

and $87 = 3 \cdot 29$.

$$\begin{array}{r|r} 3 & 27 \\ \hline 3 & 9 \\ \hline & 3 \end{array}$$

Thus $27 = 3 \cdot 3 \cdot 3$.

First, finding the prime factored form of each, we inspect the sets and find 3 to be the only common factor. Therefore, the G.C.F. is 3.

Many pairs of numbers, such as 5 and 7, have only 1 as a greatest common factor. When a set of numbers (two or more) have 1 as their highest common factor, they are considered *relatively prime*. Numbers need not be prime numbers to be relatively prime, however. For instance, $57 = 19 \cdot 3$ and $25 = 5 \cdot 5$. Neither 57 nor 25 are prime numbers. Yet none of the prime factors of 57 are the same as any of the prime factors of 25. Therefore, 25 and 57 have a G.C.F. of 1 and are considered relatively prime.

1.13 Least Common Multiple

When one number is contained as a factor in another number, the second number is said to be a multiple of the first. For example, 96 contains 3 as a factor ($96 = 3 \cdot 32$). Therefore, 96 is a multiple of 3. For the same reason, 25, 30, and 35 are multiples of 5, 36 is a multiple of 12, and so on.

In the consideration of two numbers, we often need the smallest number which is a multiple of each of the two, that is, the smallest number into which each of the numbers will divide. When the smallest such number is found, it is called the *least common multiple* or *L.C.M.* of the two numbers.

Example 1. Consider the L.C.M. of 24 and 18.

$$24 = 2 \cdot 2 \cdot 2 \cdot 3 \quad \text{and} \quad 18 = 2 \cdot 3 \cdot 3.$$

Since 2 appears as a factor at most three times and 3 appears as a factor at most two times, the L.C.M. would be $2 \cdot 2 \cdot 2 \cdot 3 \cdot 3 = 72$. Notice that 72 is $3 \cdot 24$ and therefore is a multiple of 24, and that 72 is also $4 \cdot 18$ and therefore a multiple of 18. 72 is also the smallest number containing both 24 and 18 as factors.

Notice that if the numbers in question are relatively prime, then the least common multiple of the two is the product of the two numbers.

Example 2. Find the L.C.M. of 96 and 35

$$96 = 2 \cdot 2 \cdot 2 \cdot 2 \cdot 2 \cdot 3, \text{ and } 35 = 7 \cdot 5.$$

The L.C.M. would then be

$$\begin{aligned} 2 \cdot 2 \cdot 2 \cdot 2 \cdot 2 \cdot 3 \cdot 7 \cdot 5 &= 96 \cdot 35 \\ &= 3360. \end{aligned}$$

1.14 Fractions

A portion of a quantity is called a *fraction* of the quantity. In this book fractions are written:

$$\frac{1}{2}, \ 1/2, \text{ or } \tfrac{1}{2}.$$

The lower number is called the *denominator* of the fraction, and the upper number is called the *numerator*. Thus in the fraction 7/15 the 15 is the denominator, indicating that the quantity has been divided into 15 equal parts, and the 7 is the numerator, indicating that we are considering 7 of these equal parts. The numerator and denominator are called the *terms* of the fraction.

1.15 Equivalent Fractions

Notice that in Example 1 below the numerator of one fraction is a multiple of the numerator of the other, and that the denominator of one is a multiple of the denominator of the other. Notice also that they are the same multiple, that is, if one numerator is 2 times the other, then one denominator is 2 times the other. This brings us to a very important observation, namely that *multiplying both the numerator and the denominator of a given fraction by the same number results in another fraction equivalent to the given fraction.*

Example 1. $\frac{5}{6} = \frac{5 \times 5}{6 \times 5} = \frac{25}{30}.$

Similarly, dividing both the numerator and the denominator of a given fraction by the same number results in another fraction equivalent to the given fraction. On the other hand, if the numerator and denominator of a fraction are relatively prime, then the fraction is said to be *reduced to lowest terms.*

Example 2. Reduce $\frac{575}{680}$ to lowest terms.

By factoring numerator and denominator and writing the fraction as

$$\frac{5 \cdot 5 \cdot 23}{5 \cdot 2 \cdot 2 \cdot 2 \cdot 17},$$

we can see that the only common factor is 5; thus by dividing numerator and denominator by 5 the fraction can be reduced to

$$\frac{5 \cdot 23}{2 \cdot 2 \cdot 2 \cdot 17} = \frac{115}{136}.$$

1.16 Comparison of Fractions

Fractions can be compared easily when they have the same denominator. For instance, suppose we have a set of wrenches whose sizes increase by 1/32 inch, and we need to know what the next wrench in the set larger that 5/8 inch would be, we would convert 5/8 to a fraction having a denominator of 32 by multiplying numerator and denominator by 4; that is,

$$\frac{5 \cdot 4}{8 \cdot 4} = \frac{20}{32}.$$

The next larger wrench then can easily be seen to be 20/32.

The larger of two fractions can also be determined by changing both fractions to equivalent fractions having the same denominator. Their relative sizes can then be determined by their numerators.

When two fractions are written with the same denominator, they are said to have a *common denominator*. When this denominator is the smallest one possible, it is called the *least common denominator* or *L.C.D.*

A least common denominator contains all the factors of each denominator. If we recall our previous discussion of the least common multiple of two numbers, we find that the L.C.D. of two fractions is merely the least common multiple of the individual denominators.

Example 1. Change 13/84 and 17/63 to equivalent fractions having a least common denominator.

The prime factored form of 84 is $2 \cdot 2 \cdot 3 \cdot 7$, and the prime factored form of 63 is $3 \cdot 3 \cdot 7$. The prime factored form of the least common multiple of these two numbers is therefore $2 \cdot 2 \cdot 3 \cdot 3 \cdot 7$. Therefore, the least common denominator is $2 \cdot 2 \cdot 3 \cdot 3 \cdot 7 = 252$. 13/84 would be written as

$$\frac{13 \cdot 3}{84 \cdot 3} \text{ or } \frac{68}{252},$$

and 17/63 would be written as

$$\frac{17 \cdot 2 \cdot 2}{63 \cdot 2 \cdot 2} \text{ or } \frac{68}{252}.$$

1.17 Adding and Subtracting Fractions

The properties discussed in an earlier section for whole numbers also apply to the operations with fractions. However, before fractions can be added or subtracted, they must have the same denominator. This is done by changing each fraction into an equivalent fraction having the *least common denominator* of all the fractions.

Example 1. Combine 5/8 and 2/3.

$$\frac{5}{8} = \frac{15}{24} \quad \text{and} \quad \frac{2}{3} = \frac{16}{24}$$

$$\frac{5}{8} + \frac{2}{3} = \frac{15}{24} + \frac{16}{24} = \frac{31}{24}.$$

When the numerator of a fraction is smaller than the denominator, the fraction is called a *proper fraction.* When the numerator is larger than the denominator, the fraction is called an *improper fraction.* When the number is written as a whole number combined with a fraction, it is called a *mixed number.* When adding fractions involving mixed numbers, it is often advantageous to convert the mixed number to an improper fraction. A rule for changing a mixed number into an improper fraction can be formulated as follows.

Multiply the whole number part of the mixed number times the denominator part of the fraction, and add the product to the numerator of the fraction. Place this sum over the denominator of the fraction.

Example 2. Change $7\frac{3}{8}$ to an improper fraction.
$7 \cdot 8 = 56$, $56 + 3 = 59$. Therefore, as an improper fraction, $7\frac{3}{8} = 59/8$.

Example 3. Find the sum $5\frac{3}{8} + 4\frac{3}{16} + 7\frac{5}{24}$.

$$\begin{aligned} 5\frac{3}{8} + 4\frac{3}{16} + 7\frac{5}{24} &= \frac{43}{8} + \frac{67}{16} + \frac{173}{24} \\ &= \frac{258}{48} + \frac{201}{48} + \frac{346}{48} = \frac{805}{48}. \end{aligned}$$

1.18 Multiplication of Fractions

Since multiplication is a short-cut method for adding quantities, multiplication of fractions can be generalized in much the same way as multiplication of whole numbers. That is, $3 \times 1/8$ means $1/8 + 1/8 + 1/8$ or $3/8$. Since multiplication is commutative, $1/8 \times 3$ must also be $3/8$. A general procedure for multiplying fractions can be seen if we write 3 as the fraction $3/1$ and multiply by $1/8$. Then

$$\frac{3}{1} \times \frac{1}{8} = \frac{3 \times 1}{1 \times 8} = \frac{3}{8}.$$

In other words, *the product of two fractions is the fraction formed by the product of the numerators and the product of the denominators.*

Example 1. $\frac{3}{4} \times \frac{3}{5} = \frac{3 \times 3}{4 \times 5} = \frac{9}{20}$.

In many cases a product of two fractions can be reduced. This must be done if the fraction is to be in simplest form. It is easiest if the product is left in factored form until it has been reduced as far as possible.

Example 2. $\frac{3}{4} \times \frac{5}{3} = \frac{3 \times 5}{4 \times 3} = \frac{5 \times 3}{4 \times 3} = \frac{5}{4}.$

It is usually not necessary to go through all the steps in this procedure once you have become familiar with the method. The following example illustrates a shorter method.

Example 3. $\frac{\overset{1}{\cancel{3}}}{4} \times \frac{1}{\underset{1}{\cancel{3}}} = \frac{1 \times 1}{4 \times 1} = \frac{1}{4}.$

This method is often referred to as *cancelling*. When you use it, keep in mind the justification for it—both numerator and denominator of the product are divided by the same quantity.

When performing any operations involving fractions, the answers should always be expressed in lowest terms, that is, reduced as far as possible. Recall that a fraction is in lowest terms when the numerator and denominator are relatively prime.

If the product of two numbers is 1, the two numbers are said to be *multiplicative inverses* or *reciprocals* of each other. For example, since $2 \cdot 1/2 = 1$, 2 is the reciprocal of $1/2$ and $1/2$ is the reciprocal of 2. In the same manner, $5/9$ is the reciprocal of $9/5$ and a/b is the reciprocal of b/a.

Example 4. Find the reciprocal of $5/8$

Since

$$\frac{5}{8} \times \frac{8}{5} = \frac{5 \times 8}{8 \times 5} = \frac{1}{1} = 1,$$

$8/5$ is the reciprocal of $5/8$.

Zero does not have a reciprocal. For example, $0/5 = 0$, $5/0$ is meaningless, since we cannot divide by zero.

1.19 Division of Fractions

Because multiplication and division are inverse operations, we can divide one number by another by multiplying the first number by the reciprocal of the other. In terms of fractions, we *divide one fraction by another by inverting the divisor and multiplying the fractions.*

Example 1. $\frac{1}{4} \div \frac{3}{4} = \frac{1}{4} \cdot \frac{4}{3} = \frac{1}{3}.$

Example 2. $\frac{19}{22} \div \frac{2}{11} = \frac{19}{22} \cdot \frac{11}{2} = \frac{19}{4}.$

1.20 Complex Fractions

When a fraction is formed whose numerator and denominator are both fractions, the fraction is called a complex fraction. This is nothing more than another way to write the division of one fraction by another.

Example 1. $\frac{\frac{1}{3}}{\frac{1}{4}} = \frac{1}{3} \div \frac{1}{4} = \frac{1}{3} \cdot \frac{4}{1} = \frac{4}{3}.$

Fractions can also be written with more than one number in the numerator and denominator. In this case, both the numerator and the denominator must be simplified before the division is carried out.

Example 2.

$$\frac{\frac{1}{3} + \frac{1}{4}}{\frac{5}{7} + \frac{3}{8}} = \frac{\frac{4}{12} + \frac{3}{12}}{\frac{40}{56} + \frac{21}{56}} = \frac{\frac{7}{12}}{\frac{61}{56}} = \frac{\overset{14}{\cancel{56}}}{} \frac{7}{\underset{3}{\cancel{21}}} \cdot \frac{56}{61} = \frac{7}{12} \cdot \frac{56}{61} = \frac{98}{183}.$$

Notice that the sums in the numerator and denominator were simplified first. The result of the division, the fraction 98/183, is considered the simplified form of the complex fraction.

Exercise 4

Which of the following numbers are prime, which are composite, and which are neither?

1) 75 2) 49 3) 73

4) 0 5) 11 6) 1

Find the set of prime factors of the following:

7) 63 8) 45 9) 324

10) 735 11) 1848 12) 30,030

Write the following in prime factored form.

13) 195 14) 500 15) 165 16) 792

Find the G.C.F. and L.C.M. of each of the following sets of numbers.

17) 27 and 243

18) 88 and 210

19) 100 and 150

20) 15, 60, and 135

21) 36, 48, and 144

22) 1764, 2520, and 13,230

Change to improper fractions.

23) $7\frac{3}{7}$

24) $5\frac{9}{10}$

25) $46\frac{5}{8}$

26) $\frac{26}{3}$

27) $\frac{17}{5}$

28) $\frac{465}{12}$

Perform the following operations and reduce the answers to lowest terms.

29) $\frac{1}{3}+\frac{1}{4}$

30) $\frac{2}{5}+\frac{3}{10}$

31) $\frac{5}{6}-\frac{3}{6}$

32) $\frac{7}{15}+\frac{5}{6}$

33) $\frac{3}{5}+\frac{3}{10}-\frac{2}{15}$

34) $4\frac{1}{3}+2\frac{1}{2}+2\frac{1}{6}$

35) $\frac{5}{16}\times\frac{8}{35}$

36) $\frac{21}{72}\times\frac{72}{21}$

37) $\frac{3}{10}\times\frac{5}{9}$

38) $9\frac{1}{6}\times\frac{3}{16}$

39) $6\frac{1}{8}\times 3\frac{5}{7}$

40) $2\frac{5}{9}\times 27\frac{2}{3}$

41) $\frac{9}{15}\div\frac{2}{3}$

42) $\frac{2}{7}\div\frac{6}{14}$

43) $\frac{7}{8}\div\frac{1}{4}$

44) $\frac{3}{5}\div\frac{1}{2}$

45) $\frac{3}{11}\div 7$

46) $\frac{27}{92}\div\frac{81}{46}$

47) $\frac{51}{28}\div\frac{34}{7}$

48) $\dfrac{\frac{1}{3}+\frac{3}{9}}{\frac{2}{17}+\frac{8}{51}}$

49) $\dfrac{\frac{1}{3}+\frac{2}{5}}{\frac{1}{8}+\frac{7}{16}}$

50) $\dfrac{\frac{3}{11}+\frac{2}{33}}{\frac{3}{5}+\frac{2}{3}}$

Solve each of the following problems:

51) A ten-foot board has lengths of $2\frac{1}{2}$ feet, 3 feet, and $2\frac{1}{4}$ feet cut from it. How much remains if the sawblade is 1/8 inch thick?

52) What is the perimeter (distance around) of the room below?

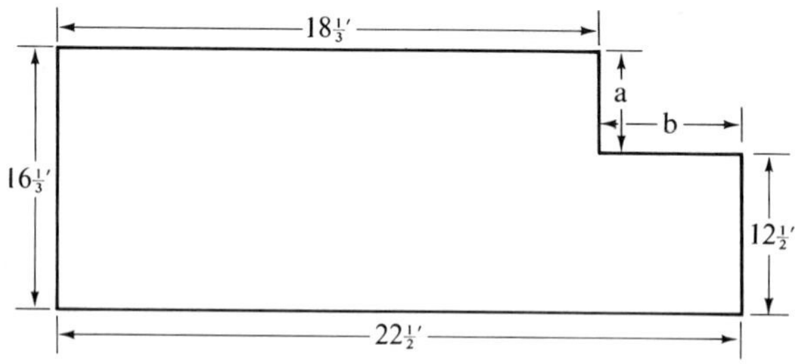

53) What are the lengths of the sides labeled *a* and *b* in Problem 52?

54) Determine the length of the bolt shown below.

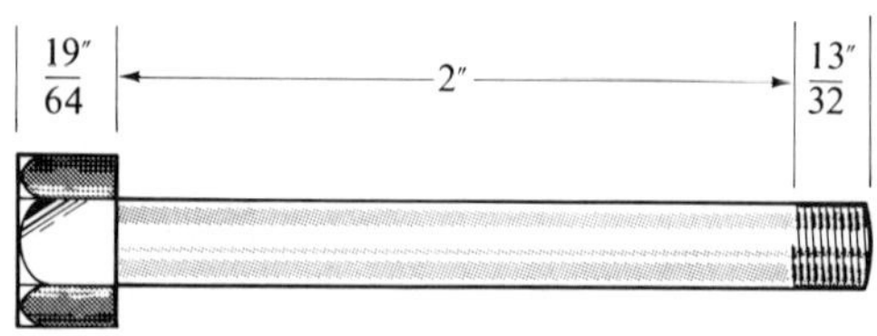

55) If machine bolts of a certain size cost 76 cents a dozen, how much do 3/4 of a dozen cost?

56) Several rectangles must be cut from a sheet of plastic 10 ft. by 10 ft. If the rectangles are to be $2\frac{1}{2}$ ft. by $3\frac{1}{3}$ ft., how many can be cut from the sheet?

57) A service station has a stock of 110 cases of oil. During the first week 3/5 of it was sold, during the following week 1/2 of the remainder was sold. How many quarts of oil remain if each case contains 24 quarts?

58) On the blueprint of a building, 1/8 inch represents 1 foot. How many inches represent the following dimensions: $15\frac{2}{3}$ ft., $27\frac{9}{10}$ ft., $14\frac{5}{8}$ ft., $96\frac{3}{5}$ ft., $126\frac{13}{16}$ ft.?

59) A journeyman painter can paint a room in 5 hours, and an apprentice can paint the same room in 8 hours. What part of the room can they paint in 1 hour working together? How long will it take both of them working together to paint the entire room? (Hint: The journeyman can paint 1/5 of the room in 1 hour, and an apprentice can paint 1/8 of the room in 1 hour. The painted room is considered as 1 job finished.)

60) A tank contains 4800 gallons of gasoline. Person A uses 1/24 of the gasoline; person B uses 1/8 of it; person C uses 1/16 of it; and person D uses 1/12 of it. How much gasoline remains in the tank?

1.21 Decimal Fractions

Just as whole numbers may be written by means of the place-value system, so may the set of fractions. If a *decimal point* (or period) is placed to the right of the last digit of a whole number (the ones place), each digit placed to the right of this decimal point also represents a place value. The first digit to the right is the tenths place and represents the number of one-tenths; the next digit to the right represents the number of one-hundredths; and so on. Each place to the right of the decimal is determined by increasing the denominator by a multiple of ten greater than the previous one.

Example 1. 27.3 means 2 tens + 7 ones + 3 tenths

$$\begin{aligned} 2 \text{ tens} &= 2 \times 10 = 20 \\ 7 \text{ units} &= 7 \times 1 = 7 \\ 3 \text{ tenths} &= 3 \times \tfrac{1}{10} = \tfrac{3}{10} \\ & \qquad 27\tfrac{3}{10} \text{ or } 27.3. \end{aligned}$$

Any decimal number having a last digit, that is, any terminating decimal can be written as a fraction. We will also see later that if a decimal number does not terminate but repeats some portion of itself over and over indefinitely, it also represents a fraction.

Example 2. $3.03 = 3\tfrac{3}{100}$.

Adding a zero behind the last digit of a decimal number does not change the value of the number. This is equivalent to multiplying numerator and denominator of a fraction by 10. That is, just as 33/100 = 330/1000, so .33 = .330.

1.22 Addition and Subtraction of Decimals

Adding or substracting decimals is performed exactly like the operation of addition or subtraction of whole numbers. Before the operation is performed the decimal points are placed one below the other. This can be done easily if zeroes are added until all the numbers have exactly the same number of decimal places.

Example 1. $50.7321 + 421.33$

$$\begin{array}{r} 50.7321 \\ +421.3300 \\ \hline 472.0621 \end{array}$$

Example 2. $497.61 - 362.851$

$$\begin{array}{r} 497.610 \\ -362.851 \\ \hline 134.759 \end{array}$$

1.23 Multiplying Decimals

Except for the handling of the decimal point, multiplication of decimal fractions is carried out in exactly the same manner as multiplication of whole numbers. *The position of the decimal point in the product is determined by counting the total number of decimal places in both factors; this sum determines the number of decimal places in the product.*

Example 1. 54.33×41.2

$$\begin{array}{r} 54.33 \\ \times\ 41.2 \\ \hline 10\ 866 \\ 54\ 33 \\ 2173\ 2 \\ \hline 2238.396 \end{array}$$

The product has three decimal places, since there are two places and one place respectively in the two factors.

1.24 Division of Decimals

Just as in multiplication, the numerical part of division of decimals is carried out in the same manner as division of whole numbers. However, we must be careful in placing the decimal point.

It is always easier to divide by a whole number. Thus if we remember that a fraction is a form of division, we can always make the denominator of the fraction (the divisor) a whole number by multiplying the numerator and the denominator by some multiple of ten. For instance, let us consider 21.85/2.3 as a complex fraction involving decimal fractions. This is equivalent to

$$\frac{21.85}{2.3}\cdot\frac{10}{10}=\frac{218.5}{23}.$$

Putting this in the form of a division problem, the decimal point in the quotient is placed directly above the decimal point in the dividend.

$$
\begin{array}{r}
9.5 \\
2.3\overline{)21.85} = 23\overline{)218.5} \\
207 \\
\hline
11\ 5 \\
11\ 5 \\
\hline
0
\end{array}
$$

Zeros can be added to the dividend, and the process of long division can be continued until a remainder of zero is obtained or until desired accuracy is reached in the quotient.

Example 1. $157.53 \div 53.4$

$$
\begin{array}{r}
2.95 \\
53.4\overline{)157.53} = 534\overline{)1575.30} \\
1068 \\
\hline
507\ 3 \\
480\ 6 \\
\hline
26\ 70 \\
26\ 70 \\
\hline
0
\end{array}
$$

In any operation involving decimal numbers, the decimal placement can be checked by approximating the answer to the problem. For instance, 96.9/47.5 is nearly 100/50 or 2. Therefore, when we find that the digits of the answer are 204, we know that the decimal point logically should follow the 2, since an answer of 204 or 20.4 would not be expected from our approximation.

1.25 Changing the Form of Fractions

To change a decimal fraction to an equivalent arithmetic fraction, we must consider the meaning of the decimal fraction. For instance, .531 is read, "five-hundred thirty-one thousandths." It can be written in fractional form as $\frac{5}{10}+\frac{3}{100}+\frac{1}{1000}$. Combining these fractions, the fractional form of .531 becomes

$$\frac{5}{10}+\frac{3}{100}+\frac{1}{1000}=\frac{500}{1000}+\frac{30}{1000}+\frac{1}{1000}=\frac{531}{1000}.$$

In this case the arithmetic fraction is in lowest terms.

Since arithmetic fractions are a form of division, they can be changed to decimal fractions by writing them in division form and carrying out the division following the procedure for division of decimals.

Example 1. Change 3/8 to a decimal.

$$\begin{array}{r} .375 \\ 8\overline{)3.000} \\ 2\;4 \\ \hline 60 \\ 56 \\ \hline 40 \\ 40 \\ \hline 0 \end{array}$$

Therefore, 3/8 = .375.

Since problems involving conversion of common fractions to decimals are common in any technical field, tables are often used to speed up this procedure. In this book, see the table provided on the end papers, but remember the procedure above for use when a table is not available or when the desired conversion is not found in the table.

1.26 Rounding

In many instances, only limited accuracy is desired in a decimal fraction. If the division does not result in a zero remainder before such accuracy is reached, the decimal is rounded off. A suitable rounding procedure is: if the digit of the place value immediately to the right of the desired accuracy is larger than 5, the preceding digit is increased by one and the remaining digits are dropped; if the digit is less than 5, the preceding digit is left the same and the following digits are dropped. If the digit to the right of the desired place value is exactly 5 and is the only nonzero digit to be dropped, and if the digit preceding is even, the digit remains unchanged and the following digits are dropped; if, however, the digit preceding the 5 is odd, the digit is increased by one and the remaining digits are dropped.

Example 1. 3.14159 rounded to five digits is 3.1416. The number has been rounded to the nearest ten-thousandth.

Example 2. 5.7715 rounded to four digits is 5.772. The number has been rounded to the nearest thousandth.

Example 3. 8.143 rounded to three digits is 8.14. The number has been rounded to the nearest hundredth.

Example 4. 44.3325 rounded to five digits is 44.332. The number has been rounded to the nearest thousandth.

In carrying out the division of decimals, it is advisable to obtain one more decimal place in the quotient than is desired and then round off to the desired accuracy.

1.27 Percent

The word *percent* means "by the hundred." Therefore, percentage refers to specific parts of one hundred. The symbol used to represent percent is %. Since our number system is based on multiples of 10, and 100 is (10)(10), operations with percent are directly related to our work with decimals.

The expression 5% means 5 parts out of 100, or the comparison of 5 to 100. As in our previous discussion, this can be written 5/100 or .05. Therefore, 5% of a quantity is the same as .05 of the quantity. For example, 5% of 450 is (.05)(450) = 22.50 = 22.5.

In general *percent can be changed to decimal form by replacing the percent symbol by* 1/100 *or* .01 *and multiplying, or simply by moving the decimal point two places to the left and removing the percent symbol.*

Example 1. 45% of a number is what decimal part of a number?

$45\% = 45 \cdot \frac{1}{100}$ or $(45)(.01) = .45$. Thus 45% of a number is .45 of the number.

Example 2. The interest rate on an investment is 6% per annum (per year). Find the amount of interest paid on an investment of \$5100 for one year.

6% of \$5100 = (.06)(\$5100) = \$306.00.

Example 3. A certain metal alloy is composed of 8% aluminum, 14% nickel, 24% cobalt, and 3% copper. If a casting of this metal alloy weighs 85 lb., what weight of each component does the casting contain?

Aluminum	= 8% of 85 lb.	= (.08)(85)	= 6.8	lb.
Nickel	= 14% of 85 lb.	= (.14)(85)	= 11.9	lb.
Cobalt	= 24% of 85 lb.	= (.24)(85)	= 20.4	lb.
Copper	= 3% of 85 lb.	= (.03)(85)	= 2.55	lb.

The percent accounted for so far is

$$8\% + 14\% + 24\% + 3\% = 49\%.$$

Since the total percent in any object is always 100%, the remaining metal not specified must be 100% − 49% = 51% of 85 or (.51)(85) = 43.35 lb. Our work may be checked by adding the component weights for a total of 85 lb.

$$6.8 \text{ lb.} + 11.9 \text{ lb.} + 20.4 \text{ lb.} + 2.55 \text{ lb.} + 43.35 \text{ lb.} = 85.00 \text{ lb.}$$

Decimal numbers can be changed to percent by multiplying the number by 100 *and placing % after the number, or simply by moving the decimal point two places to the right and placing % after the number.*

Example 4. .09 of a number is what percent of the number?
.09 = (.09)(100) or 9% of the number.

From the information given, we can write a general expression for finding percentage if we refer to the total number as the base.

$$\text{percentage} = (\text{percentage rate}) \times (\text{base})$$

or

$$\text{percentage rate} = \frac{\text{percentage}}{\text{base}}.$$

Example 5. Find 6% of 585.
In this case, the answer will be in percentage; 6% is the percentage rate, and 585 is the base. The expression then becomes percentage = (.06)(585) = 35.1. Notice that the percentage rate must be written in decimal form before the product can be found.

Another situation that arises frequently is finding the percent increase or decrease of a quantity over a period of time. For instance, if the yield of a field producing grain was 74 bushels per acre last year and 80 bushels per acre this year, we may wish to indicate the percent increase in yield. To do this, we first find the actual increase of one over the other: 81 − 75 = 6 bushels per acre increase. Since we started with a yield of 75 bushels per acre, this is our base and 6 is the percentage. That is,

$$\begin{aligned} \text{percentage rate} &= \frac{\text{percentage}}{\text{base}} \\ &= \frac{6}{75} \\ &= .08 = 8\%. \end{aligned}$$

Therefore, the percent increase is 8%.

Example 6. Joe earned \$4200.00 last year and \$4914.00 this year. What percent increase in salary did he receive?

$$\$4914.00 - \$4200.00 = \$714.00.$$

$$\text{Percentage rate} = \frac{714}{4200} = .17.$$

$$\text{Percent increase} = 17\%.$$

Example 7. A company hired 150 employees last year, and this year due to automation needs only 120 employees. What percent decrease in the staff is this?
Since we started with 150 employees, this is considered the base.

$$150 - 120 = 30$$

$$\text{Percentage rate} = \frac{30}{150} = .20.$$

$$\text{Percent decrease} = 20\%.$$

Fractional percents are found in exactly the same manner as that explained above. $8\frac{1}{4}\%$ is $8\frac{1}{4}$ or 8.25 parts out of a hundred or 825 parts out of 10,000. The decimal form of $8\frac{1}{4}\%$ is therefore .0825. Similarly, $6\frac{1}{2}\%$ in decimal form is .0650 or .065

Example 8. What is $7\frac{1}{4}\%$ of 9000?

$$\begin{aligned} 7\tfrac{1}{4}\% \text{ of } 9000 &= 7.25\% \text{ of } 9000 \\ &= .0725 \times 9000 \\ &= 652.5. \end{aligned}$$

Exercise 5

Write the following in decimal form.

1) Five-hundred five and fifty-five thousandths
2) Five-hundred twenty hundred-thousandths
3) Four-hundred fifty and seven-tenths
4) $263\frac{973}{1000}$
5) $\dfrac{475}{100{,}000}$

Write the following in fractional form and simplify:

6) .456
7) 9.375
8) .625
9) 450.7
10) .1275

Perform the following operations.

11) $8.296 + 3.21 + .297$
12) $865 - 262.91$
13) $26.975 - 12.888$
14) $57.9244 + 260.7 + 854.3$
15) $45{,}654.32 - 33{,}154.98$

Find the following sums after putting the problems into the proper form.

16)
```
  4.375
 26.28
 19.364
438.2
```

17)
```
  97.2
 145.38
9675.0
 452.175
```

18)
```
   667.4453
 1,475.3328
26,987.332
 7,445.6771
49,567.98
76,236.7092
 4,567.7785
```

Perform the following operations.

19) 6.5×7.2
20) 3.6×7.4
21) 7.77×88.8
22) 21.25×15.84
23) 7.21×6.392
24) $.217 \times .519$
25) $25.4 \times 26.3 \times 11.9$
26) $3.92(6.54 + 4.27)$
27) $7.2 \div 3.6$
28) $37.52 \div 1.2$
29) $102.003 \div 8.43$
30) $(2.6 \times 3.1) \div .002$

Divide the following and round your answer to the indicated accuracy:

31) $.84 \div .06$ to the nearest hundredth
32) $.0200 \div .28$ to the nearest thousandth
33) $.00059 \div .0127$ to the nearest tenth
34) $5.217729 \div .00326$ to the nearest thousandth
35) $45.67 \div 17.3$ to the nearest unit

Change each of the following to decimal form

36) $\frac{26}{1000}$
37) $3\frac{276}{1000}$
38) $\frac{2}{5}$
39) $\frac{5}{8}$
40) $\frac{7}{16}$

Calculate the following.

41) 5% of 97.3
42) 80% of 20% of 63
43) 130% of 34
44) $5\frac{3}{4}$% of 100

45) $7\frac{1}{2}\%$ of 550

46) What percent of 75 is 15?

47) 19 is what percent of 152?

48) If a quantity increases from 75 to 90, what percent increase is represented?

49) If an item costs 55 cents, and if a profit of 20% is desired, what should the selling price of the item be?

50) A car has a cylinder diameter of 3 in. If when the cylinder is rebored the diameter is increased by .060, what is the percent increase in the diameter?

51) Union scale for a certain job is $3.88 per hour. If the scale is raised 5% because of a cost-of-living increase, what is the new rate per hour?

52) The total precipitation last year on this date was 19.7 in. for the season. This year to date the rainfall has been 30.2 in. What is the percent increase in rainfall?

53) ABR wholesale sells lighting fixtures to contractors at list price less 40%, less 30%, less 3%, where each percentage is computed on the previous price. How much would a contractor expect to pay for a fixture listed at $80.00?

54) The 1970 Federal Income Tax tables list a tax rate of $1630.00 plus 28% of the amount over $8000.00 for a single taxpayer. How much would a man with an adjusted income of $9263 expect to pay?

1.28 Measurement Conversion

One very common application of operations with decimals in technical areas of study is converting from one system of measurement to another. In most areas of study in the United States the *English system* of measurement is used. In some technical areas and in the majority of the countries of the world the *metric system* of measurement is in use. A knowledge of the ways to convert from one system to another, as well as from one dimension to another within each system, is essential. In this text a complete table of English-to-metric and metric-to-English conversions is given in the tables inside the front cover. Their use is discussed below.

As an example of converting from one dimension to another, consider the problem of expressing 47,586 ft. in miles. Since 1 mi. = 5280 ft., then 1 ft. = 1/5280 mi. Therefore, 47,586 ft. = (47,586)(1/5280) = 9.0125 mi.

The process of converting from one dimension to another is much easier in the metric system, because each larger unit is a multiple of 10 times the smaller unit. For instance, suppose we wish to express 7.85

kilometers in centimeters; since 1 kilometer = 1000 meters and 1 meter = 100 centimeters, 7.85 km. = (7.85)(1000)(100) cm. = 785,000 cm.

When converting from one system to another, we must get the proper conversion factor from the table and multiply (or divide in some cases) by the given value. The following examples will illustrate the process.

Example 1. What is the size of a 16 millimeter wrench to the nearest hundredth inch?

$$\begin{aligned} 1 \text{ mm.} &= .0394 \text{ in.} \\ 16 \text{ mm.} &= 16(.0394) \text{ in.} \\ &= .6304 \text{ in.} \\ &= .63 \text{ in. to the nearest hundredth inch.} \end{aligned}$$

Example 2. It is 295 kilometers from Paris to Calais. How far is this distance in miles?

$$\begin{aligned} 1 \text{ km.} &= .621 \text{ mi.} \\ 295 \text{ km.} &= 295(.621) \text{ mi.} \\ &= 183.2 \text{ mi. to the nearest tenth mile.} \end{aligned}$$

Example 3. If a gas tank holds 5 liters, how many gallons does it hold?

$$\begin{aligned} 1 \text{ l.} &= 1.057 \text{ qt.} \\ 5 \text{ l.} &= 5(1.057) \text{ qt.} \\ &= \frac{5 \times (1.057)}{4} \text{ gal.} \\ &= 1.32 \text{ gal.} \end{aligned}$$

Example 4. Convert 750 cubic inches to cubic centimeters.

$$\begin{aligned} 1 \text{ in.} &= 2.54 \text{ cm.} \\ 1 \text{ cu. in.} &= (2.54)^3 \text{ cu. cm.} = 16.39 \text{ cu. cm.} \\ 750 \text{ cu. in.} &= 750(16.39) = 12{,}290 \text{ cu. cm. (also cc.)} \end{aligned}$$

Exercise 6

Convert the following to the indicated units of measure.

1) 4.62 mi. to inches
2) 75 in. to feet
3) 17 mi. to inches
4) 52,800 ft. to miles
5) .75 rd. to inches
6) 785 m. to kilometers

7) 4.56 cm. to millimeters
8) .30 m. to centimeters
9) 50 cm. to meters
10) 57.6 km. to decimeters
11) 456 oz. to pounds
12) 49 qt. to liters
13) 5 oz. av. to 5 ounces troy
14) 46 lb. to kilograms
15) 96 qt. to liters
16) 445 g. to ounces
17) 916 l. to gallons
18) 45 lb. to kilograms
19) 5 kg. to ounces
20) 75 l. to quarts
21) 15 sq. in. to square centimeters
22) 252 sq. cm. to square inches
23) 321 cu. cm. to cubic inches
24) 450 cu. cm. to cubic inches
25) 35 sq. rd. to square meters
26) 460 cu. in. to cubic yards
27) 27 cu. yd. to cubic feet
28) 85 acres to square meters
29) 975 cu. m. to cubic feet
30) 74 sq. mi. to square kilometers
31) In the Olympic games, one race is the 100-meter dash. How many yards is this race?
32) A certain weapon has a bore of 57 mm. What is the bore size in inches?
33) The thickness of a certain metal is 41.6 mm. What is the thickness of the metal in inches?
34) What is the size in inches of a 12 cm. wrench?
35) If 1 gal. of some material weighs 7.5 lb., how much does 1 kl. weigh?
36) Suppose you are in France and are driving an American-made automobile whose tank holds 25 gal. of gasoline. How many liters of gasoline will be required to completely fill the tank?

37) If a certain casting contains 15 kg. of iron, 450 g. of lead, and 3 kg. of copper, how many pounds does the casting weigh?

38) How much paint would be required to paint a rectangular room 25 ft. × 62 ft. if the walls are 8 ft. high? One gallon covers 520 sq. ft. and two coats are required.

39) How many quarts of oil would be required to fill a hydraulic reservoir manufactured in France with dimensions shown? Write the answer to the nearest quart.

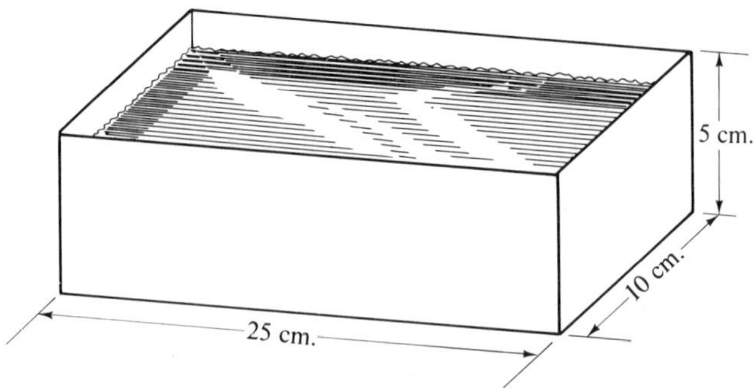

40) If a city lot has dimensions of 75 ft. by 150 ft., what is the area of the lot in acres?

Review Exercise Chapter 1

Perform the following additions:

1) $17 + 26 + 15 + 19$

2) $451 + 796 + 325$

3)
$$\begin{array}{r} 57921 \\ 72817 \\ 29455 \\ +54719 \\ \hline \end{array}$$

4)
$$\begin{array}{r} 526718 \\ 391457 \\ 913871 \\ +412685 \\ \hline \end{array}$$

Perform the following subtractions.

5) $\begin{array}{r} 457 \\ -398 \\ \hline \end{array}$

6) $\begin{array}{r} 9641 \\ -7894 \\ \hline \end{array}$

7) $\begin{array}{r} 657971 \\ -231418 \\ \hline \end{array}$

8) $\begin{array}{r} 77912 \\ -19799 \\ \hline \end{array}$

Perform the following multiplications.

9) $\begin{array}{r} 47 \\ \times 51 \\ \hline \end{array}$

10) $\begin{array}{r} 875 \\ \times 19 \\ \hline \end{array}$

11) $\begin{array}{r} 352 \\ \times 796 \\ \hline \end{array}$

12) $\begin{array}{r} 9,843 \\ \times 751 \\ \hline \end{array}$

Perform the following divisions.

13) $790 \div 12$

14) $37,500 \div 2500$

15) $478,964 \div 981$

Combine the following.

16) $1/3 + 1/4 + 5/36$

17) $71/72 - 7/80 - 11/45$

18) $13\frac{3}{8} - 12\frac{11}{18}$

Perform the following operations with decimals.

19) 93.93×87.58

20) $1743.298 + 875.42 - 96.851$

21) $457.8 \div 26.3$

22) $47.3 \times 18.5 \div 92.5$

23) 25% of a number is 48. The number is ______.

24) What percent of 325 is 91?

25) Find $5\frac{7}{8}\%$ of 29.5

Perform the following operations.

26) $(-3.5) + (-7.2) - (-6.3)$

27) $17 + (-6)(-4) - 7$

28) $\frac{-55}{11} + [(-3)(-2)] - \left(\frac{-4}{-8}\right)$

29) $(-65.3) \div (17.2)$

Convert the following.

30) 67 mi. to yards

31) 4 A. to square ft.

32) 27 gm. to ounces

33) 18 km. to miles

34) 975 cu. m. to cubic feet

35) 15 cm. to inches

2 Exponents and Radicals

2.1 Exponents

To write that a number is to be multiplied by itself many times could be quite tedious using the notation we have learned up to this point. To simplify this as well as other areas of mathematics *exponents* have been devised.

An exponent is a natural number* placed slightly above and to the right of another number indicating how many times the number is used as a factor. For instance $5 \cdot 5 \cdot 5$ is written 5^3, where 3 is called the exponent and 5, the number used as a factor, is called the *base* of the exponential expression. We say that 5^3 is "5 raised to the third power" or "5 *cubed.*"

100 or $10 \cdot 10$ can be written as 10^2, 1000 or $10 \cdot 10 \cdot 10$ can be written as 10^3, and so on. In literal number form, a^5 means $a \cdot a \cdot a \cdot a \cdot a$ and $(ab)^3$ means $ab \cdot ab \cdot ab$.

Example 1. $3 \cdot 3 \cdot 3 \cdot 3 = 3^4$ or three raised to the fourth power. 3 is the base, 4 the exponent.

Example 2. $6 \cdot 5 \cdot 5 \cdot 5 \cdot 5 = 6 \cdot 5^4$ or six times five raised to the fourth power. 5 is the base, 4 the exponent.

If a negative number is raised to a power, the sign of the product is determined by observing the exponents. If the exponent is odd, the product has an odd number of negative factors and the product is negative. If the exponent is even, the product has an even number of negative factors and the product is positive.

* We will restrict our definition of exponents to natural numbers for the present time. This will be modified later.

Example 3. $(-3)^3 = (-3)(-3)(-3) = -3[(-3)(-3)]$

$$= -3[9] = -27.$$

Example 4. $(-3)^4 = (-3)(-3)(-3)(-3)$

$$= [(-3)(-3)][(-3)(-3)] = 9 \cdot 9 = 81.$$

2.2 Multiplying with Exponential Numbers

When we multiply two exponential numbers having the same base, the product is as follows:

$$\begin{aligned} 5^2 \cdot 5^3 &= (5 \cdot 5)(5 \cdot 5 \cdot 5) \\ &= 5 \cdot 5 \cdot 5 \cdot 5 \cdot 5 \\ &= 5^5. \end{aligned}$$

or 5 used as a factor 5 times. In general, when two exponential numbers having the same base are multiplied, the exponent of the product is the sum of the exponents of the two numbers; the base remains the same.

Example 1. $2^2 \cdot 2^3 = 2^{2+3} = 2^5 = 2 \cdot 2 \cdot 2 \cdot 2 \cdot 2 = 32.$

Example 2. $a^3 \cdot a^4 = a^{3+4} = a^7.$

2.3 Dividing with Exponential Numbers

In much the same manner that we arrived at a general method for multiplying two exponential numbers, we can arrive at a general method of division for exponential numbers. The division of two exponential expressions having like bases, such as $4^3/4^2$, can be expressed as

$$\frac{4 \cdot 4 \cdot 4}{4 \cdot 4}.$$

Reducing the fraction gives us

$$\begin{aligned} \frac{4 \cdot 4 \cdot 4}{4 \cdot 4} &= \frac{4}{1} \\ &= 4 \text{ or } 4^1. \end{aligned}$$

The quotient involves the difference between the number of times the base is used as a factor in the numerator and the number of times it is used as a factor in the denominator. We can therefore generalize the division of numbers expressed in exponential form. That is, *when two exponential numbers having the same base are divided, the exponent of the quotient is found by subtracting the exponent of the divisor from the exponent of the dividend.* (The case in which the exponent of the numerator is less than the exponent of the denominator is discussed in Section 2.5.)

Example 1. $7^5 \div 7^3 = \dfrac{7^5}{7^3} = 7^{5-3} = 7^2 = 49.$

Example 2. $5^7 \div 5^3 = \dfrac{5^7}{5^3} = 5^{7-3} = 5^4 = 625.$

Example 3. $M^3 \div M = \dfrac{M^3}{M} = M^{3-1} = M^2.$

A further extension of this property is the situation where the exponents of the numerator and the denominator are the same. We know that if the base is any nonzero number, the following quotient must be one:

$$\frac{3^4}{3^4} = 1.$$

However, if we follow the rules of division for exponents established above, we get

$$\frac{3^4}{3^4} = 3^{4-4} = 3^0,$$

resulting in an exponent of zero. So

$$\frac{3^4}{3^4} = 3^0 = 1.$$

This leads to the definition that *any number (except zero) raised to the zero power is equal to one.*

Example 4. $\dfrac{8^3}{8^3} = \dfrac{8 \cdot 8 \cdot 8}{8 \cdot 8 \cdot 8} = 1$, or $\dfrac{8^3}{8^3} = 8^{3-3} = 8^0 = 1.$

Example 5. $\dfrac{a^5}{a^5} = a^0 = 1$, where $a \neq 0$. ($\neq$ means "is not equal to.")

Example 6. $(5280)^0 = 1.$

Example 7. $(x^2 + y^2)^0 = 1$, where x and y both are not zero.

2.4 Combined Operations with Exponential Numbers

Consider for a moment the product of two exponential expressions with different bases, both raised to the same power, such as $4^3 \cdot 3^3$. By using the definition of exponents and the associative and commutative properties of multiplication this number can be written as:

$$\begin{aligned} 4 \cdot 4 \cdot 4 \cdot 3 \cdot 3 \cdot 3 &= 4 \cdot 3 \cdot 4 \cdot 3 \cdot 4 \cdot 3 \\ &= (4 \cdot 3)(4 \cdot 3)(4 \cdot 3) \\ &= (4 \cdot 3)^3 \text{ or } 12^3. \end{aligned}$$

In general, if two numbers raised to the same power are multiplied, the product is the same as the product of the bases raised to the same power.

Example 1. $2^3 \cdot 3^3 \cdot 4^3 = (2 \cdot 3 \cdot 4)^3 = 24^3.$

Example 2. $a^5 \cdot b^5 \cdot c^5 = (abc)^5.$

The same reasoning may be used to arrive at an expression for the quotient of two numbers to the same power. That is, the quotient of two numbers of the same power is equal to the quotient of their bases raised to that power.

Example 3.
$$\begin{aligned} 3^5 \div 4^5 &= \frac{3^5}{4^5} \\ &= \frac{3 \cdot 3 \cdot 3 \cdot 3 \cdot 3}{4 \cdot 4 \cdot 4 \cdot 4 \cdot 4} \\ &= \frac{3}{4} \cdot \frac{3}{4} \cdot \frac{3}{4} \cdot \frac{3}{4} \cdot \frac{3}{4} \\ &= \left(\frac{3}{4}\right)^5. \end{aligned}$$

Combined multiplication and division problems are done in a similar way.

Example 4. $\dfrac{2^3 \cdot 3^3}{4^3} = \left(\dfrac{2 \cdot 3}{4}\right)^3 = \left(\dfrac{6}{4}\right)^3 = \left(\dfrac{3}{2}\right)^3.$

An exponential number raised to a power is another method of writing a repeated factor where the entire exponential number is used as a factor. $(a^2)^4$ would, therefore, be $a^2 \cdot a^2 \cdot a^2 \cdot a^2$, and the rules for multiplication of exponential numbers could be used to write the product as $a^{2+2+2+2} = a^8$. Since the addition of exponents can be shortened by multiplication, this product can be written as $a^{4 \cdot 2} = a^8$. In general, *the result of raising an exponential number to a power can be found by multiplying the exponents.*

Example 5. $(5^2)^3 = 5^{2 \cdot 3} = 5^6 = 15{,}625$.

Notice that the base of the exponent 3 is 5^2.

Example 6. $\left(\dfrac{2^2}{3^2}\right)^3 = \dfrac{2^6}{3^6} = \dfrac{64}{729}$.

Exercise 7

In Problems 1—6 generalize the discussion of exponents in the text. Example: $a^m \cdot a^n = a^{m+n}$.

1) $(a^m)^n$
2) a^0
3) $\left(\dfrac{a}{b}\right)^m$
4) $(a \cdot b)^m$
5) $\dfrac{a^m}{a^n}$; if $m > n$
6) $\dfrac{a^m}{a^n}$; if $m < n$

Evaluate each of the following using the rules of exponents:

7) 2^5
8) 3^3
9) 5^4
10) 3^5
11) $(-1)^6$
12) $(-2)(-3)^3$
13) $[(-2)(-3)]^3$
14) $(2.12)^2$
15) $(2.3)^2$
16) $2^2 \cdot 3$
17) $(2 + 3)^2$
18) $2^2 + 3^2$
19) $2 \cdot 5^2 + 2^2$
20) $[3(4 - 2)]^2$
21) $[3(4 - 2)^2]^2$
22) $(5601)^0$
23) $(5601 - 2632)^0$
24) $\left(\dfrac{5}{9}\right)^3$
25) $(.1)^3$
26) $(.02)^5$
27) $(.06 - .04 - .01)^3$
28) $(.215)^3 - (.301)^2$
29) $\left(-\dfrac{2}{3}\right)^4$
30) $-\left(\dfrac{3}{2^2}\right)^3$
31) $\left(-\dfrac{2}{3}\right)^2\left(\dfrac{3}{4}\right)$
32) $(2^2 \cdot 3^2)^2$
33) $[(2^2)^2]^2$
34) $\left(\dfrac{3}{5}\right)^2$

35) $\left[\left(\frac{2^2}{3}\right)^2\right]^2$

36) $2^2 \cdot 3^2 \cdot 4^2 \cdot 5^2$

37) Express 4^2 in terms of the base 2 and the appropriate exponent.

38) Express $4^2(8)$ in terms of the base 2 and the appropriate exponent.

Write each of the following without exponents:

39) xy^4

40) x^2y^3

41) $(xy)^4$

42) $\left(\frac{x^2}{y^2}\right)^2$

43) $(2x)^2$

Apply the rules of exponents to rewrite each of the following:

44) $x^2 \cdot x^3 \cdot x^4$

45) $x^2 \cdot x^3 \cdot x^4 \cdot x^5$

46) $x^7 \cdot x^6 \cdot x$

47) $(x^2y)(xy)$

48) $(x^2y^2)(x^3y^3)$

49) $(xy)^2(x^3y^3)$

50) $\frac{x^3}{x^6}$

51) $\frac{x^2 \cdot x^3 \cdot x^4 \cdot x^6}{x^2 \cdot x^3}$

52) $\frac{x^6}{x^2}$

53) $\frac{[(x^4)^2]^2}{(x^2)^2}$

54) $\left(\frac{x^4}{x^2}\right)^2$

55) $\frac{[(x^4)^2]^3}{[(x^3)^2]^4}$

2.5 Negative Exponents

To this point we have restricted exponents to the set of natural numbers. Let us now go back for a moment to the division of exponential numbers, and consider the case where the exponent of the denominator is larger than the exponent of the numerator. Following our original rule for division of exponential numbers,

$$4^3 \div 4^5 = \frac{4^3}{4^5} = 4^{3-5} = 4^{-2}.$$

Now

$$\frac{4^3}{4^5} = \frac{4 \cdot 4 \cdot 4}{4 \cdot 4 \cdot 4 \cdot 4 \cdot 4} = \frac{1}{4 \cdot 4}$$

$$= \frac{1}{4^2},$$

so we have

$$\frac{4^3}{4^5} = 4^{-2} = \frac{1}{4^2}.$$

We can therefore still maintain the consistency of our system of exponents by defining *a number with a negative exponent to be the reciprocal of the same number with a positive exponent.* That is,

$$5^{-2} = \frac{1}{5^2} = \frac{1}{25} \qquad \text{and} \qquad \frac{1}{3^{-2}} = \frac{1}{\frac{1}{3^2}} = 3^2 = 9.$$

The division of exponential numbers can now be done without regard to which exponent is larger.

In fact, with this extension of the properties of exponents, any integer can be used as an exponent and operations with these exponents follow all the rules for operations with integers.

Example 1. $\dfrac{a^3}{a^5} = a^{3-5} = a^{-2} = \dfrac{1}{a^2}.$

Example 2. $3^{-2} \cdot 4^{-2} = (3 \cdot 4)^{-2} = (12)^{-2} = \dfrac{1}{12^2} = \dfrac{1}{144}.$

Example 3. $(-5)^{-2} = \dfrac{1}{(-5)^2} = \dfrac{1}{(-5)(-5)} = \dfrac{1}{25}.$

Example 4. $[(3)^{-2}]^{-1} = \left[\dfrac{1}{3^2}\right]^{-1} = \left[\dfrac{1}{9}\right]^{-1} = 9.$

An exponential expression is in simplest form when all possible operations have been carried out and all exponents are positive. The reader should become completely familiar with the procedures of changing from one form to another and changing negative exponents to positive exponents and conversely.

Example 5. $\dfrac{(-3)^{-2}}{(3)^2} = \dfrac{\frac{1}{(-3)^2}}{3^2} = \dfrac{\frac{1}{9}}{9} = \dfrac{1}{9} \cdot \dfrac{1}{9} = \dfrac{1}{81}.$

Example 6. $\dfrac{x^5y^{-3}}{y^{-2}x^{-2}} = \dfrac{x^5 \cdot \frac{1}{y^3}}{\frac{1}{y^2} \cdot \frac{1}{x^2}} = \dfrac{\frac{x^5}{y^3}}{\frac{1}{x^2y^2}} = \dfrac{x^5}{y^3} \cdot \dfrac{x^2y^2}{1}$

$$= \frac{x^5 \cdot x^2 \cdot y^2}{y^3} = \frac{x^7}{y}$$

or

$$\frac{x^5y^{-3}}{y^{-2}x^{-2}} = x^{5-(-2)}y^{-3-(-2)}$$

$$= x^{5+2}y^{-3+2} = x^7y^{-1} = \frac{x^7}{y}.$$

2.6 Scientific Notation

Exponents are useful for writing very large numbers or very small numbers. Since each digit in our place value system represents a power of ten, a number such as 4,821 can be written as follows:

$$\begin{aligned} 4821 &= 4000 + 800 + 20 + 1 \\ &= (4 \times 1000) + (8 \times 100) + (2 \times 10) + (1 \times 1) \\ &= (4 \times 10^3) + (8 \times 10^2) + (2 \times 10^1) + (1 \times 10^0). \end{aligned}$$

This procedure can be extended to decimal numbers as well. That is,

$$\begin{aligned} .4821 &= \left(4 \times \frac{1}{10}\right) + \left(8 \times \frac{1}{100}\right) + \left(2 \times \frac{1}{1000}\right) + \left(1 \times \frac{1}{10000}\right) \\ &= (4 \times 10^{-1}) + (8 \times 10^{-2}) + (2 \times 10^{-3}) + (1 \times 10^{-4}). \end{aligned}$$

In this manner, a large number such as 484 billion, or 484,000,000,000, can be written in shortened notation as 4.84×10^{11}. 484 billionths or .000,000,484 can be written as 4.84×10^{-7}. In each case, the power of 10 is determined by counting the number of places to the original decimal from where the new decimal is placed. This method of writing a number as a number between one and ten times some integral power of ten is called *scientific notation*. Computation with the slide rule, logarithms, and the electronic computer are routinely handled in scientific notation.

Example 1. Put 457,000,000 in scientific notation.

$$457{,}000{,}000 = 4.57 \times 10^8.$$

Example 2. Put 0.0000000023 into scientific notation.

$$0.0000000023 = 2.3 \times 10^{-9}.$$

Example 3. Simplify by putting the numbers into scientific notation and then performing the operations:

$$\frac{32{,}000 \times 930{,}000 \times .00015}{.0008 \times 310 \times 450{,}000}$$

$$= \frac{3.2 \times 10^4 \times 9.3 \times 10^5 \times 1.5 \times 10^{-4}}{8.0 \times 10^{-4} \times 3.1 \times 10^2 \times 4.5 \times 10^5}$$

$$= \frac{\overset{.4}{\cancel{3.2}} \times \overset{\overset{1}{\cancel{3}}}{\cancel{9.3}} \times \overset{1}{\cancel{1.5}} \times 10^{4+5-4}}{\underset{1}{\cancel{8.0}} \times \underset{1}{\cancel{3.1}} \times \underset{\underset{1}{\cancel{3}}}{\cancel{4.5}} \times 10^{-4+2+5}}$$

$$= \frac{.4 \times 10^5}{10^3}$$

$$= .4 \times 10^2 = 40.$$

Example 4. Find the product of the following numbers and write the result in scientific notation.

$$\begin{aligned} 485{,}000 \times .0000028 &= (4.85 \times 10^5)(2.8 \times 10^{-6}) \\ &= (4.85)(2.8)(10^5)(10^{-6}) \\ &= (13.58)(10^{5-6}) \\ &= 13.58 \times 10^{-1} \\ &= 1.358. \end{aligned}$$

This process of scientific notation gives us a method for writing either very large or very small numbers as numbers between one and ten times some exponential number with 10 as a base. Scientific notation simplifies computation, as the reader can see if he does the problems of Examples 3 and 4 *without* using scientific notation.

Exercise 8

Write each of the following with positive exponents and evaluate:

1) x^{-2}
2) 5^{-2}
3) 7^{-1}
4) 3^{-4}
5) $5(5^{-2})$
6) 3^{-3}
7) xy^{-2}
8) $\dfrac{1}{2^{-4}}$
9) $(xy)^{-2}$
10) $x^{-2}y$

11) x^2y^{-2}

12) $\frac{y^{-5}}{y^{-3}}$

13) $\frac{3^{-2}}{2^{-3}}$

14) $\frac{4^{-2}}{2^{-4}}$

15) $\left(\frac{2}{3}\right)^{-2}$

16) $\left(\frac{4}{3}\right)^{-2}$

17) $x^{-6}x^4$

18) $\frac{ab^{-1}}{a^{-2}b}$

19) $(-3x^2)^{-3}$

20) $\frac{(ab)^{-1}}{a^{-2}b}$

21) $x^{-3}y^4$

22) $(-3)^{-2}(y^{-2})^2$

23) $(-2y)^{-3}$

24) $2x^{-3}(2x)^{-3}$

25) $4x^2(3x^{-2})^2$

26) $\frac{(2.36)^{-2}}{(1.20)^{-2}}$

27) $(5y^2)^{-1}(25y)$

28) $\left(\frac{1.44}{12}\right)^{-2}$

29) $3x^{-3}(2x^{-3})^2$

30) $2^{-3}x^{-2}(3x^{-2})^2$

Write each of the following using scientific notation:

31) 2641

32) 295

33) 2,621,000

34) 758,000

35) 3,927,176,000

36) 986.3

37) 297 followed by thirty zeros

38) 75 trillion

39) 176 billion

40) .000,000,000,075,300

41) 24 hundred-thousandths

42) .00000864

43) .000000297

44) .000035

45) 456,000,000,000,000,000,000

Divide the following after expressing them in scientific notation:

46) $\frac{263{,}000}{292{,}000{,}000}$

47) $\frac{.00373}{476{,}000}$

48) $\frac{(.0021)(.952)}{(.327)(.326)}$

Solve the following:

49) If a shim is .125 inch thick, use scientific notation to express the thickness of 5000 of these shims.

50) Wrought iron expands .00001166 inch for each degree Centigrade the temperature increases. If the temperature increases by 12.4° C., find the expansion of the metal and express the result in scientific notation.

51) The limitations of measurement of precision gage blocks are from .000002 inch to .0000008 inch. Express these measurements in scientific notation.

52) One year a large corporation reported profits of $375,000, which represented 1.25% of their total sales. What were their total sales? Express this in scientific notation.

2.7 Square Root

The *square root* of a number, indicated by using the *radical symbol* $\sqrt{\ }$, is one of two equal factors of a number. That is, to find the square root of a number we must find a number which when multiplied times itself, or squared, results in the given number. For example, $2 \cdot 2 = 4$, so 2 is one of two equal factors of 4 and $\sqrt{4} = 2$. In the same manner, $\sqrt{9} = 3$ and $\sqrt{16} = 4$. The number whose root is being found is called the *radicand.* The product of a number times itself is called the *square* of that number. For example, the square of 4 is 16.

If the numbers are large the square roots of the numbers cannot be found simply by observation. One of the more common methods for calculating square roots is illustrated by the following example.

Example 1. Find the square root of 625.

$$\begin{array}{r|l} & \ 2\ \ \ 5 \\ & \sqrt{6\ 25} \\ & \ 4 \\ 40 & 2\ 25 \\ 45 & 2\ 25 \\ & \qquad 0 \end{array}$$

First, beginning at the decimal point, we divide the number into groups of two digits each. The group furthest left need not have two digits in it. We then find the largest number which when squared will be smaller than or equal to the given group. In this example that number is 2. This number is placed above the radical, then squared, and the square is subtracted from the group of numbers. Here, $6 - 4 = 2$. Bring down the next pair of numbers. We then double the partial root ($2 \times 2 = 4$), multiply it by 10, and use it as a trial divisor by dividing it into the new number (40 into 225). Since 40 is contained in 225 five times, the true divisor is 45. When 45 is multiplied by 5, the product is placed under 225 and the difference is taken again. In this case, the difference is zero and the desired root is 25.

In the above example, if the difference had not been zero, the decimal would have been placed in the root directly above the decimal in the radicand and groups of two zeros would have been brought down and the process continued until a remainder of zero was obtained or until the desired accuracy was reached.

Example 2. Find $\sqrt{1049.76}$.

```
            3  2. 4
         √10 49.76
            9
 60     | 1 49
 62     | 1 24
    640   |25 76
    644   |25 76
                0
```

So $\sqrt{1049.76} = 32.4$.

Example 3. Find $\sqrt{25.149}$.

```
              5. 0  1  4
           √25.14 90 00
             25
 100       | 0 14
 1000      | 0 00
     1000    |14 90
     1001    |10 01
     10020     |4 89 00
     10024     |4 00 96
                  88 04
```

Therefore 5.01 is the square root of 25.149 to the nearest hundredth. This process can be continued until further accuracy is reached.

2.8 Square Root Tables

Square roots are so commonly used that it is often advisable to have tables of the roots already calculated for our use. Such a table of numbers from 1 to 50 is included in the Appendix (Table I). To find the square root of a number, find the desired number under the column headed n. The square root is found directly to the right of the number under the column headed $\sqrt{n}$. For instance, $\sqrt{2}$ can be found directly to the right of 2 as 1.414. These tables are accurate to the nearest thousandth.

The tables can be extended somewhat by using the column headed $\sqrt{10n}$. For instance, $\sqrt{200} = 14.14$ can be found to the right of 20 under the column headed $\sqrt{10n}$.

Exercise 9

Without the use of tables, find the square root of each of the following numbers:

1) 529
2) 144
3) 441
4) 12.96
5) 17.64
6) 625
7) 44,944
8) 1024
9) 841
10) 961
11) 8464
12) 48,841
13) 10.304 to the nearest tenth
14) 46,656 to the nearest tenth
15) 27.391 to the nearest tenth
16) 54.813 to the nearest hundredth
17) 271.2521 to the nearest hundredth

Use the tables in the Appendix to find the indicated root in each of the following:

18) $\sqrt{32}$
19) $\sqrt{220}$
20) $\sqrt{160}$
21) $\sqrt{180}$
22) $\sqrt{480}$
23) $\sqrt{490}$
24) $\sqrt{350}$
25) $\sqrt{17}$
26) $\sqrt{450}$
27) $\sqrt{260}$
28) $\sqrt{500}$
29) $\sqrt{410}$
30) $\sqrt{330}$
31) $\sqrt{48}$
32) $\sqrt{9}\sqrt{16}$
33) $\sqrt{9 \cdot 16}$
34) $\sqrt{144}$

35) Is there any apparent relationship between the three previous problems? Can you generalize the statement?

Evaluate the following:

36) $(\sqrt{4})^3$
37) $(\sqrt{36})^2$
38) $-\sqrt{16}$
39) $(-\sqrt{16})^2(\sqrt{9})^2$
40) $(\sqrt{144})^3$
41) $(\sqrt{25})^6$
42) $[(\sqrt{4})^2]^3$
43) $[(\sqrt{4})^2(\sqrt{9})^3]^2$
44) $[\sqrt{(4)^3}]^3$

45) In a later chapter it will be shown that the diagonal of a square of side a is given by $\sqrt{2a^2}$. How long is the diagonal of a square 5 inches on a side? Write the answer to the nearest hundredth.

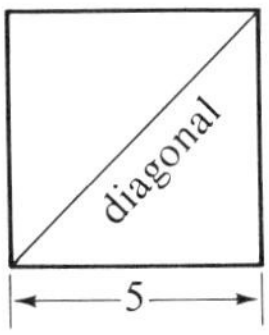

2.9 Irrational Numbers

In many cases the square root of a number will never result in a terminating decimal, nor will the decimal approximation ever repeat itself. $\sqrt{2}$ is such a number. To nine decimal places, the square root of 2 is 1.414213562. Many numbers other than square roots fall in this category, that is, they cannot be written as a terminating or repeating decimal. If this is so, then they cannot be written as the quotient of integers and hence they are not rational numbers. The set of all numbers which are not rational, that is, which cannot be written as the quotient of two integers, is called the set of *irrational numbers.*

In the following sections, many of the numbers we will work with are irrational. The set consisting of the set of rational numbers together with the set of irrational numbers is called the set of *real numbers.* Every real number corresponds to a point on the number line, and every point on the number line can be represented by a real number. All the properties discussed so far apply to the set of real numbers. All the sets of numbers discussed so far are subsets of the set of real numbers.

2.10 Radicals

We have already discussed a square root as one of two equal factors of a number. When we consider one of three equal factors of a number, this factor is called the *cube root* of a number. (Recall that $a \cdot a \cdot a$ or a^3 is "a cubed.") The cube root is written $\sqrt[3]{\ }$. The small 3 is called the *index* of the radical expression. Similarly, one of four equal factors of a number is called the fourth root of the number, written $\sqrt[4]{\ }$. In general, one of n equal factors of a number is called the *nth root* of the number. Thus, the nth root of A is written $\sqrt[n]{A}$, where n is the *index*, $\sqrt{\ }$ is the *radical* and A is the *radicand*. If the index is 2, indicating a square root, the index is not written but is understood. Some basic properties of radical expressions will be discussed briefly in this section.

To this point, all the square roots that we have considered have been positive. We see, however, that -2, as well as $+2$, is one of two equal factors of 4. That is, $(-2)(-2) = 4$ and $(+2)(+2) = 4$. Unless otherwise marked, the radical symbol indicates the positive root. This is known as the *principal root* of a number. If the negative root is desired, a minus sign is used in front of the radical, and if both roots are desired, a dual symbol ($\pm$) is used. This dual symbol is read "plus or minus."

Example 1. $\sqrt{9} = 3.$

$$-\sqrt{9} = -3.$$

$$\pm\sqrt{9} = \pm 3.$$

Recall that any number to an even power is positive. For this reason, within the set of real numbers we cannot find the square root, or for that matter, any even root, of a negative number. On the other hand, because a negative number to an odd power is negative, an odd root of any negative number can be found in the set of real numbers.

Example 2. $\sqrt{-16}$ does not exist in the set of real numbers since $(-4)(-4) = +16$ and $(+4)(+4) = +16$. That is, there is no real number that can be multiplied by itself and result in a product of -16.

Example 3. $\sqrt[3]{-27} = (-3)$ since $(-3)(-3)(-3) = -27$.

Let us formalize the connection between powers and roots. If the square root of a number is squared, the result is the radicand. That is, $(\sqrt{4})(\sqrt{4}) = (2)(2) = 4$. Also, if the cube root of a number is cubed it results in the radicand, and in general, when the nth root of a number is raised to the nth power the result is the radicand. Symbolically this could be written as $(\sqrt[n]{A})^n = A$. For this to be true A must be positive when n is even.

Example 4. $(\sqrt{450})(\sqrt{450}) = (\sqrt{450})^2 = 450.$

Example 5. $(\sqrt[4]{26})^4 = (\sqrt[4]{26})(\sqrt[4]{26})(\sqrt[4]{26})(\sqrt[4]{26}) = 26.$

This concept can be used to simplify radical expressions, using the properties of numbers previously discussed.

Example 6. $(\sqrt{6})^5 = [(\sqrt{6})(\sqrt{6})][(\sqrt{6})(\sqrt{6})] \cdot \sqrt{6}$

$$= 6 \cdot 6 \cdot \sqrt{6} = 36\sqrt{6}.$$

We know that

$$(\sqrt{5})(\sqrt{5}) = 5 = \sqrt{25}.$$

Similarly, as long as the index of the expressions are the same, the product of any two real roots is equal to the root of the product of the radicands.

Example 7. $\sqrt{5} \cdot \sqrt{3} = \sqrt{15}$.

Example 8. $\sqrt[5]{8} \cdot \sqrt[5]{5} = \sqrt[5]{40}$.

Using the same reasoning, the quotient of two real roots is equal to the root of the quotient of the radicands, where the divisor is not zero. Like the others, this property may be used to simplify radicals.

Example 9. $\dfrac{\sqrt{10}}{\sqrt{5}} = \sqrt{\dfrac{10}{5}} = \sqrt{2}$.

Example 10. $\sqrt{\dfrac{4}{5}} = \dfrac{\sqrt{4}}{\sqrt{5}} = \dfrac{2}{\sqrt{5}}$.

Example 11. $\dfrac{\sqrt{a}}{\sqrt{b}} = \sqrt{\dfrac{a}{b}}$, where $b \neq 0$.

The properties illustrated here are very useful in simplifying radical expressions such as $\sqrt{175}$, a root which cannot be found from Table I of the Appendix, for example. First factoring the radicand, we get

$$\begin{aligned}\sqrt{175} &= \sqrt{5 \cdot 5 \cdot 7} \\ &= \sqrt{5} \cdot \sqrt{5} \cdot \sqrt{7} \\ &= 5\sqrt{7}.\end{aligned}$$

We can find $\sqrt{7}$ in the table and multiply this decimal approximation by 5 to find the approximate square root of 175. That is,

$$\sqrt{7} = 2.646$$

and

$$5(2.646) = 13.230.$$

Example 12. Find $\sqrt{578}$ using Table I.

$$\begin{aligned}\sqrt{578} &= \sqrt{17 \cdot 17 \cdot 2} \\ &= \sqrt{17} \cdot \sqrt{17} \cdot \sqrt{2} \\ &= 17\sqrt{2} \\ &= 17(1.414) \\ &= 24.038.\end{aligned}$$

Example 13. Find $\sqrt{750}$ using Table I.

$$\sqrt{750} = \sqrt{5 \cdot 5 \cdot 5 \cdot 3 \cdot 2} = \sqrt{5 \cdot 5} \cdot \sqrt{5 \cdot 3 \cdot 2}$$
$$= 5\sqrt{30} = 5(5.477) = 27.385.$$

Example 14. Find $\sqrt{256}$.

$$\sqrt{256} = \sqrt{2 \cdot 2 \cdot 2 \cdot 2 \cdot 2 \cdot 2 \cdot 2 \cdot 2}$$
$$= \sqrt{2^4 \cdot 2^4}$$
$$= 2^4$$
$$= 16.$$

Exercise 10

Use the properties of radicals discussed in the preceding sections to evaluate each of the following. Use tables where necessary.

1) $\sqrt{216}$ 2) $\sqrt{288}$ 3) $\sqrt{255}$

4) $\sqrt{135}$ 5) $\sqrt{384}$ 6) $\sqrt{364}$

7) $\sqrt{296}$ 8) $\sqrt{513}$ 9) $\sqrt{180}$

Simplify each of the following using the properties of radicals.

10) $\dfrac{\sqrt{50}}{\sqrt{2}}$ 11) $\sqrt{15} \cdot \sqrt{3}$ 12) $\sqrt{15}\sqrt{5}$

13) $\sqrt{14}\sqrt{7}$ 14) $\dfrac{\sqrt{2}}{\sqrt{200}}$ 15) $\sqrt{7}\sqrt{343}$

16) $\dfrac{\sqrt{75}}{\sqrt{3}}$ 17) $\dfrac{4\sqrt{80}}{\sqrt{20}}$ 18) $\sqrt{1250}$

19) $\dfrac{\sqrt{27}}{\sqrt{3}}$ 20) $\sqrt{125}$ 21) $\dfrac{4}{\sqrt{400}}$

22) $\dfrac{\sqrt{875}}{25}$ 23) $\dfrac{\sqrt{88}\sqrt{2}}{\sqrt{11}}$ 24) $\dfrac{\sqrt{90}}{\sqrt{10}}$

Evaluate the following. Round the answers to the nearest tenth.

25) $\dfrac{\sqrt{7} - \sqrt{243}}{6}$ 26) $\dfrac{\sqrt{2} + \sqrt{5}}{\sqrt{3}}$ 27) $\dfrac{\sqrt{2} - \sqrt{5}}{\sqrt{6}}$

2.11 Fractional Exponents

If we observe some of the basic properties of exponents, we notice that $x^a \cdot x^a = x^{a+a} = x^{2a}$.

If $x^{2a} = x^1$, this implies that $a = 1/2$. Therefore, $x^{1/2} \cdot x^{1/2} = x$. We have previously defined one of two equal factors of x to be the $\sqrt{x}$. That is $(\sqrt{x}) \cdot (\sqrt{x}) = x$. Also $(x^{1/3})^3 = x^{1/3 \cdot 3} = x$ and $(\sqrt[3]{x})^3 = x$ and in general $(x^{1/n})^n = x^{1/n \cdot n} = x$ and $(\sqrt[n]{x})^n = x$. Since the factors of a number are unique, the need for consistency in our number system makes the following definition essential.

A number to the $1/n$ power is the nth root of the number. In symbols, $x^{1/n} = \sqrt{x}$.

Example 1. $4^{1/2} = \sqrt{4} = 2$.

Example 2. $3^{1/3} = \sqrt[3]{3}$.

Example 3. $27^{1/3} = \sqrt[3]{27} = 3$, since $3 \cdot 3 \cdot 3 = 27$.

Example 4. $16^{1/4} = \sqrt[4]{16} = 2$, since $2 \cdot 2 \cdot 2 \cdot 2 = 16$.

Now, since raising an exponential number to some power requires a product of the exponents, $(a^{1n})^m = a^{m/n}$. This means that $(\sqrt[n]{a})^m = a^{m/n}$. If a is positive, $a^{1/n \cdot m} = a^{m \cdot 1/n}$ and $a^{m \cdot 1/n} = \sqrt[n]{a^m}$.

Example 5. $4^{3/2} = (\sqrt{4})^3 = (2)^3 = 8$ also, since $4 > 0$,

$$4^{3/2} = \sqrt{(4)^3} = \sqrt{64} = 8.$$

Example 6. $3^{3/4} = (\sqrt[4]{3})^3$ or $\sqrt[4]{3^3} = \sqrt[4]{27}$.

2.12 Imaginary Numbers

We have previously required that the radicand of any square root be greater than zero, since the square root of a negative number does not exist in the set of real numbers. In certain areas of mathematics and particularly in the field of electronics, solutions to equations often result in negative radicands. Because of the occurence of negative radicands, the *imaginary numbers* were developed which resulted in the set of *complex numbers*.

A *pure imaginary number* is one which involves the square root of a negative number such as $\sqrt{-4}$. In order to operate with imaginary numbers, $\sqrt{-1}$ is defined as equal to i. (This numeral is sometimes called j.) Since $\sqrt{-1} = i$, $(\sqrt{-1})^2 = i^2 = -1$. Also, $i^3 = (-1)i = -i$ and $i^4 = i^2 \cdot i^2 = (-1)(-1) = 1$. With this definition along with the laws for working with radicals we can now work with negative square roots. Consider the following examples.

Example 1. $\sqrt{-4} = \sqrt{-1}\sqrt{4} = i\sqrt{4} = i \cdot 2 = 2i.$

Example 2.
$$\begin{aligned}\sqrt{-6}\sqrt{-3} &= i\sqrt{6} \cdot i\sqrt{3} = i^2 \cdot \sqrt{3 \cdot 3 \cdot 2}\\ &= i^2 \cdot 3\sqrt{2} = -3\sqrt{2}.\end{aligned}$$

Example 3. $$\frac{\sqrt{-27}}{\sqrt{-3}} = \frac{i\sqrt{27}}{i\sqrt{3}} = \sqrt{\frac{27}{3}} = \sqrt{9} = 3$$

Example 4. $$\frac{1}{\sqrt{-2}} = \frac{1}{i\sqrt{2}} = \frac{i\sqrt{2}}{i\sqrt{2}} \cdot \frac{1}{i\sqrt{2}} = \frac{i\sqrt{2}}{i^2 \cdot 2} = \frac{i\sqrt{2}}{-2} = \frac{-i\sqrt{2}}{2}$$

A radical expression is considered in simplest form when the denominator contains no radical expressions and no imaginary numbers. The process shown in Example 4 is called *rationalizing the denominator*.

2.13 Complex Numbers

When the imaginary numbers are combined with the set of real numbers, the result is a number of the form $a + bi$, where a is an element of the set of real numbers and bi is an element of the set of pure imaginaries. Since a and bi belong to different sets, the addition can only be indicated.

Example 1. $5 + 3i$, $2 - i$, $4 + \frac{5}{2}i$ are all examples of complex numbers. Complex numbers are added or subtracted by combining their real parts and combining their imaginary parts.

Example 2. $(3 + 5i) + (2 - i) = (3 + 2) + (5 - 1)i = 5 + 4i.$

Example 3. $(2 - i) - (5 + 2i) = (2 - 5) + (-1 - 2)i = -3 - 3i.$

Example 4. $(6 - 2i) - (3 - 5i) = (6 - 3) + [-2 - (-5)]i = 3 + 3i.$

The product of two complex numbers is found by using the distributive law as shown in the following examples.

Example 5. $(2 + 3i)(1 - i) = 2(1 - i) + 3i(1 - i) =$
$2 - 2i + 3i - 3i^2 = 2 + i - 3(-1) = 5 + i.$

Example 6. $(5 + 2i)(3 + 4i) = 5(3 + 4i) + 2i(3 + 4i) =$
$15 + 20i + 6i + 8i^2 = 15 + 26i + 8(-1) = 7 + 26i.$

Further operations with the set of complex numbers will be discussed in Chapter 4.

Exercise 11

Write each of the following in the form $a + bi$ or $a - bi$

1) -9
2) -25
3) $\dfrac{2 - \sqrt{-4}}{2}$
4) $\dfrac{4\sqrt{-9}}{6}$
5) $2 + \sqrt{-18}$
6) $3 - \sqrt{-48}$
7) $\dfrac{6 + \sqrt{-72}}{12}$
8) $\dfrac{5 - \sqrt{-50}}{5}$
9) $\dfrac{2 + \sqrt{-64}}{3}$
10) $\dfrac{4 + \sqrt{-63}}{3}$

Carry out the following operations. Write the answers in the form $a + bi$ or $a - bi$.

11) $(3 + i) + (6 + 4i)$
12) $(2 - 7i) + (2 + 5i)$
13) $(6 + 3i) + (2 + 4i) + (5 + 2i)$
14) $(3 + 6i) - (3 + 2i)$
15) $(2 + 7i) - (5 + 3i)$
16) $[(5 + 2i) + (3 + 4i)] - (6 + 2i)$
17) $(2 + \sqrt{-9}) + (6 + \sqrt{-4})$
18) $(5 - \sqrt{-11}) + (2 + \sqrt{-44})$
19) $(3 + \sqrt{2}i) + (7 + \sqrt{-8})$
20) $(7 + \sqrt{19}i) + (-7 - \sqrt{76}i)$
21) $\dfrac{3 - \sqrt{-27}}{3} + \dfrac{12 - \sqrt{-48}}{3}$

Review Exercise Chapter 2

1) In an exponential expression such as $5x^3$, x is called the _____ and 3 is called the _____.

2) When exponential numbers with like bases are multiplied the exponents are _____; when such numbers are divided the exponents are _____.

3) What does A^{-3} mean?

4) When a very large or a very small number is written as a number between one and ten times some integral power of ten, the procedure is called _____ _____.

5) To raise an exponential number to a power, the exponents are _____.

6) In the expression $\sqrt[3]{6}$, 3 is called the _____, $\sqrt{\ }$ is called the _____, and 6 is called the _____.

7) The product of the nth root of 2 numbers is the nth root of the _____ of their radicands.

Perform the following operations and simplify.

8) $x^5 \cdot x^3 \cdot x^{-2}$ 9) $5^2 \cdot 5^{-6}$ 10) $4 \cdot 8^2$

11) $\dfrac{x^4}{x^3}$ 12) $\dfrac{(xy)^3}{x^2}$ 13) $\dfrac{(x^2y)^{-3}}{y^{-2}}$

Find the following without tables.

14) $\sqrt{729}$ 15) $\sqrt{1024}$

Simplify the following numbers, leaving them in radical form.

16) $5\sqrt{75}$ 17) $\dfrac{\sqrt{80}}{\sqrt{20}}$ 18) $\sqrt{\dfrac{7}{25}}$

19) $\sqrt{27}\sqrt{3}$ 20) $\sqrt{60}\sqrt{30}$

Using Table I, find decimal approximations for the following.

21) $\sqrt{400}$ 22) $\sqrt{73}$ 23) $\sqrt{780}$

24) $\sqrt{560{,}000}$ 25) $\sqrt{4800}$

Simplify using the laws of exponents.

26) $9^{1/2}$ 27) $27^{2/3}$ 28) $8^{3/4}$

29) $16^{3/2}$ 30) $81^{1/4}$

Write in the form $a + bi$.

31) $3 + \sqrt{-9}$

32) $\dfrac{10 - \sqrt{-75}}{5}$

33) $\dfrac{2 + \sqrt{-8}}{\sqrt{2}}$

Carry out the following operations.

34) $(4 - i) + (2 + i)$

35) $(6 - 2i) + (-4 - i)$

36) $(1 + i) - (4 - 2i)$

37) $4 + \sqrt{-8} - 3 - \sqrt{-2}$

38) $\dfrac{3 + \sqrt{-9}}{3} - \dfrac{10 - \sqrt{-75}}{5}$

3
Slide Rule

3.1 Parts of the Slide Rule

The slide rule is a mechanical device for mathematical calculations. It is not used in addition or subtraction, but greatly simplifies finding approximate answers for many other operations. In this section we will discuss the use of the slide rule for multiplication, division, and square roots and cube roots of numbers.

The slide rule consists of three parts: the *stator* or *body* of the rule, the *slide*, and the *cursor* or *indicator*. The line on the indicator is called the *hairline*. (See Figure 3.1.)

Figure 3.1

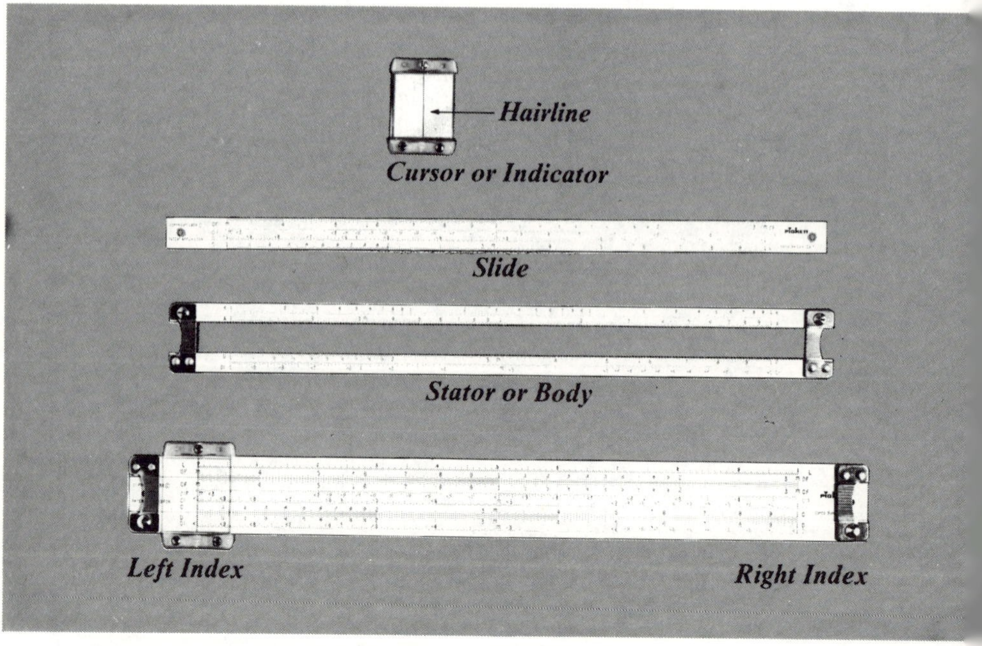

Slide Rule courtesy of Pickett Industries, Santa Barbara, California.

Each scale on the rule is labeled with a letter or letters and will be referred to by these letters. Both ends of each scale are labeled with the numeral 1. These are called the *left index* and *right index* of the scale, respectively.

Notice that some scales are made up of 2 or 3 identical, repeating parts. The A scale, for instance, is composed of 2 identical sections. The K scale is made up of three identical sections.

In order to use the rule effectively, the user must know how to read the scales, how to set the slide and the indicator for each operation, and where to position the decimal point in the answer.

3.2 Reading the Scales

Slide rule scales are read much like the scales on an ordinary ruler. The large divisions of the scale, called *primary divisions*, are indicated by lines numbered from 1 to 10. (See Figure 3.2.) *Secondary divisions* are marked by smaller lines, and depending on the scale and position on the scale, may or may not be numbered. Still smaller divisions called *tertiary divisions* are marked by smaller lines between the secondary divisions.

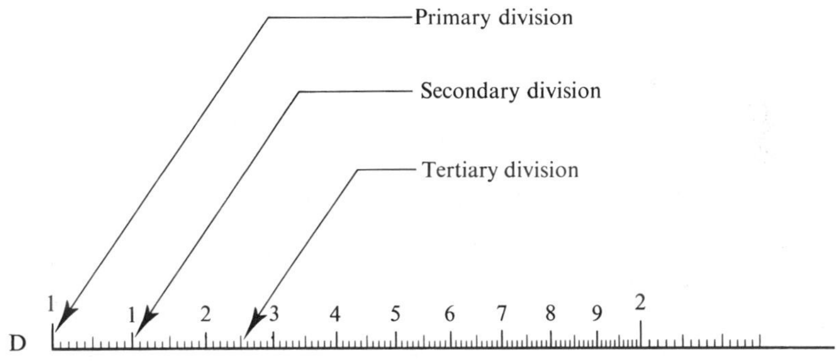

Figure 3.2

To mark the position of a number on any scale, we can either slide the hairline over the desired position, or move the slide so that either the left or the right index lies at the desired position.

Using primary, secondary, and tertiary divisions, we can find a point on a scale corresponding to any one-, two-, or three-digit number.

Example 1. Find 95 on the C and D scales, using the hairline.

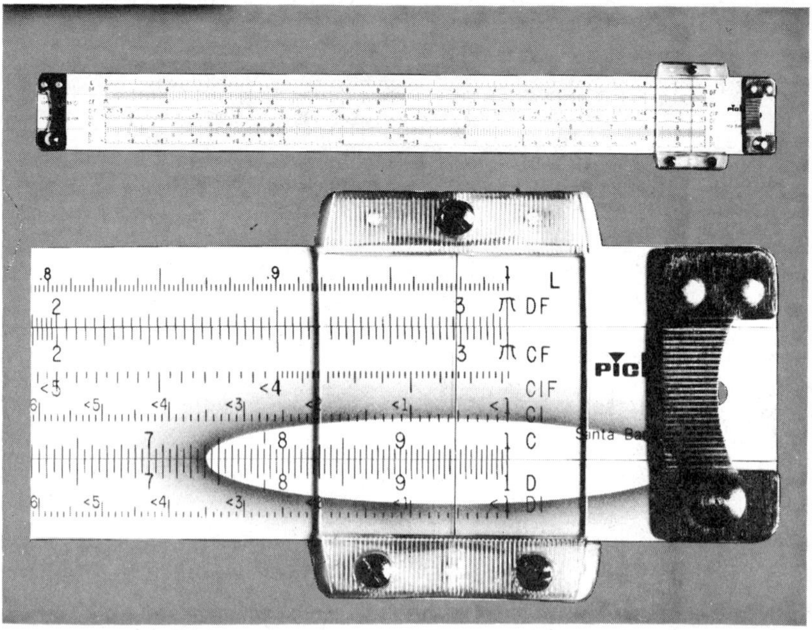

Example 2. Find 20 on the D scale, using the left index of the C scale.

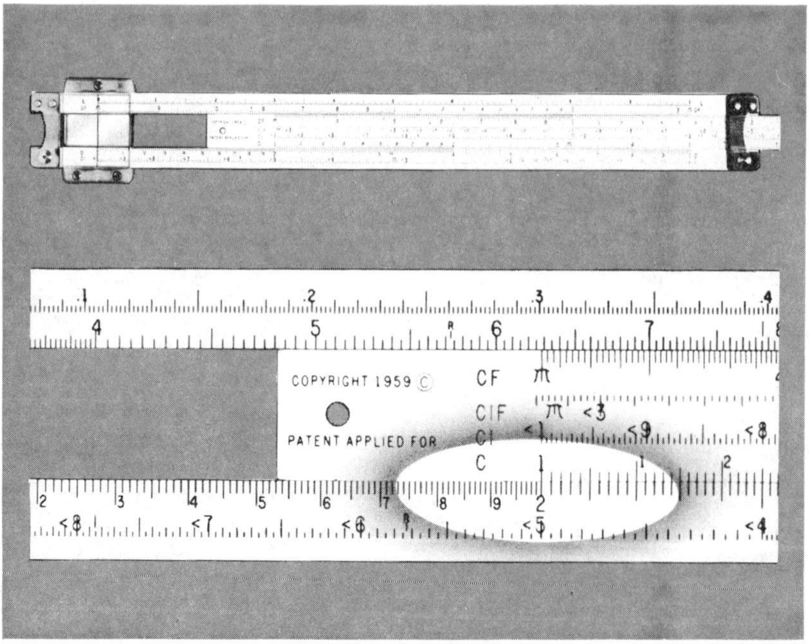

Notice in Example 2 that the settings for 20 and 2 are identical. In fact, the settings for 2000, .02, and 2,000,000 are all at the same position on the scale. In other words, the position of the decimal point cannot be indicated using the slide rule.

Numbers of more than two or three digits must be set by approximation, since the scales cannot be marked into continually shrinking divisions.

Example 3. Find 1725 on the C scale, using the right index of the D scale.

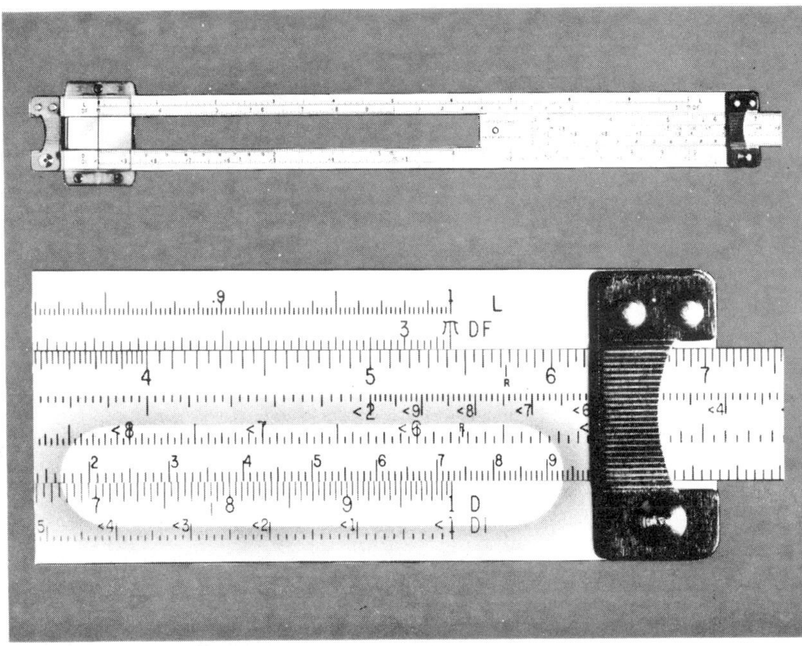

Notice that we can mark 1720 precisely on the scale; to set the fourth digit, 5, we approximate half-way between the tertiary 2 and 3.

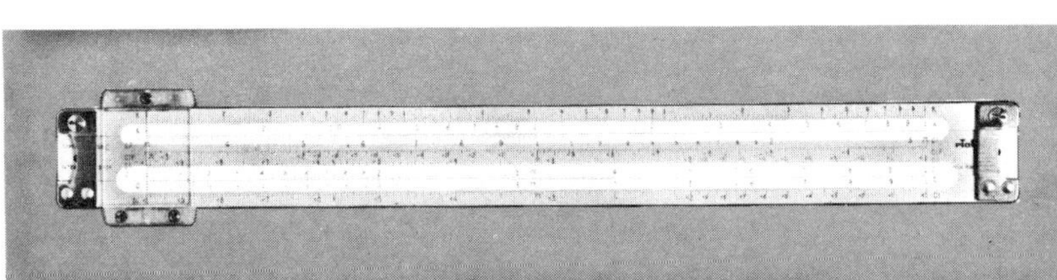

Example 4. Find 843 on the C and D scales, using the hairline.

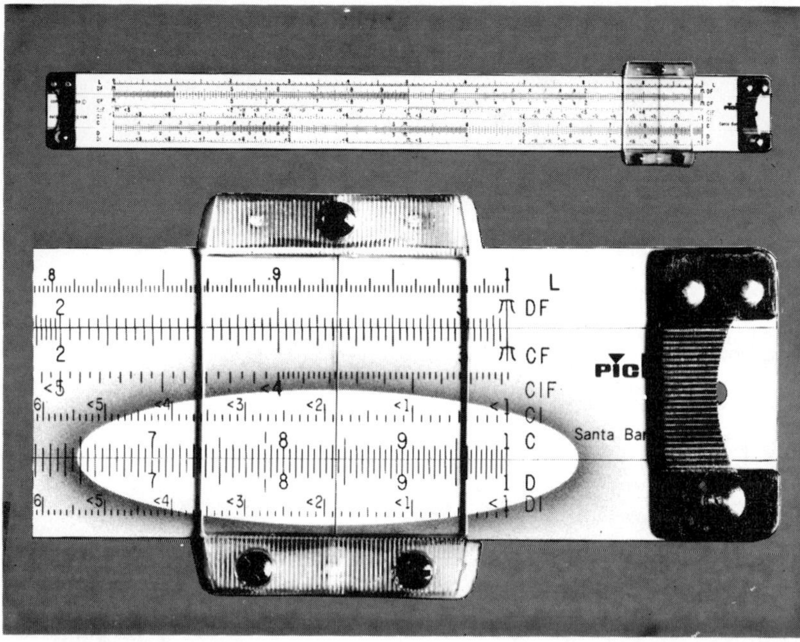

Notice that the hairline, once set at a particular number on one scale, crosses other numbers on other scales.

Example 5. If the hairline is set on 420 on the D scale, what three-digit number is under the hairline on the A scale?

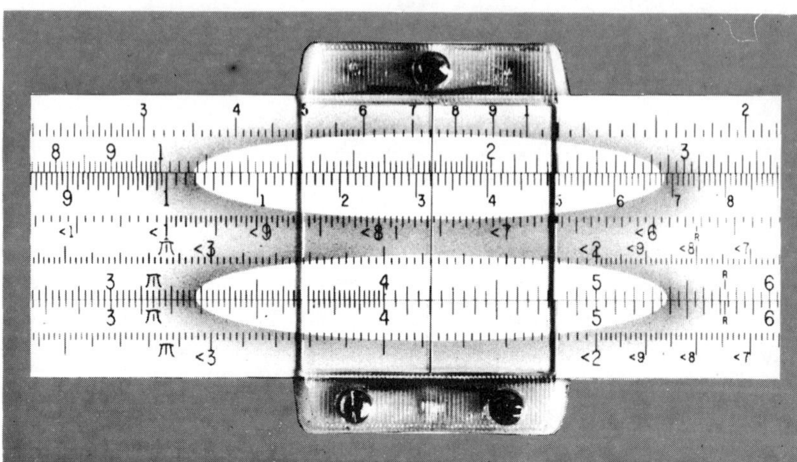

On the D scale, each secondary division indicates one unit, so the hairline is set on the second secondary division line to the right of 4. On the A scale the hairline is between primary divisions 1 and 2, so the first digit of our number is 1. Here each secondary division indicates one unit, and since the hairline is between the seventh and the eighth secondary division, the second digit of our number is 7. Finally, each tertiary division indicates 2/10 units, and since the hairline is over the third tertiary division to the right of the seventh secondary division, the third digit of the number is 6. The number on the A scale is 176. Notice that since both the A and D scales are immediately next to the slide, we could just as easily use one of the indices to locate corresponding numbers on the two scales.

Verify the following correspondences on your own slide rule.

Example 6. 400 on the D scale corresponds to 160 on the A scale.

Example 7. 300 on the D scale corresponds to 900 on the A scale.

Example 8. 300 on the D scale corresponds to 2700 on the K scale.

Example 9. 1.85 on the D scale corresponds to 342 on the A scale.

Example 10. 764 on the D scale corresponds to 445 on the K scale.

Exercise 12

Fill in the following table by putting the hairline on the values and scales given, and reading corresponding *three-digit* values on the other scales.

	Value on D scale	Value on A scale	Value on K scale
1)	717	—	—
2)	—	551, right half	—
3)	362	—	—
4)	1755	—	—
5)	—	—	7, left section
6)	499	—	—
7)	—	—	836, middle section
8)	1991	—	—
9)	—	665, left half	—
10)	423	—	—

	Value on D scale	Value on A scale	Value on K scale
11)	—	—	185, left section
12)	—	—	419, right section
13)	—	991, left half	—
14)	54.1	—	—
15)	—	—	651, left section
16)	201	—	—
17)	210	—	—
18)	—	—	104, middle section
19)	—	724, right half	—
20)	—	155, left half	—

21) What happens to the accuracy with which you can read the D scale as you move from left to right?

22) How much does each tertiary division represent if the number in question is located on the D scale between 1 and 2?

23) How much does each secondary division represent if the number in question is located on the D scale between 8 and 9?

24) How much does each tertiary division represent if the number in question is located on the D scale between 2 and 3?

25) How much does each tertiary division represent if the number in question is located on the A scale between 1 and 2?

26) How much does each tertiary division represent if the number in question is located between 5 and 6 on the K scale?

27) What observations have you made about the two halves of the A scale and the three sections of the K scale?

28) What is another name for the body of your slide rule?

29) The numeral 1 found on the left and right hand sides of the D scale is referred to by what name?

30) Can you think of any place on the slide rule where you could read 5 digits accurately?

3.3 Finding the Decimal Point

No decimal points are given on the slide rule, so the operator must determine the proper position of the decimal point in the result after the digits have been found on the slide rule. Scientific notation is an easy method of handling the decimal when calculating on the slide rule. This is done in much the same manner as approximate answers were found in Section 1.18.

Example 1. $58.5 \times 49.7 = ?$

58.5 is approximately 60 and 49.7 is approximately 50, so the position of the decimal point in the product can be found by

$$\begin{aligned} 60 \times 50 &= 6 \times 10 \times 5 \times 10 \\ &= 6 \times 5 \times 10^2 \\ &= 30 \times 10^2 \\ &= 3000. \end{aligned}$$

Therefore the decimal will follow the fourth digit in the answer.

Example 2. $\dfrac{971 \times .00366}{492} = ?$

Think of 971 as approximately $1000 = 1 \times 10^3$, .00366 as approximately $.004 = 4 \times 10^{-3}$, and 492 as approximately $500 = 5 \times 10^2$. Then

$$\begin{aligned} \frac{1000 \times .004}{500} &= \frac{1 \times 10^3 \times 4 \times 10^{-3}}{5 \times 10^2} \\ &= \frac{1 \times 4 \times 10^{3-3-2}}{5} \\ &= .8 \times 10^{-2} = .008^{-3}. \end{aligned}$$

Therefore there will be two zeros in front of the first digit of the answer.

Exercise 13

The first three or four digits of the answer are given for each of the following problems. By approximation, place the decimal point in the answer.

1) $\dfrac{9.63 \times 4.17}{1.234} = 325$

2) $500 \times 600 = 300$

3) $250 \times 2500 = 625$

4) $14 \times 360 = 504$

5) $\dfrac{70 \times 45}{30} = 1050$

6) $\dfrac{8500 \times 41}{720} = 484$

7) $5.06 \times 6.72 = 340$

8) $\dfrac{498 \times .563}{89.3} = 314$

9) $\dfrac{7.4 \times 28}{9} = 230$

10) $\dfrac{.874}{3.95 \times .707} = 313$

11) $\dfrac{.772 \times 386}{2.85} = 1046$

12) $\dfrac{147.5 \times 8.76}{3260} = 396$

13) $2.43 \times 4.17 \times 3.24 \times 6.92 = 2272$

14) $\dfrac{41.6 \times 8.08}{113 \times 2.31} = 1287$

15) $\dfrac{1.62 \times 2.71}{3.43 \times 7.92} = 1617$

16) $\dfrac{4240}{7.55 \times 11.2 \times 27.4} = 1830$

17) $\dfrac{3.72 \times 13.34 \times 4.07}{1.005} = 201$

18) $23.6 \times .721 \times 5.03 = 856$

19) $\dfrac{14.1 \times 9.37}{4.11 \times 2.93} = 1097$

20) $1.776 \times 1.625 \times 1.880 = 543$

21) $\dfrac{86 \times 70.8}{2.85} = 214$

22) $\dfrac{.797 \times 5.96}{.502} = 946$

23) $\dfrac{37 \times 86}{75.7} = 420$

24) $\dfrac{17.9 \times 587}{2580} = 407$

25) $\dfrac{7.59 \times 5.49}{.876} = 476$

26) $\dfrac{333}{1.084 \times 12.35 \times 2.05} = 1213$

27) $7.07 \times .771 \times 76.7 = 418$

28) $\dfrac{217 \times .0983}{72.1 \times .870} = 340$

29) $\dfrac{152.3}{8.53} = 1785$

30) $\dfrac{17.33 \times 31.6}{111.5} = 491$

31) $\dfrac{18 \times 45 \times 37}{23 \times 29} = 449$

32) $.785 \times 4.27 \times 3.76 = 1260$

33) $\dfrac{1375 \times .0642}{76{,}400} = 1155$

34) $\dfrac{218}{4.23 \times 50.8} = 1014$

35) $\dfrac{65.7 \times .835}{3.58} = 1532$

36) $.535 \times 1.47 \times 3.44 = 271$

37) $\dfrac{24.7}{8.11 \times 2.02} = 1508$

38) $6.72 \times 2.76 \times .885 \times 11.23 = 184$

39) $\dfrac{28{,}300}{42.1 \times 3.72 \times 5.05} = 358$

40) $\dfrac{17.6 \times 2.84}{27.3} = 183$

3.4 Multiplication

The C and D scales are used to multiply two numbers on the slide rule. The procedure is as follows:

1) locate one factor on the D scale with the index of the C scale;
2) move the hairline over the second factor on the C scale;
3) the product is found below the hairline on the D scale.

If using the left index to locate the first factor on the rule results in the second factor being beyond the range of the indicator on the C scale, then the right index is used. The remaining procedure is the same.

Example 1. Multiply 2×2 on the slide rule.

Find 2 on the D scale with the left index of the C scale. Locate 2 on the C scale with the hairline. Read 4 under the hairline on the D scale.

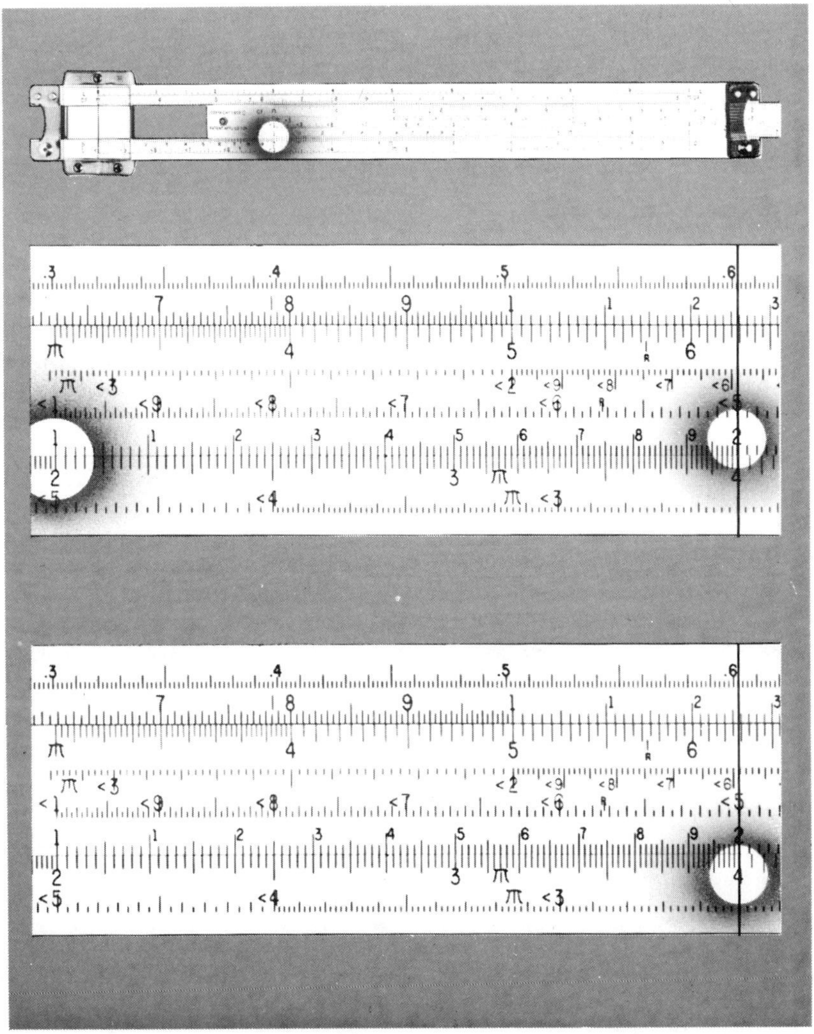

Example 2. Multiply 8×5.

Locate 8 on the D scale with the left index. Notice that 5 on the C scale cannot be located by the hairline, since it is off the rule. So, we find 8 on the D scale with the right index of the C scale. Locate 5 on the C scale with the hairline. Read 40 below the hairline on the D scale.

Example 3. 58.5×49.7

Find 585 on the D scale with the right index of the C scale. Locate 497 on the C scale with the hairline. Read 291 below the hairline on the D scale. Use the method of Example 1, Section 5.3, to locate the decimal point to obtain 2910 as the approximate answer.

Longer problems can be solved by using the same procedure repeatedly. After the product of the first two factors has been found, move the appropriate index to this product on the D scale and find the next product on the C scale with the hairline. This can be repeated until the desired product has been found.

Example 4. $5 \times 4 \times 3 \times 2$

Find 5 on the D scale with the right index of the C scale. Move the hairline to 4 on the C scale, reading 20 below the hairline on the D scale. Without moving the hairline slide the left index of the C scale under the hairline. Move the hairline to 3 on the C scale, reading 60 on D. Again without moving the hairline, slide the right index of the C scale under the hairline. Move the hairline to 2 on the C scale, reading 120 under the hairline on the D scale.

The procedure can be continued indefinitely. Remember, however, that you must determine the decimal placement yourself, using the procedure of the previous section.

Exercise 14

Perform the following operations on the slide rule.

1) 3×5	2) 7×2
3) 13×21	4) 72×4
5) 23×8	6) 81×9
7) 25×42	8) 321×7
9) 643×8	10) $.4 \times 17$
11) 44×11	12) $.3 \times 1.6$

13) $.9 \times 1.8$

14) 5.6×7.1

15) 3.2×6.9

16) $.54 \times 8.2$

17) 6.71×5.24

18) 2.9×3.27

19) 5.06×6.72

20) 5.63×1.820

21) 17.63×61.7

22) 146×4.37

23) 57.6×1.845

24) 582×13.1

25) $192.1 \times .421$

26) 75.2×74.3

27) 3.16×2.17

28) $86{,}000 \times 74{,}300$

29) 132.1×62.1

30) $6.17 \times 8.62 \times 4.16$

31) $57.2 \times .0810$

32) $212 \times 3.16 \times 8.12$

33) $3.21 \times 4.51 \times 9.25$

34) $2.78 \times 6.42 \times 3.8$

35) $11.26 \times 5.31 \times 7.65$

36) 6.14×8.36

37) $521 \times .860 \times 7.17$

38) 12.2×9.71

39) $4.5 \times 7.6 \times 8.7 \times .93$

40) $2.97 \times 41.7 \times 17.1 \times 64.1$

Use your slide rule to compute each of the following:

41) If the thickness of one shim is .125 inch, find the thickness of a stack of 475 shims.

42) The volume of a box is found by multiplying the length times the width times the height. Find the volume of a box whose dimensions are 4.13 feet, 6.21 feet, and 3.45 feet.

43) A contractor figures that it will cost him \$8500 to complete a job. If his profit is $12\frac{1}{2}\%$ of the cost, how much will his profit be?

44) A board foot of lumber is a piece 1 inch thick by 12 inches wide by 12 inches long. This same measure is constituted by a board 1 inch thick, 24 inches long and 6 inches wide. Some examples are shown below. Notice that the volume of a board foot is always length $\times$ width $\times$ height = 144 cubic inches. Find the number of board feet in a stack of lumber containing 480 pieces of 1 inch by 12 inches by 8 feet lumber.

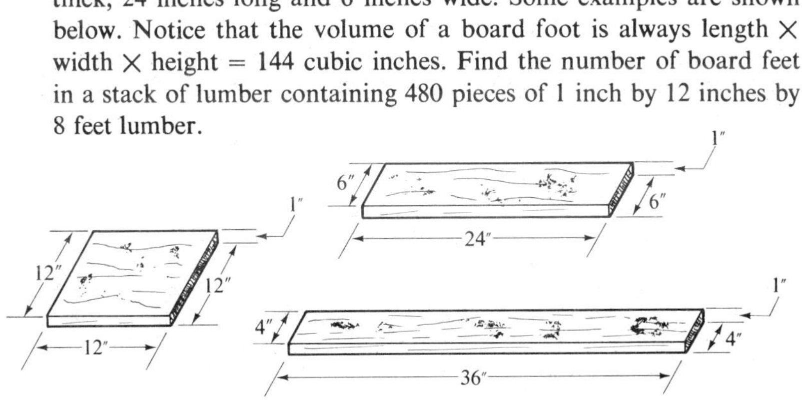

45) The cost per board foot for lumber is \$.17. Find the cost of a stack of 115 (2 × 6) boards each containing 18 board feet.

46) Find the number of board feet in a stack of lumber containing 320 pieces of 1 inch by 12 inches by 12 feet and 247 pieces of 1 inch by 6 inches by 12 feet.

47) Assuming that the price per board foot of each piece of lumber in Problem 46 is the same, what is the total cost of the stack to the nearest dollar if the cost per board foot is \$.27? Use care in reading the scales.

48) Find the total weekly wages paid 45 men earning \$3.17 per hour if each man works 42 hours per week.

49) A certain television tube costs \$2.45. What will be the total cost of a gross (144) of these tubes?

50) When punching holes through sheet metal or steel, punching tonnage is calculated by multiplying the hole circumference by the material thickness by the tensile strength in tons. How many tons of force would be required to punch each of the following:

	Hole Diameter	Hole Circumference	Material	Thickness	Tensile Strength
a)	$\frac{11}{16}$ in.	2.159 in.	Low carbon steel	.25 in.	25 tons
b)	$\frac{1}{2}$ in.	1.5708 in.	Common bronze	.1875 in.	26 tons
c)	$\frac{1}{4}$ in.	.7854 in.	Annealed stainless steel	.125 in.	55 tons

51) 22-gauge steel is approximately .0299 inches thick. How much shelf clearance would be necessary for a stack of 52 sheets?

3.5 Division

Since division is the inverse operation of multiplication, division on the slide rule is done by a method which is the inverse of multiplication. To divide one number by another,

1) the dividend is located on the D scale with the hairline;
2) the divisor on the C scale is moved under the hairline;

3) the quotient is then read on the D scale opposite the index on the C scale.

Example 1. Divide 56 by 7.

56 is located on the D scale with the hairline. 7 on the C scale is moved under the hairline. Opposite the right index of the C scale, the quotient is read as 8 on the D scale.

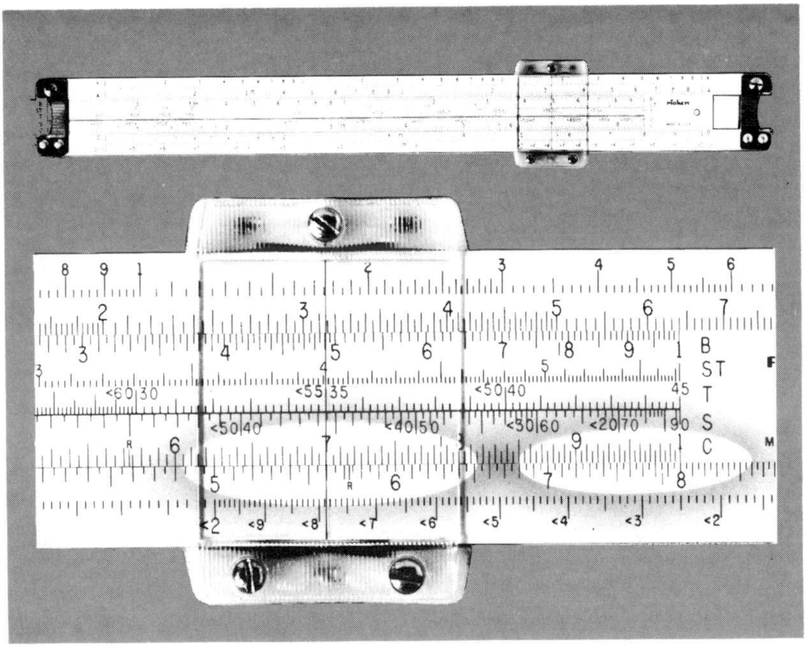

Example 2. Divide 729 by 185.

Locate 729 on the D scale with the hairline. Slide 185 on the C scale under the hairline. The answer, 3.93, is read on the D scale opposite the left index of C.

3.6 Combined Operations

Longer combination problems can be done easily on the rule by combining the procedures for multiplication and division. To do longer problems using the C and D scales only, one should try to obtain a partial answer in the most convenient position. For instance, if we have $\frac{5 \times 4}{3}$ as a problem, we can begin by multiplying 5×4 in the usual manner, obtaining 20 on the D scale as partial answer. If the hairline is then moved to this partial answer, the divisor on the C scale can be moved under the hairline and the digits 667 will be opposite the index on the C scale. The answer is then 6.67.

Alternatively, we could solve the same problem by dividing 5 by 3 first, locating 5 on the D scale with the hairline and moving the 3 on the C scale below the hairline, giving the partial result opposite the index on the C scale. Without moving the scale, the hairline could be moved directly to 4 on the C scale giving the product 667 or 6.67 on the D scale. This method involves moving the scale only once for the entire problem, and is therefore the most convenient method of solving the problem.

Example 1. $\dfrac{4 \times 18}{7} = (4 \div 7) \times 18.$

Move the hairline to 4 on the D scale. Slide 7 on the C scale below the hairline. Slide the hairline to 18 on C. Read 1023 below the hairline on the D scale. Therefore

$$\frac{4 \times 18}{7} = 10.23.$$

Example 2. $\dfrac{75 \times .045 \times 1.8}{7.4 \times 15} = 75 \div 7.4 \times 1045 \div 15 \times 1.8.$

Move the hairline to 75 on the D scale. Slide 74 on the C scale under the hairline. Move the hairline to 45 on the C scale. Slide 15 on the C scale under the hairline. Move the hairline to 18 on the C scale. Read 548 on the D scale. By the procedure discussed previously locate the decimal point and find the answer as .0548.

The digits for which the reader supplied the decimal points in Exercise 11 were all arrived at by solving the problems with a slide rule. If you need practice, solve the problems in Exercise 11 on your own slide rule, then check your answers against the ones given. Try to arrive at the result with the minimum movement of the slide.

Exercise 15

Find the following quotients to three places using a slide rule.

1) $\dfrac{417}{6.72}$ 2) $\dfrac{45}{750}$ 3) $\dfrac{81.7}{22.3}$

4) $\frac{18}{95}$ 5) $\frac{10.42}{0.316}$ 6) $\frac{43.3}{21.2}$

7) $\frac{43.7}{6.21}$ 8) $\frac{8.50}{.0085}$ 9) $\frac{8.32}{14.64}$

10) $\frac{47.1}{576}$ 11) $\frac{557}{2.17}$ 12) $\frac{27.3}{.271}$

13) $\frac{63.7}{19.2}$ 14) $\frac{.371}{59.2}$ 15) $\frac{1463}{362}$

16) $\frac{155}{75}$ 17) $\frac{7428}{2.93}$ 18) $\frac{14.3}{21.3}$

Convert the following to three-place decimal fractions using a slide rule.

19) $\frac{1}{7}$ 20) $\frac{1}{9}$ 21) $\frac{5}{6}$

22) $\frac{1}{3}$ 23) $\frac{3}{8}$ 24) $\frac{5}{8}$

25) $\frac{19}{32}$ 26) $\frac{7}{16}$ 27) $\frac{15}{47}$

28) $\frac{5}{9}$ 29) $\frac{9}{16}$ 30) $\frac{15}{32}$

31) $\frac{87}{615}$ 32) $\frac{18}{43}$ 33) $\frac{896}{1144}$

34) $\frac{95}{117}$ 35) $\frac{37}{53}$ 36) $\frac{77}{125}$

37) $\frac{9}{11}$ 38) $\frac{17}{32}$ 39) $\frac{16}{17}$

40) $\frac{12}{87}$

Compute the following to three places using a slide rule.

41) $\frac{9.63 \times 4.17}{1.234}$ 42) $\frac{41.6 \times 8.08}{113 \times 2.31}$

43) $\frac{24.7}{8.11 \times 2.02}$ 44) $\frac{14.1 \times 9.37}{4.11 \times 2.93}$

45) $\frac{14.55 \times 9.04}{31.6 \times 2.33}$ 46) $\frac{1.62 \times 2.71}{3.43 \times 7.92}$

47) $\frac{4240}{7.55 \times 11.2 \times 27.4}$ 48) $\frac{3.72 \times 13.34 \times 4.07}{1.005}$

49) $\frac{2.17 \times .0983}{72.1 \times .870}$ 50) $\frac{4.21 \times 1.170}{8.08 \times .312}$

51) $\dfrac{41.3 \times 26.2}{9.4}$

52) $\dfrac{7.2 \times 8.5}{16.3}$

53) $\dfrac{455}{81.6 \times 22.3}$

54) $\dfrac{458 \times 796}{854 \times 13}$

55) $\dfrac{912 \times 8.91}{.751 \times 84.3}$

56) $\dfrac{11.12 \times 17.93}{8.47}$

57) $\dfrac{875 \times 14 \times 86}{75 \times 11.3 \times 75}$

58) $\dfrac{34.3 \times 84.2 \times 19.1}{45.3 \times 19.4}$

59) $\dfrac{347 \times .432 \times 4.97}{.00045 \times 86.1}$

60) $\dfrac{99.7 \times 87.3 \times 19.2 \times 451 \times 21.4}{26.3 \times 14.5 \times 906}$

61) $\dfrac{.905 \times 111 \times 4.26}{.529}$

3.7 The CI and DI Scales

The CI and DI scales are identical scales as are C and D and A and B. For this reason, some operations which can be done on A and B or C and D can alternatively be done on CI and DI. However, the CI and DI scales have a very unique property. If n is found on the D scale with the hairline, $1/n$ or n^{-1} will be under the hairline on the DI scale. Likewise, if n is found on the C scale with the hairline, $1/n$ or n^{-1} will be under the hairline on the CI scale. This also implies that if n is found on CI or DI, the reciprocal of n will be found on C or D respectively. For this reason, CI and C and DI and D are called reciprocal or inverse scales.

Notice that the CI and DI scales are read from right to left instead of from left to right. Also, to avoid error in reading reciprocals, if the number is found on D, its reciprocal should be found on DI; if the number is located on C, the reciprocal should be read from CI.

Example 1. Find the value of $\dfrac{1}{3.71}$.

Locate 3.71 on D with the hairline. Read $1/3.71 = .269$ on the DI scale ($1/3.71$ is close to $1/3$ or $.3$).

Example 2. $\dfrac{1}{.521} = 1.919 \qquad \left(\dfrac{1}{.5} = \dfrac{10}{5} = 2.0\right).$

Example 3. $\frac{1}{65.2} = .01533 \qquad \frac{1}{100} < \frac{1}{65} < \frac{1}{10}$

$$.01 < \frac{1}{65} < .1.$$

Further use of the CI and DI scales can be made in cases where multiplication and division by using C and D scales alone would result in having the slide extended or otherwise not convenient. For instance, 85 × 1.15 can be found on the C and D scale to be 9.78. However, 8.5 × 1.15 = $\frac{8.5}{1/1.15}$. We can therefore perform the operation by locating 8.5 on the D scale with the hairline and sliding 1.15 on CI under the hairline. (1/1.15 is on C under the hairline). The answer 9.78 is located on D opposite the index of C.

Example 4. Use reciprocal scales to find 9.2 × 1.07.
$9.2 \times 1.07 = \frac{9.2}{1/1.07}$. Locate 9.2 on D with the hairline. Slide 1.07 on CI under the hairline. The answer 9.84 is opposite the right index of C on the D scale.

By the same reasoning, the reciprocal scales can be used to find a quotient. Instead of dividing by a number we can multiply by its reciprocal using CI or DI. For instance 1.5/2 can be thought of as 1.5 × 1/2. If 1.5 is located on the D scale with the left index of C and the hairline is placed over 2 on the CI scale (it would be over 1/2 on the C scale), the answer .75 will be under the hairline on the D scale.

Example 5. Using reciprocal scales find 1.762/4.85.
Locate 1.762 on D with the left index of C. Move the hairline to 4.85 on the CI scale. The answer .363 is under the hairline on D.

This procedure can be extended with longer problems to allow for minimum movement of the slide.

Example 6. Using reciprocal scales, find 4.52 × 8.73 × 1.42 × 6.28.
One method would be to locate 4.52 on D with the right index of C. Move the hairline to 8.73 on D finding the first product on C and D only. Next, instead of changing indices to perform the next product, divide by 1/1.42 by moving 1.42 on CI under the hairline. The second product {(4.52 × 8.73) × 1.42} is opposite the right index of C on the scale. The final product can be found by moving the hairline to 6.28 on the C scale and finding the result 351. under the hairline on the D scale.

Example 7. Find $\frac{7.31 \times 4.26}{1.65 \times 2.73}$.

Locate 7.31 on the D scale with the right index of C. Move the hairline over 1.65 on CI. Move the right index of C under the hairline. Move the hairline to 4.26 on the C scale. Move 2.73 on C under the hairline. The answer 6.91 is opposite the right index of C on the D scale.

Other procedures can be used for the two previous examples using the same scales. The convenience of the reciprocal scales makes learning their use worthwhile.

Further use of the reciprocal scales will be discussed in the chapter on trigonometry.

3.8 Square Root

The A and B scales are identical, as are the C and D scales. Notice, however, that the A and B scales are each divided into two smaller identical scales. The A and B scales are used with the C and D scales for finding square root on the slide rule.

To find square root on the slide rule, no movement of the slide is necessary. The radicand is located on the A scale with the hairline. The square root of the radicand will be found on the D scale below the hairline. The same procedure can be used with the B and C scales of the slide rule.

The most important thing to remember when finding square root on the slide rule is to locate the radicand on the proper part of the A or B scale. For radicands greater than one, if there are an even number of digits to the left of the decimal point, the number is located on the right half of the A or B scale; if the number has an odd number of digits to the left of the decimal point, it is then located on the left half of the A or B scale.

Example 1. Find the square root of 16.5.

Since the number has 2 digits left of the decimal point, it is located on the right half of the A scale with the hairline. The square root is then read on the D scale under the hairline as 4.06.

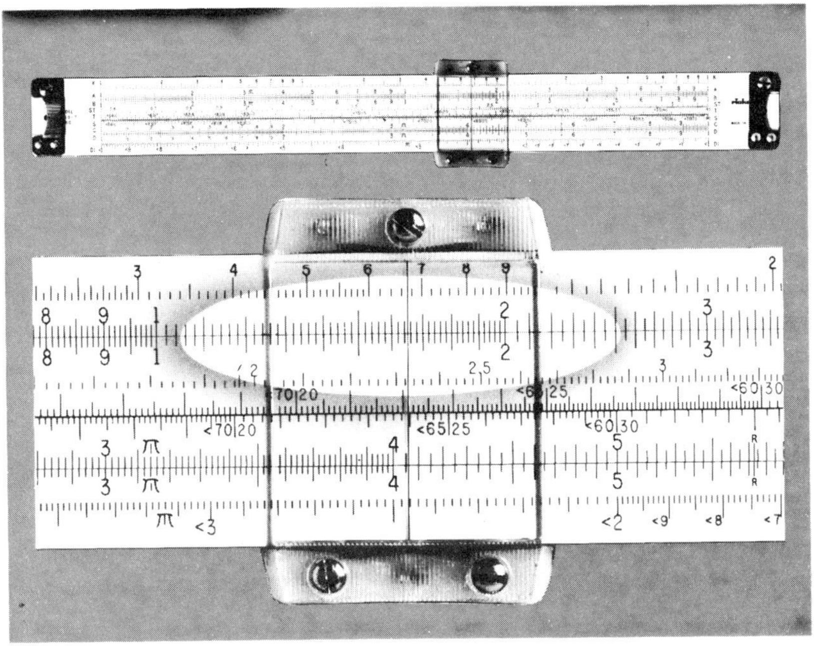

Example 2. Find the square root of 165.0.

Use the left half of the A scale to locate 165 with the hairline. The answer, 1284, is read under the hairline on the D scale. Since (12)(12) = 144 and (13)(13) = 169, the decimal is to the right of the 12 (Section 2.5), and the answer is 12.84.

Example 3. Find the square root of 4.63.

Use the left half of the A scale. $\sqrt{4.63} = 2.15$.

Another way to determine which half of the A scale to use is to write the number in scientific notation. (See Section 2.6.) If the exponent of 10 is even, the number will be found on the left half of the A scale; if the exponent of 10 is odd, the number will be found on the right half of the A scale. Try this rule on Examples 1—3 to convince yourself that it works.

Using this rule we can now find square roots of numbers less than 1. Recall that when a decimal number is multiplied by another decimal number, the number of decimal places in the answer is the sum of the decimal places in the factors. The placement of the decimal point can be determined by mentally calculating the approximate root, and comparing that with the actual digits obtained on the slide rule.

Example 4. $\sqrt{.0035} = \sqrt{3.5 \times 10^{-3}}$

In scientific notation the exponent of 10 is odd, so we use the right half of the A scale. We read 591 on the D scale. Since (.06)(.06) = (.0036), the root is .0591.

Example 5. $\sqrt{.000875} = \sqrt{8.75 \times 10^{-4}}$.

In scientific notation the exponent of 10 is even, so we use the left half of the A scale. We read 296 on the D scale. Since (.03)(.03) = .0009, the root is .0296.

3.9 Squaring a Number

Squaring a number is finding the product of the number multiplied by itself. For this reason, a number can be squared by using the C and D scales. However, using the A and D scales a number can be squared without movement of the slide. This is the inverse of the process of finding square root. That is, we need only locate the number on the D scale with the hairline. The square of the number is read under the hairline on the A scale.

Example 1. Find 7^2.

Move the hairline over 7 on the D scale. The square of 7 or $7^2 = 49$ is found under the hairline on the A scale.

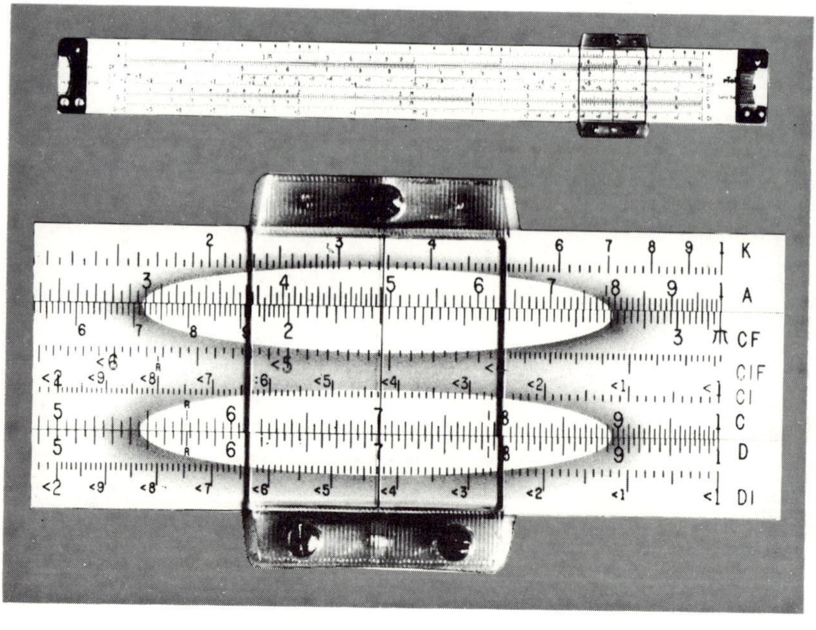

Exercise 16

Find the following square roots on the slide rule.

1) $\sqrt{75}$	2) $\sqrt{412}$	3) $\sqrt{5.19}$
4) $\sqrt{.116}$	5) $\sqrt{.441}$	6) $\sqrt{4.41}$
7) $\sqrt{44.1}$	8) $\sqrt{.033}$	9) $\sqrt{23.7}$
10) $\sqrt{26,400}$	11) $\sqrt{.875}$	12) $\sqrt{31.2}$
13) $\sqrt{26.1}$	14) $\sqrt{.00352}$	15) $\sqrt{2.716}$
16) $\sqrt{.000292}$	17) $\sqrt{.871}$	18) $\sqrt{7.31}$
19) $\sqrt{.125}$	20) $\sqrt{6.25}$	21) $\sqrt{62.5}$
22) $\sqrt{.00625}$	23) $\sqrt{.0341}$	24) $\sqrt{7,740,000}$

Square the following numbers using the A and D scales.

25) 47.5	26) 3.4	27) 19
28) 8.25	29) .0217	30) 29.6
31) 10.3	32) 91.3	33) 22.4
34) 8.13	35) 445	36) .731
37) .0625	38) 7.37	39) 426
40) 9.91	41) 4.32	42) 17.2

3.10 Cube Root

Just as the A and B scales are divided into two identical sections, the K scale is divided into three identical sections. The K scale is used with the D scale to find cube roots. Again, no movement of the slide is necessary. Note, however, that the K scale has no duplicate scale on the slide, as do the A and D scales.

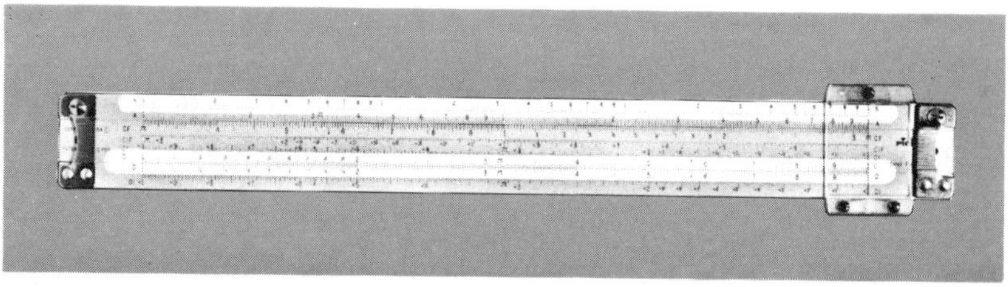

Figure 3.3

To find the cube root of a number on the slide rule, the radicand is located on the K scale with the hairline. The cube root of the number is read below the hairline on the D scale. As in square roots, the key to success lies in using the correct portion of the K scale and correct decimal placement.

The following applies to radicands greater than one. If the radicand contains 1, 4, 7, 10, . . . digits to the left of the decimal point, the left section of the K scale is used. If the number has 2, 5, 8, 11, . . . digits to the left of the decimal, the middle section is used. If the number contains 3, 6, 9, 12, . . . digits to the left of the decimal point the right section of the K scale is used. In short, divide the number into groups of three digits. If the left group contains one digit, use the left portion of the scale; if it contains two digits use the middle of the scale; and if it has three digits use the right part of the scale.

Example 1. $\sqrt[3]{49.3}$

Use the middle section of the K scale. Read 367 from the D scale. Since

$$3^3 = 27 \text{ and } 4^3 = 64$$

it follows that

$$\sqrt[3]{49.3} = 3.67.$$

Example 2. $\sqrt[3]{1.31}$

Use the left section of the K scale. Read 1094 from the D scale. Since $1^3 = 1$,

$$\sqrt[3]{1.31} = 1.094.$$

Example 3. $\sqrt[3]{596}$

Use the right section of the K scale. Read 841 from the D scale. Since $(10)^3 = 1000$, the root is less than 10 and

$$\sqrt[3]{596} = 8.41.$$

If the number for which the cube root is being found is less than one, divide the number into groups of three digits starting with the decimal point. If the group containing the first nonzero digit contains only one digit, the left section of the K scale is used. If this group has two digits (in this case a zero following the first nonzero digit is counted), the middle section is used, and if this group contains three digits, the right section is used.

Example 4. $\sqrt[3]{.00851} = \sqrt[3]{.008\ 510}$

The first group of 3 digits contains one nonzero digit; therefore, the left section of the K scale is used. The digits read from the D scale are 204. Since $(.2)^3 = .008$, the answer is .204.

Example 5. $\sqrt[3]{.000\ 000\ 056\ 2}$

The first group containing a nonzero digit has two nonzero digits; therefore the middle section of the K scale is used. The digits of the answer are 383. Since $(.004)^3 = .000000064$, the answer is .00383.

Finding cube roots can be further simplified if we recall from our discussion of radicals that $\sqrt[n]{ab} = \sqrt[n]{a}\sqrt[n]{b}$. This law is used to simplify $\sqrt[3]{1.59 \times 10^{-14}}$ by writing it as $\sqrt[3]{15.9 \times 10^{-15}} = \sqrt[3]{15.9}\sqrt[3]{10^{-15}}$. The cube root of 15.9 is between one and ten and the cube root of 10^{-15} is 10^{-5}. This also enables us to find the decimal easily. This simplification process works equally well with square and cube roots.

Example 6. $\sqrt[3]{75{,}000{,}000} = \sqrt[3]{75} \cdot \sqrt[3]{10^6} = \sqrt[3]{75} \times 10^2.$

Since 75 has two digits, from the middle section of K we read 422 on D as the digits of the answer. The answer is $4.22 \times 10^2 = 422$.

Example 7. $\sqrt[3]{.000000475} = \sqrt[3]{475} \cdot \sqrt[3]{10^{-9}} = \sqrt[3]{475} \times 10^{-3}.$

Since 475 has three digits, we put the hairline over the right section of the K scale and read 780 on scale D. So

$$\begin{aligned}\sqrt[3]{475} \times 10^{-3} &= 7.80 \times 10^{-3}\\ &= .00780.\end{aligned}$$

3.11 Cubing a Number

A number can be cubed by multiplying it times itself three times using the C and D scales. However, a procedure which is the inverse of finding cube roots can be used to cube a number without movement of the slide. This is done by locating the number on the D scale with the hairline and reading the cube of the number on the K scale under the hairline.

Example 1. Find $(18)^3$.

Set the hairline over 18 on the D scale. Read the digits 580 on the K scale. Using scientific notation we can express $(18)^3$ as

$$(1.8 \times 10)^3 = 5.8 \times 10^3 = 5800.$$

Exercise 17

Find the cube roots of the following:

1) $\sqrt[3]{5.37}$	2) $\sqrt[3]{.00351}$	3) $\sqrt[3]{452}$
4) $\sqrt[3]{45.2}$	5) $\sqrt[3]{715}$	6) $\sqrt[3]{1.91}$
7) $\sqrt[3]{27,000}$	8) $\sqrt[3]{8.04}$	9) $\sqrt[3]{26.5}$
10) $\sqrt[3]{614}$	11) $\sqrt[3]{44.4}$	12) $\sqrt[3]{4.44}$
13) $\sqrt[3]{.444}$	14) $\sqrt[3]{925}$	15) $\sqrt[3]{.0849}$
16) $\sqrt[3]{750}$	17) $\sqrt[3]{45}$	18) $\sqrt[3]{26.2}$
19) $\sqrt[3]{43.1}$	20) $\sqrt[3]{450,000}$	21) $\sqrt[3]{75,000}$
22) $\sqrt[3]{465}$	23) $\sqrt[3]{.142}$	24) $\sqrt[3]{.0035}$
25) $\sqrt[3]{.76}$	26) $\sqrt[3]{7.88}$	27) $\sqrt[3]{87.1}$

Cube each of the following using the D and K scales:

28) 14	29) 4.25	30) 16.3
31) 43.7	32) 77.3	33) 7.73
34) 195	35) 36	36) 412
37) .0334	38) .41	39) .107

The following example may be of help in determining the order in which the following operations should be done on the slide rule.

Example: $5\sqrt[3]{75}$ Find $\sqrt[3]{75} = 4.22$ on the D scale. Move the right index under the hairline. Move the hairline to 5 on the C scale. Read the answer 21.1 under the hairline on the D scale.

40) $17\sqrt[3]{9}$	41) $(\sqrt[3]{16})(7)$	42) $27.3\sqrt[3]{450}$
43) $95\sqrt[3]{-95}$	44) $\sqrt{5}\sqrt[3]{6}$	45) $\sqrt[3]{7}\sqrt{3}$
46) $\dfrac{\sqrt[3]{7}\sqrt[3]{15}}{\sqrt{12}}$	47) $\sqrt[4]{75}$	

48) $\sqrt[6]{18}$ (Hint: $\sqrt[6]{18} = \sqrt[2]{\sqrt[3]{18}}$.)

49) If the volume of a cube is known, the length of an edge can be found by finding the cube root of the volume. Find the length of the edge of a cube whose volume is 45 cubic inches.

50) The volume of a sphere is found by taking 4/3 times 3.14 times the radius cubed. How long is the edge of a cube if the cube has the same volume as a sphere with a 3-inch radius?

Review Exercise Chapter 3

Use the slide rule for all of these problems.

Convert the following to three-digit decimals:

1) $\frac{1}{8}$ 2) $\frac{3}{8}$ 3) $\frac{5}{16}$ 4) $\frac{9}{32}$ 5) $\frac{19}{64}$

Perform the following operations:

6) 5.6×8.3

7) 96.2×41.9

8) $65{,}800 \times .00472$

9) 550×8.65

10) $\frac{4.36}{3.92}$

11) $\frac{5.82 \times 4.21}{20.5}$

12) $\frac{111.2 \times 1.75}{14.3 \times 10.5}$

13) $\frac{765 \times 876 \times .00342}{14.2 \times 93.1}$

14) $\frac{86.5 \times 4.38}{1.25 \times 78.5 \times 9.62}$

15) $\frac{1}{8} \times \frac{5}{16} \times \frac{3}{5}$

16) $\sqrt{485}$

17) $\sqrt{4.85}$

18) $\sqrt{48.5}$

19) $\sqrt[3]{485}$

20) $\sqrt[3]{48.5}$

21) $\sqrt[3]{4.85}$

22) If a piece of cardboard is 0.025 inch thick, how thick is a stack of 4,560 sheets of the cardboard?

23) If a case of bolts contains 48 cartons, each containing a gross (144) of individual bolts, how many bolts does the case contain?

24) If a man's salary is \$4.83 per hour, and if he received a check for \$176.20, approximately how many hours had he worked?

25) If water weighs approximately 8.3 pounds per gallon and one cubic foot contains approximately 7.5 gallons, how much does a container holding 475 cubic feet of water weigh? How many gallons does it contain?

4

Algebra

4.1 The Meaning of Algebra

Algebra is an extension of arithmetic in which letters are used to represent numbers. Probably you have already encountered formulas which are algebraic in nature and have worked successfully with them without too much difficulty. For example, to find the area of a rectangle, we multiply the number of units in the length by the number of units in the width. If we let l represent the number of units in the length and w represent the number of units in the width, then the area, which we will represent by A, is given by the algebraic relation

$$A = lw.$$

4.2 Literal Numbers and Variables

In section 1.4 the concept of a literal number was introduced. In the example in Section 4.1 above, l, w, and A are literal numbers, since the letters are used to represent quantities of length, width, and area. The formula $A = l \cdot w$ will allow us to find the area of any rectangle of which both the length and the width are known. If the length were 4 inches and the width 5 inches, the area would be 20 square inches. If the length were 6 feet and the width 2 feet, the area would be 12 square feet. In other words, if the dimensions vary, the area varies. This observation suggests that the A, l, and w might also be referred to as *variables*, since their units of measure vary from case to case. If the width and length do not vary, they are called *constants*.

4.3 Fundamental Operations

The fundamental operations of algebra are essentially the same as those of arithmetic. They are indicated by the symbols $+$, $-$, $\times$, and $\div$, which have the same meaning as in arithmetic. If we let a and b represent two numbers, then the following are examples of these operations.

Example 1. The sum of a and b is represented by

$$a + b.$$

Example 2. The difference of a and b is represented by

$$a - b.$$

Example 3. The product of a and b is represented by

$$a \times b \quad \text{or} \quad a \cdot b \quad \text{or} \quad ab.$$

Example 4. The quotient of a and b is represented by

$$a \div b \quad \text{or} \quad \frac{a}{b} \quad \text{or} \quad b\overline{)a}.$$

Example 5. Four more than three times a is represented by

$$3a + 4.$$

One times a is written simply as a.

Example 6. The difference between a and b divided by the sum of c and d is shown by the expression

$$(a - b) \div (c + d),$$

where c and d also represent numbers.

The four operations discussed above are not the only operations used in algebra. Extracting square roots such as $\sqrt{a}$, raising quantities to a power using exponents such as a^5, and others have already been encountered in the arithmetic sections of this book. Additional operations will be introduced at appropriate points.

4.4 Algebraic Terminology

Notation is central to the study of algebra and one should be particularly aware of its meaning and application. For example, any combination of arithmetic and literal numbers written according to a specified set of rules and operations is called an *algebraic expression.* $a + b$, $a \cdot b$, $a \div b$, $a + b - c + 2d$, and $a^2 - 3ab + bd$ are all algebraic expressions.

An *algebraic term* or simply, a *term,* is an expression in which the parts are not separated by an addition or subtraction sign, that is, an expression in which the only operations involved are multiplication, division, or both. Thus $9xyz$, $7x \div y$, and $3a^2b$ all qualify as terms while $9 - x$, $a + b + 2z$, and $3x - 2y + 4z \div y$ do not.

Any algebraic expression of one term, such as $9xyz$, $7x \div y$, or $3a^2b$, is called a *monomial.* An expression made up of two terms, such as $3x - 2y$ or $5xy + 3z$, is called a *binomial.* An expression made up of three terms is called a *trinomial.* In general, any expression having one or more terms is called a *polynomial,* though the word is usually used to refer to expressions of four or more terms.

Example 1. Classify the following as monomials, binomials, trinomials or polynomials:

a) $5xy^2 + 3$ b) $\dfrac{12xy}{5}$

c) $xyz - x^2 + z^2$ d) $5x^4 + 3x^3 - 2x^2 - 4x + 5$

Answers:

a) is a binomial; $5xy^2$ is one term and 3 is another.
b) is a monomial since the only operations are multiplication and division.
c) is a trinomial; the terms are xyz, x^2, and z^2.
d) is a polynomial; it has the five terms $5x^4$, $3x^3$, $2x^2$, $4x$, and 5.

If we consider the expression $6ab$, 6 is said to be the *numerical coefficient* of ab and ab is said to be the *literal coefficient* of 6. We can also say that $6a$ is the coefficient of b or that $6b$ is the coefficient of a. In short, the coefficients of any factor or group of factors of a term are the remaining factors of the term.

If two terms have identical literal coefficients, the terms are said to be *like terms* or *similar terms.* Note here that not only do all the variables of the literal coefficients have to be the same, but they must be raised to identical powers to be considered like terms.

Example 2. $3a$ and $6a$ are similar terms since they both have a as a literal coefficient.

Example 3. $2a^2$, $3a^2$, and $7a^2$ are similar terms since they all have a^2 as a literal coefficient.

Example 4. $3ab$, $4ac$, and $5ad$ are not similar terms since their literal coefficients are not the same.

4.5 The Degree of a Polynomial

The degree of a polynomial in one variable is determined by the greatest exponent that appears on the variable in any term. If no variable occurs in an expression, the degree is zero. For the purposes of this section we will restrict the exponents to natural numbers and will not consider the degree of polynomials in more than one variable.

Example 1. $x^2 + 6x + 9$ is a trinomial of degree 2.

Example 2. $x^5 + 2x^2 + 6x + 1$ is a polynomial of degree 5.

Example 3. $x^6 + 2x^2 + 5x + 1$ is a polynomial of degree 6.

Exercise 18

Write an algebraic expression for each of the following:

1) the sum of 3, a, d, and f
2) z subtracted from s
3) 9 more than t
4) t decreased by f
5) the sum of x and y added to their product
6) the quotient of x and y subtracted from their product
7) 6 times the difference between x and y (Hint: use symbols of grouping.)
8) the sum of x and y decreased by their product
9) t decreased by 5
10) the product of x and y decreased by their sum
11) the product of 2 times x and 7 times b
12) 5 more than 6 times x
13) c less than t
14) the product of x and y divided by their sum
15) twice a number increased by 7
16) a number added to 8 times itself
17) $\frac{2}{3}$ of a number plus twice the same number

18) a number minus the number divided by 6

19) 5 times the sum of a number and 9

20) a number times 3 times itself plus the number divided by twice itself

Classify each of the following as a monomial, binomial, trinomial, or polynomial. More than one classification may apply.

21) $9xy$

22) $a + 2b - c$

23) $2a - b + c - d$

24) $\frac{3ab}{c}$

25) $6abc \div 2ad$

26) $6abc(3cde)$

27) $(5x)(7y) + 9xy$

28) $25a + 16b - 4c$

29) $a + b + 5d + 6e$

30) $(ab)(cd)(abc)$

Name the similar terms within each group below.

31) $2a, 3a, 6b, 9a^2$

32) $ab, ac, 3ab, a^2b$

33) $2ab, 7ab, 9a^2b, 2ab^2$

34) $4a^2b^2, 6a^2b^2, ca^2b^2, d^2a^2b^2$

35) xyz, xy, x^2y^2

36) $5(m + n), 3(n + m), 2(mn)$

37) $x + y, x - y, xy, \frac{x}{y}$

38) $5x^2y^2, 6xy^2, 4(xy)^2$

39) $(a + b), 3a + 3b, a + 3b$

40) $x + y, x^2y^3, (xy)^2y$

Answer the following:

41) If it takes x pounds of iron and y pounds of copper to form an alloy to pour a casting, indicate the amount of metal needed to pour 24 such castings.

42) A certain automobile has c cubic inches displacement. By working the cylinders and heads the displacement is increased by 3%. Indicate the new displacement.

43) One man earns n dollars per hour. Another man earns five dollars per hour more than half that amount. Show algebraically how much the second man earns.

44) A set of hand tools costs x dollars. A tune-up scope, timing light, and compression tester cost twice as much as the hand tools. A brake drum lathe and valve grinding machine cost three times as much as the others combined. Show the total cost of all the tools listed.

45) Given a rectangle with dimensions as shown:
   a) write an expression for the perimeter of the rectangle;
   b) write an expression for the area.

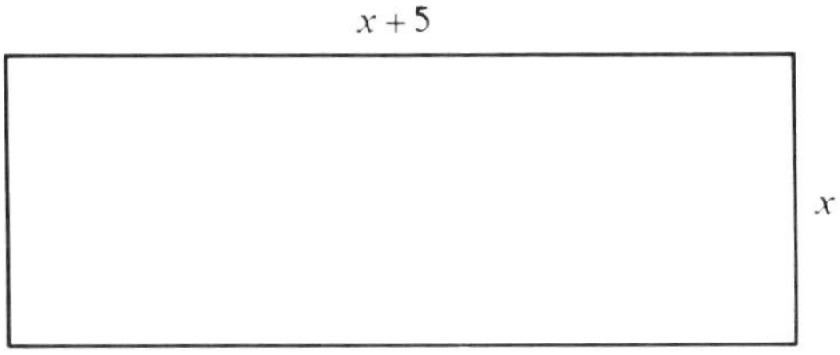

46) If union scale for a certain job is \$3.85 per hour, and if due to recent negotiations the rate increases by 7% beginning next month, indicate an expression for the new hourly rate.

47) There are x feet of 5/32-inch welding rod per pound. How many feet of rod are in a carton weighing 150 pounds?

48) If a casting weighs p pounds before machining and 22% of the metal is removed by machining, how much does the finished casting weigh?

49) A man earns x dollars per hour, and time and one-half on time over 40 hours per week. How much does the man make if he works 48 hours in one week?

50) An alloy contains 61% copper, 3% lead, 1% tin, and the rest zinc. If a casting of the alloy weighs x pounds, how much zinc does it contain?

51) The property tax on a house was n dollars in a given year. The following year the tax increased by 10%. Indicate the new tax.

52) If the width of a rectangle is w feet and the length is 4 feet more than the width, show the area of the rectangle.

53) The area of a square is given by s^2 where s is the length of a side. If the length of a side is increased by 10 inches, show the new area algebraically.

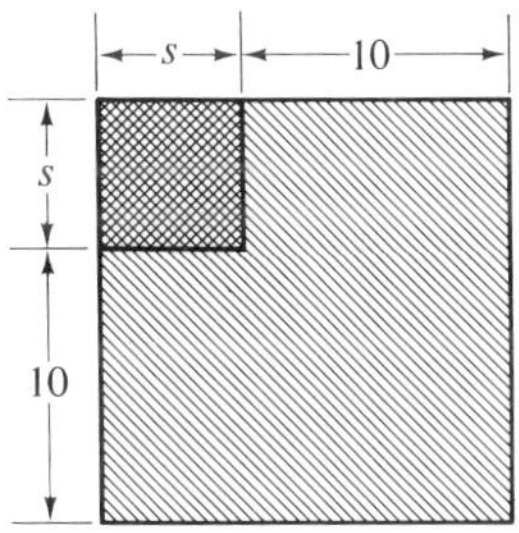

4.6 Operations with Algebraic Expressions

Algebraic expressions may be evaluated if the value of their literal coefficients is known. However, they may be added, subtracted, multiplied, and divided whether or not their literal coefficients are assigned an arithmetic value. The following examples will serve to illustrate some of these operations.

Example 1. Evaluate $x + 2y + z$ if $x = 2$, $y = 3$, and $z = 4$.

Since we know what x, y, and z represent, we can substitute these quantities for the literal numbers.

$$2 + 2(3) + 4 = 2 + 6 + 4 = 12.$$

Example 2. Evaluate $3x - y + 2z$ if $x = 2$, $y = -1$, and $z = 3$.

$$3(2) - (-1) + 2(3) = 6 + 1 + 6 = 13.$$

The student will recall that subtraction is the process of adding the additive inverse.

Example 3. Evaluate $3(y + z)$ if $y = 2$ and $z = 4$.

$$3(2 + 4) = 3(6) = 18.$$

Notice, however, that we could have used the distributive law first:

$$\begin{aligned} 3(y + z) &= 3y + 3z \\ &= 3 \cdot 2 + 3 \cdot 4 \\ &= 6 + 12 = 18. \end{aligned}$$

Example 4. Evaluate $2a + 3a - 4a$ when $a = 17$.

If we substitute directly, we have

$$\begin{aligned} 2 \cdot 17 + 3 \cdot 17 - 4 \cdot 17 &= 34 + 51 - 68 \\ &= 85 - 68 = 17. \end{aligned}$$

Notice how much easier the problem is if we simplify the polynomial first:

$$2a + 3a - 4a = a = 17.$$

Example 5. Simplify $5a^2 + 2ab + 6a + 3a^2 - 7a$.

$$\begin{aligned} 5a^2 + 2ab + 6a + 3a^2 - 7a \\ &= (5a^2 + 3a^2) + 2ab + (6a - 7a) \\ &= 8a^2 + 2ab - a. \end{aligned}$$

The second step is permissible because of the commutative and associative laws.

The student should observe in each of the above examples that similar terms are added by adding their numerical coefficients. Notice in Example 5 that only similar terms have been added and that further addition and subtraction can only be indicated.

When we encounter longer addition or subtraction problems, the work can be simplified by writing the terms in column form.

Example 6. Add $2x^2 + 6x + y^2$, $3x^2 + 2x - y^2$, and $4x^2 + 6x + 3y^2 + z$.

$$\begin{array}{l} 2x^2 + \ \ 6x + \ \ y^2 \\ 3x^2 + \ \ 2x - \ \ y^2 \\ \underline{4x^2 + \ \ 6x + 3y^2 + z} \\ 9x^2 + 14x + 3y^2 + z \end{array}$$

Notice that like terms are placed in the same vertical column. The sum is then the sum of the numerical coefficients multiplied by the literal coefficient of each column.

Example 7. Subtract $x^2 + y^2 - 2z^2 + w$ from $3x^2 + y^2 - z^2 - 3w$.

$$\begin{array}{r} 3x^2 + y^2 - \ \ z^2 - 3w \\ \underline{-(\ x^2 + y^2 - 2z^2 + \ \ w)} \end{array}$$

or

$$\begin{array}{l} \ \ 3x^2 + \ y^2 - \ \ z^2 - 3w \\ \underline{-\ \ x^2 - \ y^2 + 2z^2 - \ \ w} \\ \ \ 2x^2 + 0y^2 + 1z^2 - 4w = 2x^2 + z^2 - 4w \end{array}$$

Recall that subtraction is the same as adding the additive inverse, so that subtracting x^2 from $3x^2$ is the same as adding $-x^2$ to $3x^2$, and so on.

Example 8. Subtract $(-x^2 - 2y^2z + z^2 - 3k^2 - w)$ from $(6x^2 + 2y^2z + 3z^2 - k^2 + w)$.

This is the same as $(6x^2 + 2y^2z + 3z^2 - k^2 + w) + (x^2 + 2y^2z - z^2 + 3k^2 + w)$ or

$$\begin{array}{l} 6x^2 + 2y^2z + 3z^2 - \ \ k^2 + \ \ w \\ \underline{\ \ x^2 + 2y^2z - \ \ z^2 + 3k^2 + \ \ w} \\ 7x^2 + 4y^2z + 2z^2 + 2k^2 + 2w \end{array}$$

Exercise 19

Evaluate the following if $a = 2$, $b = -3$, $c = 4$, and $d = -1$.

1) $a + b + c$

2) $a - b - c$

3) $2a + b - 3c$

4) $5abc$

5) $abc - d$

6) $3(a + b)$

7) $2(a + b + c)$

8) $(a + b) + (d - c)$

9) $2(a + b) + 3(c - d)$

10) $(a + b)(c + d)$

11) $(a + b)(c - d)$

12) $\dfrac{5a + b}{7}$

13) $\dfrac{3(a + b) + 2(c + d)}{6}$

14) $\dfrac{2(a + b) + 3(c + d)}{6(a + b)}$

15) $\dfrac{5abc + 2cd + d}{cd}$

16) $\dfrac{3a + b - 4d}{a + d}$

17) $a^2 + b^2$

18) $a^3 + b^2 + d^3$

19) $a^2 + c^2 + (2d)^2$

20) $a^3 + b^2c^2$

21) $a^3b^2 + a^2b^3 + (a^2b)^2$

Add the following:

22)
$$\begin{array}{r} 3a \\ \underline{4a} \end{array}$$

23)
$$\begin{array}{r} -5b \\ \underline{3b} \end{array}$$

24)
$$\begin{array}{r} -4xy^2 \\ \underline{xy^2} \end{array}$$

25)
$$\begin{array}{r} ac \\ \underline{6ac} \end{array}$$

26)
$$\begin{array}{r} 3mt \\ 6mt \\ 2mt \\ -8mt \\ \underline{23004mt} \end{array}$$

27)
$$\begin{array}{r} 3xy^2 \\ 4xy^2 \\ 5xy^2 \\ \underline{-2xy^2} \end{array}$$

28)
$$\begin{array}{r} u + v \\ 6u - 2v \\ \underline{2u + 6v} \end{array}$$

29)
$$\begin{array}{r} x + y \\ x + 2y \\ \underline{3x - y} \end{array}$$

30)
$$\begin{array}{r} 2x - y + z \\ x - y + z \\ \underline{3x + y - 2z} \end{array}$$

31)
$$\begin{array}{r} a + b + c^2 \\ 2a + b + c^2 \\ \underline{a + 2b + 2c^2} \end{array}$$

32)
$$\begin{array}{rrrr} 2m + & r + & s + & t \\ m - & r + & 2s - & t \\ -3m + & 2r & & - 3t \\ \hline & & - 3s + & 3t \end{array}$$

33)
$$\begin{array}{r} 3a + 4b - c \\ 6a + 2b + 3c \\ -2a + b - 5c \\ \underline{- a + 2b + 2c} \end{array}$$

34)
$$\begin{array}{r} 2x^2 - 3xy + 7y^2 \\ 3x^2 - 2xy - 7y^2 \\ \underline{2x^2 + 5xy + y^2} \end{array}$$

35)
$$\begin{array}{r} a + b + a^2 + b^2 \\ a - b - 2a^2 + b^2 \\ \underline{-2a - 2b + a^2 - 2b^2} \end{array}$$

Subtract the following:

36)
$$\begin{array}{r} 6a \\ \underline{-(2a)} \end{array}$$

37)
$$\begin{array}{r} 2a + b \\ \underline{-(\ a + b)} \end{array}$$

38)
$$\begin{array}{r} 3a^2 + 2b^2 \\ \underline{-(\ a^2 - b^2)} \end{array}$$

39)
$$\begin{array}{r} 4a^2 - b^2 + x \\ \underline{-(\ a^2 + b^2 - x)} \end{array}$$

40)
$$\begin{array}{r} 3a + 2b + 6c \\ \underline{-(\ a + b + 2c)} \end{array}$$

41)
$$\begin{array}{r} 2a - 3b + c^2 \\ \underline{-(6a - 2b + c^2)} \end{array}$$

42)
$$\begin{array}{r} 2a - b + 3c^2 \\ \underline{-(\ a - 2b + 4c^2)} \end{array}$$

43)
$$\begin{array}{r} 2a^3 + 3a^2 - 6c \\ \underline{-(\ a^3 + a^2 - 4c)} \end{array}$$

44)
$$\begin{array}{r} a^2 - 2b^2 + c^2 \\ \underline{-(a^2 - c^2 + z)} \end{array}$$

45)
$$\begin{array}{r} 2c^2 + d^2 - 8c^2 \\ \underline{-(\ c^2 + d^2 - 4c^2)} \end{array}$$

46) Subtract $3x^2 - 2x + 4$ from $5x^2 - 6x + 9$.

47) Subtract $x^2 + 2x + 7$ from $3x^2 - 9x + 14$.

48) Add $2x^2 + x + y$ to the sum of $3x^2 + 4x + 2y$ and $x^2 + x + y$.

49) Subtract $3x^2 - 2y + z$ from the sum of $4x^2 + 6y + 3z$ and $x^2 + y - 2z$.

50) The total surface of a solid is given by the expression $2lw + 2lh + 2wh$, where l, w, and h are the length, width, and height of the solid. Find the total surface if $l = 5$, $w = 4$, and $h = 6$.

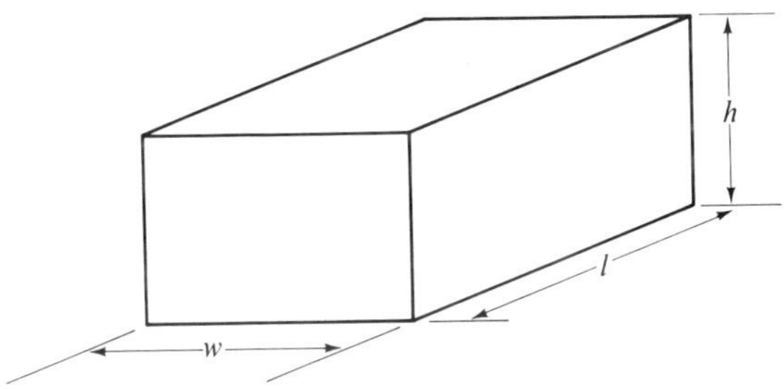

51) In discussing the energy of a moving body, the expression $mv_1^2 + mv_2^2$ is used. Evaluate the expression if $m = 25$, $v_1 = 6$, and $v_2 = 4$.

52) If two resistances are connected in parallel, the expression for the total resistance is given by the expression $\frac{1}{R_1} + \frac{1}{R_2}$. Evaluate the expression when $R_1 = 5$ ohms and $R_2 = 3$ ohms.

53) The distance a body travels in free-fall is given by the expression $v_0t + \frac{1}{2}gt^2$. Find the distance when $v_0 = 200$, $g = 32$, and $t = 5$.

54) The diagonal of a rectangular prism is given by the formula $d = \sqrt{l^2 + w^2 + h^2}$. Determine d for each set of values in the table.

l	w	h
2	4	3
4	2	2
2.4	.3	1.2

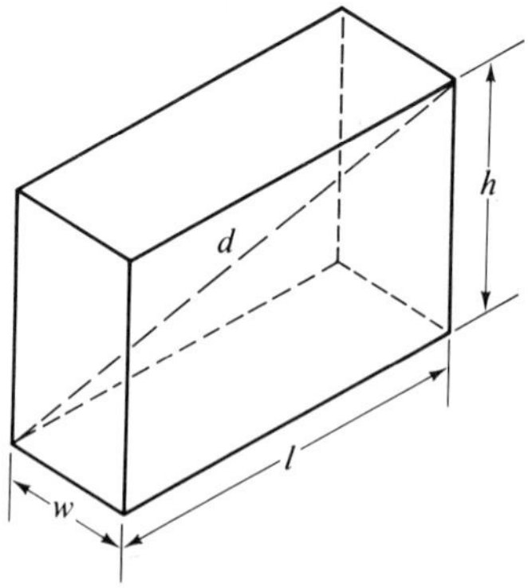

55) Shop drawings frequently give dimensions in terms of letters or symbols. This practice is frequently used when many parts of the same shape but of different sizes are to be made. In the drawing below, determine the volume of metal necessary to pour three of each of the castings whose dimensions are given in the table, where $V = \frac{1}{2}ABC$.

A	B	C
1″	2″	3″
1.2″	2.7″	3.9″
$1\frac{2}{3}$″	$1\frac{1}{8}$″	$4\frac{1}{4}$″

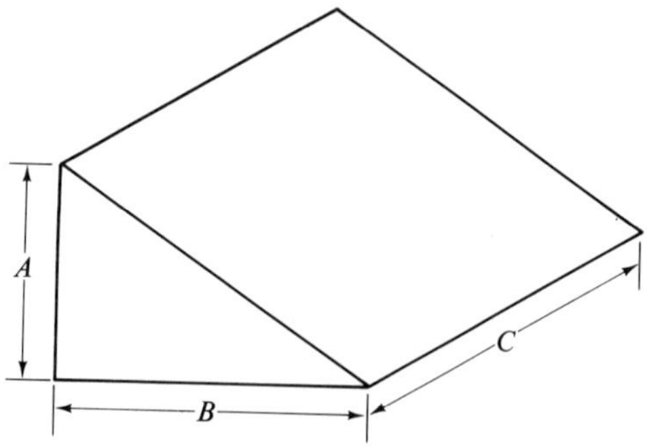

4.7 Multiplication of a Polynomial by a Monomial

Multiplication in algebra requires a knowledge of the multiplication processes that are familiar to the student from arithmetic, as well as a knowledge of the rules for multiplication of exponential and signed numbers.

The latter rules were developed in Chapter 2 and generalized in Problems 1 through 6 of Exercise 5. It is also necessary that the student remember the commutative, associative, and distributive laws.

Consider the product of a binomial and a monomial. In order to multiply $a(a^2 + ab)$ we first use the distributive law to write $a(a^2) + a(ab)$. We then use the associative law to write $(a \cdot a^2) + (a \cdot a)b$, and finally we write $a^3 + a^2b$ by using the rules of exponents. Therefore

$$a(a^2 + ab) = a^3 + a^2b.$$

The same processes that were used to multiply a binomial by a monomial can be extended to the products of monomials and polynomials. In the following examples only the work is shown; the reader should supply the reasons.

Example 1. $$\begin{aligned} 2x^2y(x + y - xy^2) &= 2x^2y(x) + 2x^2y(y) - (2x^2y)(xy^2) \\ &= 2(x^2x)y + 2x^2(y \cdot y) - 2(x^2x)(y \cdot y^2) \\ &= 2x^3y + 2x^2y^2 - 2x^3y^3. \end{aligned}$$

Example 2. $$\begin{aligned} ab^2c(a + b + c) &= (ab^2c)a + (ab^2c)b + (ab^2c)c \\ &= (a \cdot a)(b^2c) + a(b^2b)c + (ab^2)(c \cdot c) \\ &= a^2b^2c + ab^3c + ab^2c^2. \end{aligned}$$

Once the reader is certain he understands the reasons for the steps in the above examples, he should try writing the product omitting the intermediate steps.

4.8 Multiplication of a Polynomial by a Polynomial

Consider the product of $(a + b)(c + d)$.

$$\begin{aligned} (a + b)(c + d) &= (a + b)c + (a + b)d \\ &= ac + bc + ad + bd \end{aligned}$$

using the distributive law for both steps. None of the terms of the product are similar terms so their addition can only be indicated.

Very frequently products of polynomials are written in vertical form.

Example 1.

$$\begin{array}{l} a + b \\ \underline{c + d} \\ ac + bc \\ \underline{\qquad\quad ad + bd} \\ ac + bc + ad + bd \end{array}$$

To use this method, we multiply each term of the multiplicand, $(a + b)$, by the first term of the multiplier, c. Next we multiply each term of the multiplicand, $(a + b)$, by the second term of the multiplier, d. If any of the products result in similar terms, these terms are placed in the same column to facilitate their addition.

Example 2.

$$\begin{array}{l} x + y \\ \underline{2x - 3y} \\ 2x^2 + 2xy \\ \underline{\qquad - 3xy - 3y^2} \\ 2x^2 - xy - 3y^2 \end{array}$$

Example 3.

$$\begin{array}{l} x + 2y + 3z \\ \underline{x - 2y} \\ x^2 + 2xy + 3xz \\ \underline{\quad - 2xy \qquad\qquad - 4y^2 - 6yz} \\ x^2 \qquad\quad + 3xz - 4y^2 - 6yz \end{array}$$

4.9 Special Products

Certain products appear with such frequency in algebra that it is worthwhile to look at them separately. If we observe closely the procedure described above for multiplication of polynomials, we see that this method is really a method of using the distributive law over and over again and organizing the sums by writing like terms in the same column.

We can write a convenient rule for the special case of the product of two binomials. *The product of two binomials is found as the product of the first terms plus the product of the outside terms plus the product of the inside terms plus the product of the last terms.* This is often abbreviated as FOIL (First, Outer, Inner, Last).

Example 1. $(x + y)(a + b) = ax + bx + ay + by.$

The first terms are x and a, the outer terms are x and b, the inner terms are y and a, and the last terms are y and b.

Example 2.
$$\begin{aligned}(4x + y)(3x + 2y) &= 12x^2 + 8xy + 3xy + 2y^2 \\ &= 12x^2 + 11xy + 2y^2.\end{aligned}$$

Another common product is the square of a binomial, that is, the product of the type $(a + b)(a + b)$. Using the rule just described, this would be $a^2 + ab + ab + b^2$. Notice, however, that the two middle terms are alike and can be combined to give $a^2 + 2ab + b^2$. This is always the case when squaring a binomial, and the procedure can be generalized as: *the square of a binomial is the square of the first term plus twice the product of the two terms plus the square of the last.*

Example 3. $(x + y)^2 = x^2 + 2xy + y^2$.

The first term is x; the last term is y.

Example 4. $(4a + 3b)^2 = 16a^2 + 24ab + 9b^2$.

Example 5. $(2m - n)^2 = 4m^2 - 4mn + n^2$.

The third special product we will discuss is the product of two binomials having terms which are identical except for the sign between them, such as $(a + b)(a - b)$. Using the first method described in this section, this becomes $a^2 - ab + ab - b^2$. Observe that the middle terms in this case are additive inverses of one another and add up to zero. The result is simply *the square of the first term minus the square of the last term.*

Example 6. $(a^2 - b)(a^2 + b) = a^4 - b^2$.

Example 7. $(xy + 3z)(xy - 3z) = x^2y^2 - 9z^2$.

Exercise 20

Find the following products:

1) $x(x + y)$
2) $a(5 + a)$
3) $x^2(x^2 + y^2)$
4) $5(a + b)$
5) $5a(a - b^2)$
6) $ab^2(a^2b^2 + 1)$
7) $x^2(2x + 2y)$
8) $ab(a^2b + ab^2 + a)$
9) $3ab^2c(a - 2b^2 - ab^2c)$
10) $a^2bc^2(a^2 + b^2 + c^2)$
11) $-2ab^2c(a^2 - b + 2c^2)$
12) $-3a^2bc(a^2 - 2b - c^2 + ac)$
13) $(x - y)(2x - y)$
14) $(a - 2b)(a + 2b)$
15) $(x - y)(x - y)$
16) $(x - y)(x + y)$
17) $(a - b)(x + y)$
18) $(x^2 + 1)(x^2 - 1)$
19) $(a - b)^2$
20) $(x - y)(3x + 2y)$

21) $(a + b)(a^2 - ab + b^2)$
22) $(a - b)(a^2 + ab + b^2)$
23) $(a - b)(a + b)$
24) $(x^2 + x + 1)(x^2 - x - 1)$
25) $(x^2 + xy + 1)^2$
26) $(5a - 3b)(a^2 + ab + 2b^2)$
27) $(a + b)(a - b)(3a - 2b)$
28) $(x + a)(x^2 + ax + a^2)$
29) $(x + a)(x + a)$
30) $(x - a)(x - a)$
31) $(x + a)(x + a)(x - a)$
32) $(x + a)(x + a)(x + a)$
33) $(x + 2)(x - 3)(x + 4)$
34) $(x + m)^2$
35) $(a - b + c)^2$
36) $(r - s)(r + s)$
37) $(5 - 3t)(5 + 3t)$
38) $(15 + m)^2$

Find the following products by putting the factors into binomial form and using a special product rule.

39) Multiply (32)(28) by expressing 32 as $(30 + 2)$ and 28 as $(30 - 2)$.
40) (51)(49)
41) (98)(102)
42) (47)(33)
43) $(25)^2$
44) $(16)^2$
45) (25)(15)

4.10 Division of a Polynomial by a Monomial

To divide a polynomial by a monomial, divide each term of the polynomial by the monomial. Recall that division is the process of multiplying by the inverse of a quantity, so the distributive law can be used to justify this rule.

Example 1. Divide $8 + 10$ by 2, where $8 + 10$ is the polynomial and 2 is the monomial.

$$(8 + 10) \div 2 = \frac{1}{2}(8 + 10)$$

$$= \frac{1}{2} \cdot 8 + \frac{1}{2} \cdot 10$$

$$= \frac{8}{2} + \frac{10}{2}.$$

Notice that we can avoid the intermediate steps by using the rule given above.

Example 2. $(2x^3 + 6x^2 + 12x) \div 2x = \dfrac{2x^3}{2x} + \dfrac{6x^2}{2x} + \dfrac{12x}{2x}$

$$= x^2 + 3x + 6.$$

(Recall from Chapter 2 that $a^m/a^n = a^{m-n}$.)

Example 3. $\dfrac{6a^2x^3y^2 + 12x^2y^2 - 4xy}{3xy} = \dfrac{6a^2x^3y^2}{3xy} + \dfrac{12x^2y^2}{3xy} - \dfrac{4xy}{3xy}$

$$= 2a^2x^2y + 4xy - \frac{4}{3}.$$

4.11 Division of a Polynomial by a Polynomial

Division of a polynomial by a polynomial is performed in a way similar to that of long division of two arithmetic quantities. The analogy can be seen from the following examples:

Example 1.

$$\begin{array}{r} 22 \\ 11\overline{)242} \\ \underline{22} \\ 22 \\ \underline{22} \\ 0 \end{array}$$

Example 2. $242 \div 11 = [2(10^2) + 4(10) + 2] \div [1(10) + 1]$.

$$\begin{array}{r} 2(10) + 2 \\ 1(10) + 1\overline{)2(10^2) + 4(10) + 2} \\ \underline{2(10^2) + 2(10)} \\ 2(10) + 2 \\ \underline{2(10) + 2} \\ 0 \end{array}$$

Consider both the dividend and the divisor in place value notation. We first answer the question: how many times will 1(10) divide into $2(10^2)$? This result, 2(10), is then multiplied by all parts of the divisor, yielding $2(10^2) + 2(10)$. We then subtract and bring the next digit down. Now we answer the new question: how many times will 1(10) divide into $2(10) + 2$? This result, 2, is multiplied by all parts of the divisor yielding $2(10) + 2$. We subtract and find the remainder zero. The solution is then $2(10) + 2 = 22$. If the remainder were not zero we would continue until the highest power of 10 in the divisor exceeded the highest power of 10 in the dividend.

If we carefully study the processes used to carry out the division in Example 2, we can use the same ideas to complete the following division problems.

Example 3. $(x^2 + 6x + 5) \div (x + 1)$.

$$\begin{array}{rl} & x \;+5 \\ x+1\,\big) & \overline{x^2 + 6x + 5} \\ & \underline{x^2 + \;\; x} \\ & \qquad 5x + 5 \\ & \qquad \underline{5x + 5} \\ & \qquad\qquad 0 \end{array}$$

First we divide x into x^2 obtaining x as a partial quotient. This quotient is multiplied by the divisor, $(x + 1)$, and the result subtracted. We then divide x into $5x$ obtaining 5 as the quotient. Now we multiply $(x + 1)$ by 5 and subtract. Since there is no remainder, the division is complete.

Example 4. $(x^3 - 6x^2 + 7) \div (x^2 - 1)$.

$$\begin{array}{rl} & x \;\; - 6 \\ x^2-1\,\big) & \overline{x^3 - 6x^2 \qquad + 7} \\ & \underline{x^3 \qquad\quad - x} \\ & \quad - 6x^2 + x + 7 \\ & \quad \underline{- 6x^2 \qquad + 6} \\ & \qquad\qquad\quad x + 1 \end{array}$$

Notice that like terms are placed under each other and that both divisor and dividend are arranged in descending powers of x. The reader should also notice where the x and x^2 terms are missing and that adequate spacing is allowed to compensate for this in the division. The quotient with remainder is written in the same manner as in arithmetic, namely,

$$x - 6 + \frac{x + 1}{x^2 - 1}.$$

Methods of reducing this remainder will be discussed in a later section.

Exercise 21

Perform the following divisions:

1) $(2x + 4y + 6z) \div 2$
2) $(3x^2 + 6x + 9) \div 3$
3) $(8z^3 + 4z + 8) \div 2$

4) $(3x^2y + 9x^3y^2 + 12x^2) \div 3x^2$

5) $(2x^2y^2 + 3y^2 + y^4) \div y^2$

6) $(2a^5 + 4a^4 + 2a^2 + 4a) \div 2a$

7) $(16a^3 + 12a^2 - 16a) \div 4a$

8) $(12a^2b^2 + 18a^3b^3 - 24a^4b^4) \div 6a^2b^2$

9) $(18x^2y^2 + 36x^3y^5 - 54x^6y^7) \div (-6x^2y^2)$

10) $(a^2 + 2ab + b^2) \div (a + b)$

11) $(12k^2 - 18k^4 - 36k^6) \div (-2k^2)$

12) $(12x^2y^3 - 36x^3y^2) \div 4x^2y^2$

13) $(x^2 + 6x + 5) \div (x + 5)$

14) $(x^2 + 10x + 16) \div (x + 2)$

15) $(x^2 + 3x + 2) \div (x + 1)$

16) $(x^2 + 2x - 8) \div (x - 2)$

17) $(x^2 - 4x - 21) \div (x + 3)$

18) $(m^2 - m - 20) \div (m - 5)$

19) $(8x^2 - 6x - 5) \div (2x + 1)$

20) $(5a^2 + 9ab - 2b^2) \div (a + 2b)$

21) $(x^2 - x - 19) \div (x - 5)$

22) $(x^2 + x - 3) \div (x - 7)$

23) $(x^2 + 2x + 1) \div (x - 1)$

24) $(x^3 + 4x^2 + 7x + 6) \div (x + 2)$

25) $(y^3 + y^2 - 10y + 8) \div (y - 1)$

26) $(z^3 + z^2 - z - 1) \div (z - 1)$

27) $(x^3 + 7x^2 - 17x + 9) \div (x + 9)$

28) $(x^4 + x^3 + x^2 + x) \div (x + 1)$

29) $(2r^4 - 3r^3 - r^2 + 3r - 1) \div (2r^2 + r - 1)$

30) $(3b^4 + 4b^3 - 2b^2 - 4b - 1) \div (3b^2 - 2b - 1)$

31) $(x^3 - 27) \div (x - 3)$

32) $(x^3 - 6x^2 + 12x - 8) \div (x^2 - 4x + 4)$

33) $(x^3 - 3x^2 + 2x + 1) \div (x - 1)$

34) $(x^3 - 1) \div (x + 1)$

35) $(2x^3 - 3x^2 - 13x + 12) \div (x - 3)$

36) $(10x^2 + 7xy - 12y^2) \div (2x + 3y)$

4.12 Synthetic Division

In the preceding section we saw that the division of two polynomials can be carried out in a manner analogous to the long division processes of arithmetic. If the polynomial that acts as the divisor is of the form $x - a$ or $x + a$, the division can be synthesized and carried out by a much shorter and somewhat simpler process. This process is known as *synthetic division*. To illustrate the process consider the quotient

$$\frac{3x^3 + 3x^2 + 4x + 1}{x + 2}$$

which can be divided by long divisions as follows.

$$\begin{array}{r}
3x^2 - 3x + 10 \\
x + 2\overline{)3x^3 + 3x^2 + 4x + 1} \\
\underline{3x^3 + 6x^2} \qquad\qquad \\
-3x^2 + 4x \qquad \\
\underline{-3x^2 - 6x} \qquad \\
10x + 1 \\
\underline{10x + 20} \\
-19
\end{array}$$

This result would be written as

$$\frac{3x^3 + 3x^2 + 4x + 1}{x + 2} = 3x^2 - 3x + 10 + \frac{-19}{x + 2}.$$

We now rewrite the same division problem, but this time only the coefficients of the terms are shown.

$$\begin{array}{r}
3 - 3 + 10 \\
1 + 2\overline{)3 + 3 + 4 + 1} \\
\underline{3^* + 6} \qquad\qquad \\
-3 + 4^\dagger \qquad \\
\underline{-3^* - 6} \qquad \\
10 + 1^\dagger \\
\underline{10^* + 20} \\
-19
\end{array}$$

We now observe the following five points about the display.

1) The lead coefficient in the quotient will always be the same as the lead coefficient in the dividend.
2) The exponent on the variable of the lead coefficient of the quotient will always be one less than the exponent on the lead coefficient of the dividend.
3) The numbers marked with an asterisk are repetitions of the numbers written immediately above them.

4) The numbers marked with the cross are repetitions of the numbers appearing above them in the dividend.
5) The numbers marked with an asterisk are the same numbers that appear in the quotient.

If we eliminate the repetition discussed above, we can shorten the display as follows. First, we eliminate the numbers marked with an asterisk, since they are repetitions and a difference of zero is obtained when subtraction is carried out. Further, we eliminate the first number in the divisor, since it is the product of this number and the numbers in the quotient that produced the terms marked by the asterisk. We now have

$$\begin{array}{rrrrr}
 & 3 & -\,3 & +\,10 & \\ \cline{2-5}
2) & 3 & +\,3 & +\,4 & +\,1 \\
 & & 6 & & \\ \cline{3-4}
 & & -\,3 & +\,4 & \\
 & & & -\,6 & \\ \cline{3-5}
 & & & 10 & +\,1 \\
 & & & & 20 \\ \cline{5-5}
 & & & & -\,19
\end{array}$$

We next eliminate those numbers marked with a cross and move all other terms up until they are immediately under the numbers that appear above them.

$$\begin{array}{rrrrr}
 & 3 & -\,3 & +\,10 & \\ \cline{2-5}
2) & 3 & +\,3 & +\,4 & +\,1 \\
 & & 6 & -\,6 & +\,20 \\ \cline{3-5}
 & & -\,3 & +\,10 & -\,19
\end{array}$$

We now notice that we can bring the first number in the dividend down to the third line of the display creating the quotient as the first three numbers. The fourth number is the remainder.

We now remove the quotient from above the division symbol as well as removing the division symbol itself. We now have

$$\begin{array}{r|rrrrl}
\underline{2} & 3 & +\,3 & +\,4 & +\,1 & \qquad \text{row 1} \\
\multicolumn{1}{r}{} & & +\,6 & -\,6 & +\,20 & \qquad \text{row 2} \\ \cline{2-5}
\multicolumn{1}{r}{} & 3 & -\,3 & +\,10 & -\,19 & \qquad \text{row 3}
\end{array}$$

Finally, we observe that each number in the second row has been subtracted from each number in the first row. Observe also that each of these numbers were obtained by multiplying the second number in the divisor by the corresponding numbers in the quotient. If we change the sign on the second number before dividing, we could add instead of subtract. The display is now

$$\begin{array}{r|rrrr}
\underline{-2} & 3 & +\,3 & +\,4 & +\,1 \\
\multicolumn{1}{r}{} & & -\,6 & +\,6 & -\,20 \\ \cline{2-5}
\multicolumn{1}{r}{} & 3 & -\,3 & +\,10 & -\,19
\end{array}$$

A final analysis of the process shows nine steps, stated below, which achieve the desired result for the division.

1) The coefficients are detached from the variables and written in a horizontal fashion on the first line.
2) The sign is changed on the second term of the divisor, and it is written to the left of the first coefficient of the dividend.
3) The first coefficient 3 is brought down to the third line.
4) 3 is multiplied by the divisor -2 and the product -6, written under the second coefficient 3.
5) The sum of 3 and -6, or -3, is found and written on the third line.
6) -3 is multiplied by the divisor -2, and the product 6 is written under the next coefficient 4.
7) The sum of 4 and 6, or 10, is found and written on the third line.
8) 10 is multiplied by the divisor -2, and the product -20 is written under the last coefficient.
9) The sum -19 is found by completing the process.

Since the quotient has a leading term whose exponent is one less than the dividend, the result is read as $3x^2 - 3x + 10 - 19/(x + 2)$. The last term is always the remainder.

Example 1. Use synthetic division to divide $x^2 - 8x + 12$ by $x - 6$. Following the steps outlined above we have

$$\begin{array}{r|rrr} 6 & 1 & -8 & +12 \\ & & +6 & -12 \\ \hline & 1 & -2 & 0 \end{array}$$

The quotient is $x - 2$, and the remainder is 0.

Example 2. Use synthetic division to divide $x^3 + x^2 + 5$ by $x - 1$.

$$\begin{array}{r|rrrr} 1 & 1 & +1 & +0 & +5 \\ & & +1 & +2 & +2 \\ \hline & 1 & +2 & +2 & +7 \end{array}$$

Notice that 0 has been supplied as the coefficient of the 1st degree term.
The quotient is $x^2 + 2x + 2$, and the remainder is 7.

Example 3. Divide $x^4 + 3x^3 + 2x^2 - x + 1$ by $x + 2$.
At this point we need to observe that the variable is not arranged in terms of descending powers. This must be done before the division.
Rearranging the terms we have

$$\begin{array}{r|l} -2 & 1 + 3 - 1 + 2 + 1 \\ & \quad - 2 - 2 + 6 - 16 \\ \hline & 1 + 1 - 3 + 8 - 15 \end{array}$$

The quotient is $x^3 + x^2 - 3x + 8$, and the remainder is -15.

Example 4. Divide $x^5 + 2x^2 + 24$ by $x + 2$.

$$\begin{array}{r|l} -2 & 1 + 0 + 0 + 2 + 0 + 24 \\ & \quad - 2 + 4 - 8 + 12 - 24 \\ \hline & 1 - 2 + 4 - 6 + 12 + 0 \end{array}$$

The quotient is $x^4 - 2x^3 + 4x^2 - 6x + 12$, and the remainder is zero. Notice that zero has replaced each coefficient not present in the original form.

Exercise 22

Use synthetic division to carry out each of the following problems. Write the answers in the form of the quotient plus the remainder divided by the divisor.

1) $\dfrac{x^2 + x - 6}{x - 3}$ 2) $\dfrac{x^2 + 5x + 6}{x + 2}$ 3) $\dfrac{x^2 + 2x - 8}{x + 4}$

4) $\dfrac{x^2 + 4x + 4}{x + 2}$ 5) $\dfrac{x^2 + x - 12}{x - 3}$ 6) $\dfrac{x^2 - 1}{x + 1}$

7) $\dfrac{x^3 + 3x^2 + 3x + 1}{x + 1}$ 8) $\dfrac{x^3 - 4x^2 + 5x - 2}{x - 2}$

9) $\dfrac{2x^4 - 3x + 6}{x - 2}$ 10) $\dfrac{x^5 - 1}{x - 1}$

11) $\dfrac{x^4 - 2x^3 - 3x^2 + 8x - 4}{x + 2}$ 12) $\dfrac{x^7 + 2x^5 + x^2 - 6x + 11}{x + 3}$

13) $\dfrac{x^6 + 3x^3 + x}{x - 1}$ 14) $\dfrac{x^5 + 3x^4 + x^2 + 2x + 1}{x - 2}$

15) $\dfrac{2x^4 + 7}{x - 5}$ 16) $\dfrac{x^7 + 1}{x + 2}$

4.13 Factoring

One of the frequent problems encountered by the student of algebra is that of recognizing the factors of an algebraic expression. Recall from Chapter 1 that when two or more numbers are multiplied, each number is called a factor of the product. This concept can be extended to include algebraic expressions, as we shall discover shortly.

In Section 4.6 we saw that in order to find the product of a monomial and a binomial we apply the distributive law. For example.

$$2x^2(x^2 + xy + 2y^2) = 2x^4 + 2x^3y + 4x^2y^2.$$

Note that each result in the product is obtained as the product of two factors. If we were to reverse the process, we would be identifying the highest factor common to each term. The indicated product of this highest common factor and the remaining factor would satisfy the condition that the polynomial be *factored*. Each of the following problems illustrate factoring by removing the highest common factor.

Example 1. $3ax^2 + 6ax + 9a^2 = 3a(x^2 + 2x + 3a).$

Example 2. $9x^2yz + 18x^3y^3z^3 - 6x^2y = 3x^2y(3z + 6xy^2z^3 - 2).$

Example 3. $x^2y^2z^2 + xyz + x^2y^3z^2 + xy = xy(xyz^2 + z + xy^2z^2 + 1).$

At first glance, an expression such as $x^2 + 3x + 2$ does not appear to be factorable since a common monomial factor is not present. However, if we find the product of the two binomials $(x + 1)$ and $(x + 2)$ we will see that the expression $x^2 + 3x + 2$ has these two binomials as factors:

$$(x + 1)(x + 2) = x^2 + 3x + 2.$$

If we are to be able to recognize these factors, we will first need to review the means of obtaining their product.

Example 4.

$$\begin{array}{rrrr} x & - & 3 & \\ x & + & 4 & \\ \hline x^2 & - & 3x & \\ & + & 4x & - 12 \\ \hline x^2 & + & x & - 12 \end{array}$$

Example 5.

$$\begin{array}{rrrr} x & + & 1 & \\ x & + & 2 & \\ \hline x^2 & + & x & \\ & + & 2x & + 2 \\ \hline x^2 & + & 3x & + 2 \end{array}$$

Notice that in both examples the first term of the product is the product of the first terms of the binomials while the last term of the product is the product of the last terms of the binomials.

At this point we can conclude that if a trinomial of the form $ax^2 + bx + c$ (a, b, c are constants) is factorable into the product of two binomials then the following must be true:

1. The first term of each binomial is a factor of the first term of the trinomial.
2. The last term of each binomial is a factor of the last term of the trinomial.
3. The middle term is formed from the sum of the products of the first and last terms, as indicated by the arrows in the example above.

If we examine the result of Example 4 under the first two criteria, we see that

$$(x-1)(x+12)$$
$$(x+1)(x-12)$$
$$(x+2)(x-6)$$
$$(x-6)(x+2)$$
$$(x+3)(x-4)$$
$$(x+4)(x-3)$$

are all possible factors. Only the last pair, however, satisfy criterion 3 to produce the correct middle term.

As a final example, consider the factors of $6x^2 - 13x + 6$. If we write all the possible binomial factors that produce the correct first and last terms we consider each of the following:

$(2x+3)(3x+2)$	$(6x+3)(x+2)$
$(2x-3)(3x-2)$	$(6x-3)(x-2)$
$(2x+2)(3x+3)$	$(6x+2)(x+3)$
$(2x-2)(3x-3)$	$(6x-2)(x-3)$
$(2x+6)(3x+1)$	$(6x+6)(x+1)$
$(2x-6)(3x-1)$	$(6x-6)(x-1)$
$(2x+1)(3x+6)$	$(6x+1)(x+6)$
$(2x-1)(3x-6)$	$(6x-1)(x-6)$

Only the second pair of these factors produce the proper middle term, so the factors of $6x^2 - 13x + 6$ are $(2x-3)(3x-2)$.

Exercise 23

Factor each of the following by removing the highest common factor:

1) $ax + bx + cx$
2) $3x^2 + 6x + x$
3) $3x^4 - 6x^2 - 5x^9$
4) $2x^6 + 8x^{10} - 4x^6$
5) $ab + a^2b^2 + a^3b^3$
6) $xyz^2 + 2yz^2 + y^2z^3$
7) $2^2a + 2^3a^2 + 2^4a^3$
8) $a^3 + a^3b - b^3$
9) $4x^2y^2z^2 + 16x - 32$
10) $3x^2y^5 + 27x^3y^9 - 243x^3y^{10}$
11) $7x^2 + 14x^2y^2 + 98x^2z$
12) $2x^2y + 24xy + 96y$

13) $a^2bc + ab^2c + abc^2$

14) $9a^3b^2 + 27a^2b + 81ab^2$

15) $9abc - 5ab + 6bc$

16) $25 - 15x^2 + 75x^3$

17) $5(a - b) + 7a(a - b)$

18) $x(m + n) - y(m + n)$

19) $45m^2n^3y - 15m^3n^2y^2$

20) $5x(m - n) + 6x(n - m)$

Factor completely:

21) $a^2 + 2a + 1$

22) $x^2 + 5x + 6$

23) $x^2 - x - 2$

24) $x^2 - 5x - 14$

25) $x^2 + 11x + 18$

26) $x^2 - 1$

27) $x^4 - 16$

28) $3x^2 - 7x + 4$

29) $4x^2 - 9a^2$

30) $7y^2 + 5y - 2$

31) $9a - 25a^3$

32) $3a^3 + 4a^2 + a$

33) $x^2 + 16x + 64$

34) $5x^2 - 8x - 4$

35) $2x^2 - 5x - 12$

36) $3x^2 - 15x + 18$

37) $25m^2 + 10mn + n^2$

38) $15x^2 + 75x + 60$

39) $3z^3 - 4z^2 + z$

40) $a^2b^2 - x^2y^2$

41) $(m + n) - 3m - 3n$

42) $82x^2 - 41$

43) $75x^2 - 125y^2$

44) $25m^2 - 50mn + 25n^2$

4.14 Factoring by Grouping

In the previous sections we have seen how certain polynomials could be factored. In each example either a common monomial factor was present in every term or the polynomial was factored by trial and error into the product of two binomials. Polynomials of these types are not the only ones which occur frequently in technical applications, and several other types of factoring are worth consideration. One of these is called *factoring by grouping*.

Consider, for example, the product of the two binomials $(x + a)$ and $(y + b)$, namely $(x + a)(y + b)$. If we apply the distributive law twice, we get

$$(x + a)y + (x + a)b$$

on the first application and

$$xy + ay + xb + ab$$

on the second application. The product in no way resembles anything we have considered up until this point. However, by considering the steps in reverse, it does provide us with a method for factoring certain types of polynomials. To illustrate what we mean, consider the above product

$$xy + ay + xb + ab.$$

If we use the associative property for addition to group the first two terms and the last two terms, we have

$$(xy + ay) + (xb + ab).$$

We now see that y is a factor of the first grouping and b a factor of the last. By removing these factors we have

$$(x + a)y + (x + a)b$$

which has $(x + a)$ as a factor common to both terms. We can now factor this term from the expression which is then written as

$$(x + a)(y + b).$$

At this point it should be observed that more than one grouping may be used to factor any one polynomial. To illustrate this, consider once again the polynomial

$$xy + ay + xb + ab.$$

Instead of grouping the first two terms and the last two terms as we did previously, we group the first and third terms and the second and fourth terms. This yields

$$(xy + xb) + (ay + ab).$$

We can now factor an x from the first grouping and an a from the second grouping. We now have

$$(y + b)x + (y + b)a$$

which has $(y + b)$ as a common factor. If we now factor this term from the expression we have

$$(y + b)(x + a)$$

which is precisely what we had before.

The only requirement that must be met to use factoring by grouping successfully is that each group possess a factor common to the other groups.

Example 1. Factor $ax^2 + bx^2 + ay + by$.
Group $(ax^2 + bx^2) + (ay + by)$. Factoring the common monomials x^2 and y from each term, we have

$$x^2(a + b) + y(a + b).$$

Factoring the common binomial $(a + b)$ from each term, we have

$$(a + b)(x^2 + y).$$

Example 2. Factor $3m - 6n + am - 2an$.
Group $(3m - 6n) + (am - 2an)$.
Factoring 3 from the first group and a from the second group, we have

$$3(m - 2n) + a(m - 2n).$$

Factoring the common binomial $(m - 2n)$ from each term, we have

$$(m - 2n)(3 + a).$$

Example 3. $x^2 + 2xy + y^2$.
Rewrite $x^2 + xy + xy + y^2$. Group $(x^2 + xy) + (xy + y^2)$. Factor $x(x + y) + y(x + y)$. Factoring the common binomial $(x + y)$ from each term, we have $(x + y)(x + y)$.

4.15 Special Types of Polynomials

Some polynomials fit into recognizable patterns and as such can be factored by inspection. The following forms are illustrations of but a few of these.

$$a^2 - b^2 = (a + b)(a - b).$$

$$a^3 + b^3 = (a + b)(a^2 - ab + b^2).$$

$$a^3 - b^3 = (a - b)(a^2 + ab + b^2).$$

The first one is called *the difference of two squares*, while the second and third are called *the sum and difference of two cubes*, respectively. The reader is encouraged to use multiplication to verify these factorizations.

To apply these forms, one must be able to recognize patterns that exist in certain problems. For example,

$$x^2 - y^2$$

is of the first form and can be factored as

$$(x - y)(x + y).$$

Each of the following examples are factored by direct application of one of the three forms.

Example 1. Factor $4m^2 - 9x^2$.
Using the first form, we have

$$4m^2 - 9x^2 = (2m)^2 - (3x)^2 = (2m - 3x)(2m + 3x).$$

Example 2. Factor $16m^4 - 81r^4$.
Using the first form, we have

$$16m^4 - 81r^4 = (4m^2)^2 - (9r^2)^2 = (4m^2 - 9r^2)(4m^2 + 9r^2) = (2m - 3r)(2m + 3r)(4m^2 + 9r^2).$$

The third factor $4m^2 + 9r^2$ is the sum of two squares and cannot be factored within the real number system.

Example 3. Factor $8x^3 + y^3$.
Using the second form, we have

$$8x^3 + y^3 = (2x)^3 + y^3 = (2x + y)(4x^2 - 2xy + y^2).$$

Example 4. Factor $27t^3 - 8s^3$.
Using the third form, we have

$$27t^3 - 8s^3 = (3t)^3 - (2s)^3 = (3t - 2s)(9t^2 + 6ts + 4s^2).$$

Example 5. $x^2 - y^2 + ax + ay$.
Group $(x^2 - y^2) + (ax + ay)$. Factor $(x - y)(x + y) + a(x + y)$. Factoring the common binomial $(x + y)$ from each term, we have

$$(x + y)(x - y + a).$$

Exercise 24

Factor each of the following.

1) $a^2 - b^2$
2) $x^2 - y^2$
3) $4x^2 - 4y^2$
4) $x^3 - 1$
5) $16x^4 - 4m^4$
6) $x^3 + 8$
7) $m^3 + 27$
8) $27a^3 - 64b^3$
9) $x^3y^3 - r^3$
10) $x^8 - y^8$
11) $ab + a + b + b^2$
12) $ab + bc + ad + cd$
13) $4x^2 - 4 + 3(x^2 - 1)$
14) $y^3 - 125x^3$
15) $(x + y)^2 - z^2$
16) $5ab - 5ac + b - c$

17) $15x^2 - 10xy - 12x + 8y$ 18) $ab - 4a + b - 4$

19) $x^6 - 1$ Hint: Express first as $(x^3)^2 - 1$.

20) $49r^2 - 81$ 21) $24x^3 - 81y^3$

22) $2x^2 - x + 2xy - y$ 23) $-x^2 + 2xy + y^2$

24) $mn + m^2 + mn + n^2$

4.16 Complex Numbers in Algebra

In an earlier section you encountered expressions involving the square roots of negative numbers. As you recall, numbers such as $2i$ were called pure imaginary numbers while numbers of the form $2 + 3i$ were called complex numbers. In this section we consider some of the properties of complex numbers under the operations of arithmetic as well as their behavior when subjected to the rules of exponents. First we need to consider just what type of numbers the set of complex numbers contains. To do this we consider the formal definition of a complex number.

A complex number is any number of the form $a + bi$, where a and b are elements of the set of real numbers and $i = \sqrt{-1}$.

If two complex numbers differ only in the algebraic sign found between the real and imaginary parts they are said to be *complex conjugates.* Thus $2 + 3i$, $2 - 3i$, and $5 + \sqrt{7}i$, $5 - \sqrt{7}i$ are complex conjugates.

Using the definition as a basis, we see that if $b = 0$ and zero is a real number, then we are considering numbers of the form

$$a + 0i = a.$$

In other words, the set of real numbers itself is a subset of the set of complex numbers.

$$2 + 0i = 2, \quad -3 + 0i = -3, \quad \sqrt{2} + 0i = \sqrt{2}$$

are all complex as well as real. If we now consider the cases where $a = 0$, we have $0 + bi = bi$ or all of the pure imaginaries. In general, the complex number system embraces as subsets all the previous systems that you have considered, but it also contains infinitely more numbers. $2 + 3i$, $3 - 6i$, $5 + 4i$ are but a few examples of complex numbers that are neither real nor pure imaginaries.

Two complex numbers $a + bi$ and $c + di$ are said to be equal, if and only if $a = c$ and $b = d$. Thus

$$3 + 4i = (2 + 1) + (2 + 2)i,$$

but

$$3 + 4i \neq 4 + 3i.$$

The sum of two complex numbers is defined as

$$(a + bi) + (c + di) = (a + c) + (b + d)i.$$

Similarly the difference of two complex numbers is defined as

$$(a + bi) - (c + di) = (a - c) + (b - d)i.$$

There certainly is nothing mysterious in these definitions, since they parallel precisely the rules you have used for adding like terms in some of the previous sections. For example,

$$(3 + 2b) + (4 + 6b) = 7 + 8b,$$

and

$$(2 + 5i) + (3 + 2i) = 5 + 7i.$$

We might also state, without proof, that the set of complex numbers is closed, commutative, and associative with respect to addition.

We will encounter complex numbers in various forms in later sections of the book and should have the ability to write them in the form $a + bi$ or $a - bi$. Usually, all that is required is to apply the rules for simplifying radicals along with a few other algebraic operations.

Example 1. Write $\sqrt{-18}$ in the form $a + bi$.

$$\begin{aligned}\sqrt{-18} &= i\sqrt{18}\\ &= i\sqrt{9\cdot 2}\\ &= 3i\sqrt{2} \text{ or } 3\sqrt{2}i.\end{aligned}$$

It is not necessary to write the number as $0 + 3\sqrt{2}i$.

Example 2.

$$\begin{aligned}3 + \sqrt{-20} &= 3 + i\sqrt{4\cdot 5}\\ &= 3 + 2i\sqrt{5}.\end{aligned}$$

Example 3.

$$\begin{aligned}\frac{8 + \sqrt{-8}}{6} &= \frac{8 + i\sqrt{4\cdot 2}}{6}\\ &= \frac{8 + 2i\sqrt{2}}{6}\\ &= \frac{4 + i\sqrt{2}}{3}.\end{aligned}$$

4.17 Product of Complex Numbers

The product of two complex numbers is defined in a way that conforms to the product of any two binomials. That is,

$$(a + bi)(c + di) = (ac - bd) + (ad + bc)i.$$

To check the reasonableness of this definition we apply the distributive law to the left-hand member of the definition. This yields

$$(a + bi)c + (a + bi)di = ac + bci + adi + bdi^2,$$

but i^2 must be defined as -1, since $\sqrt{-1}\sqrt{-1} = -1$ and $\sqrt{-1} = i$. We now have the left-hand member equal to $ac + bci + adi - bd = (ac - bd) + (bc + ad)i$. Usually we do not formally use the definition but form the products mentally as we did in verifying the reasonableness of the definition.

Example 1. $(3 + 2i)(4 - 6i) = 12 - 18i + 8i - 12i^2$
$= 12 - 18i + 8i + 12 = 24 - 10i.$

Example 2. $6(5 + 7i) = 30 + 42i.$

Notice that it was not necessary to write 6 as $6 + 0i$ to form the product.

When i is raised to increasing integral powers, an interesting periodic property occurs. Notice that

$$\begin{aligned} i &= i \\ i^2 &= -1 \\ i^3 &= i^2 \cdot i = -i \\ i^4 &= (i^2)^2 = (-1)^2 = 1 \\ i^5 &= i^4 \cdot i = 1 \cdot i = i \\ i^6 &= i^4 \cdot i^2 = 1 \cdot (-1) = -1 \\ i^7 &= i^4 \cdot i^3 = 1 \cdot (-i) = -i \\ i^8 &= i^4 \cdot i^4 = 1 \cdot 1 = 1 \end{aligned}$$

We can use this property to simplify powers of i where the exponent is large. For example,

$$\begin{aligned} i^{67} &= (i^4)^{16} i^3 \\ &= 1 \cdot i^3 \\ &= -i. \end{aligned}$$

All that one need do is divide the exponent by 4 and record the remainder. The expression can be replaced by i, written with the remainder as the exponent.

Example 3. Simplify i^{621}.
Dividing 621 by 4 leaves a remainder of 1.
We have $i^{621} = i^1 = i$.

Example 4. Simplify i^{255}.
Dividing 255 by 4 leaves a remainder of 3.
We have $i^{255} = i^3 = -i$.

The relationship $\sqrt{a}\sqrt{b} = \sqrt{a \cdot b}$ is not valid for the system of complex numbers if both a and b are negative. To avoid such complications we always write the number in the *i form* before multiplying. For example,

$$\sqrt{-2}\sqrt{-3} = i\sqrt{2}i \cdot \sqrt{3} = i^2\sqrt{2 \cdot 3} = -\sqrt{6};$$

whereas

$$\sqrt{-2}\sqrt{-3} \neq \sqrt{(-2)(-3)} = \sqrt{6}.$$

Exercise 25

Simplify each of the following.

1) i^{10}
2) i^{27}
3) i^{141}
4) i^{629}

Find the product in each of the following and simplify.

5) $(3 + i)(2 + i)$
6) $(2 + 5i)(3 - 6i)$
7) $2i(6 + 3i + i^2)$
8) $(7 - 3i)(5 + 2i)$
9) $(2 + \sqrt{3}i)^2$
10) $(5 - 2i)(4 + 3i)$
11) $(4 - 5i)^2$
12) $(2 + 3i)(2 - 3i)$
13) $(5 - 2i)(5 + 2i)$
14) $\sqrt{-3}(\sqrt{-7} + i\sqrt{6})$
15) $i(4 - 3i)(4 + 3i)$
16) $(4 + 2\sqrt{-3})(4 - \sqrt{-3})$
17) $(2 - \sqrt{-2})(\sqrt{-2} - 2)$
18) $\left(\frac{2 - \sqrt{-2}}{4}\right)\left(\frac{3 + \sqrt{-2}}{6}\right)$
19) $\left(\frac{2}{5} + \frac{1i}{5}\right)\left(\frac{3}{5} + \frac{2i}{5}\right)$
20) $(3 + 2i)^2 + (3 - 2i)^2$
21) $i(1 + i)^2 + i^3(1 - i)$
22) $(1 + i)^2(1 - i)^2$
23) $(2 + 3i) + (2 + i)^2$
24) $(1 + i)i + (1 - i)^2i^2$

Review Exercise Chapter 4

Write the following products and simplify.

1) $(4x)(12x^3)$
2) $4m(5x + 3y)$
3) $(3n)(4m)$
4) $x^5 \cdot x^2$
5) $(.01)^2$
6) $(6x)^3$
7) $(a + b)^2$
8) $(5x + 3y)(x - 2y)$
9) $(x + y)(x^2 + 2x + 1)$
10) $3x^2(2x^3 + 4x - 1)$

Find the following quotients.

11) $(x^3 + 3x^2 + 5x) \div x$

12) $(x^2 + 3x + 6) \div (x + 1)$

13) $(55x^2y + 25xy^2 + 20x^5y^3) \div 5xy$

14) $(x^3 + 3x^2 + 3x + 1) \div (x + 1)$

Factor the following:

15) $26x^2 + 39x^3y + 52x^4$

16) $x^2 - 4$

17) $x^2 - 6x - 7$

18) $x^2 + 8x + 16$

19) $x^3 - 16x$

20) $5x^2 - 20x$

5

Linear Equations, Functions, and Graphs

5.1 Equations

A statement that two quantities are equal is called an *equation*. The symbol = is used to indicate that two quantities have the same value, for example, $a = b$. The portion of an equation to the left of an equal sign is called the *left member* and the portion to the right the *right member*.

Two types of equations are most often encountered in elementary algebra. These are the *conditional equation* and the *identical equation* or *identity*. The conditional equation is one which is true for a specific set of values of the variable. This set of values is referred to as the *solution set* and the values are *solutions of the equation*. An identity is an equation which is true for all permissible values of the variable.

Example 1. $2x + 1 = 9$ is a conditional equation, since it is true only if $x = 4$.

Example 2. $x^2 - 3x + 2 = 0$ is conditional, since it is true only if $x = 1$ or $x = 2$.

Example 3. $4/6x = 2/3x$ is an identity for all values of x except zero. Since division by zero is meaningless, zero is not a permissible replacement for x.

Example 4. $\dfrac{3x}{x^2} + \dfrac{5}{x} = \dfrac{8}{x}$ is an identity, since it is true for any permissible value which is assigned to x. (Again $x \neq 0$).

5.2 Solving Equations

To solve equations the student will find that he needs a set of rules as tools with which to work. These tools are called *axioms*. They will be accepted as true and no attempt will be made to prove any of them.

Axiom 1. *Equality is Reflexive.* $a = a$.
Axiom 2. *Equality is Symmetric. If* $a = b$, *then* $b = a$.
Axiom 3. *Equality is Transitive. If* $a = b$ *and* $b = c$, *then* $a = c$.
Axiom 4. *The same quantity may be added to or subtracted from both sides of an equation. If* $a = b$, *then* $a + c = b + c$. *If* $a = b$, *then* $a - c = b - c$.
Axiom 5. *Both sides of an equation may be multiplied or divided by the same quantity. If* $a = b$, *then* $ac = bc$. *If* $a = b$, *then* $a/c = b/c$ $(c \neq 0)$.
Axiom 6. *A quantity may be substituted for its equal in an equation.*

5.3 Equations Solved by Addition and Subtraction

The process of following mathematical steps or performing operations to obtain a numerical value for the variable which belongs to the solution set of the equation is called *solving the equation*, and the result obtained is called a *solution* or *root of the equation*. The verification of this value is called a *check*. When no number exists in the solution set, the solution set is said to be *empty*. The following examples illustrate the use of axiom 4 in solving equations.

Example 1. $x + 2 = 7$.

$x + 2 - 2 = 7 - 2$ Subtract 2 from each side.

$x = 5$.

Example 2. $x - 2 = 6$.

$x - 2 + 2 = 6 + 2$ Add 2 to each side.

$x = 8$.

Example 3. $n + 13 = 72$.

$n + 13 - 13 = 72 - 13$ Subtract 13 from each side.

$x = 59$.

In each of the preceding examples, the addition or subtraction of a single term from each side was sufficient to complete the solution. However, many times equations require some simplification before we are ready to perform the addition or subtraction.

Example 4. $(2c + 10) + (3 - c) = 18.$

We first remove symbols of inclusion and collect like terms to obtain

$$2c + 10 + 3 - c = 18$$
$$c + 13 = 18$$

or

$$c + 13 - 13 = 18 - 13 \quad \text{Subtract 13 from both sides.}$$
$$c = 5.$$

Example 5. $-32 + w - 5 = 17 - 2.$

Collecting like terms we have

$$-37 + w = 15$$
$$-37 + 37 + w = 15 + 37 \qquad \text{Add 37 to both sides.}$$
$$w = 52.$$

Since the probability of a mistake increases with each additional operation that is performed, the correctness of the solution should be checked. Such a check is accomplished by substituting the solution into the original equation. The check of the solution to Example 5 is shown below.

Check:

$$-32 + w - 5 = 17 - 2$$
$$-32 + 52 - 5 = 15$$
$$15 = 15.$$

Example 6. $4(x + 2) - 2(x + 3) = x - 1.$

We first remove symbols of inclusion and collect like terms.

$$4x + 8 - 2x - 6 = x - 1$$
$$2x + 2 = x - 1$$
$$2x - x + 2 = x - x - 1. \qquad \text{Subtract } x \text{ from each side.}$$
$$x + 2 = -1$$
$$x + 2 - 2 = -1 - 2$$
$$x = -3. \qquad \text{Subtract 2 from both sides.}$$

Check:

$$4(-3 + 2) - 2(-3 + 3) = -3 - 1$$
$$4(-1) - 2(0) = -4$$
$$-4 = -4.$$

Example 7. $6(x - 2) + 4x = 9x - 6.$

$$6x - 12 + 4x = 9x - 6$$
$$10x - 12 = 9x - 6$$
$$x - 12 = -6 \qquad \text{Subtract } 9x \text{ from both sides.}$$
$$x = 6. \qquad \text{Add 12 to both sides.}$$

Check:

$$6(6 - 2) + 4 \cdot 6 = 9 \cdot 6 - 6$$
$$6(4) + 4 \cdot 6 = 9 \cdot 6 - 6$$
$$24 + 24 = 54 - 6$$
$$48 = 48.$$

Exercise 26

Solve each of the following equations.

1) $x + 9 = 15$
2) $x - 7 = 26$
3) $x + 9 = 15 - 7$
4) $d + 31 = -11$
5) $200 = 150 + x$
6) $x + 30 = 35$
7) $180 - d = 28$
8) $2x - (x + 6) = 13$
9) $(x + 1) - 6 = 4$
10) $(4 + s) + 6 = 12$
11) $(y + 2) - 3 = 9$
12) $(3 + n) + (7 - n) + n = 6$
13) $-5 + [y - 1] = 9$
14) $(2x + 2) - x = 11 - 6$
15) $-t + 2 = -15$
16) $170 = 5 + R$
17) $3/4 = 1.25 - s$
18) $(y + 3) + 4 = 6$
19) $(6y - 2) - (y - 4) - 4y = 0$
20) $6 + H = 6$
21) $H + 9 = H + 9$
22) $H + 4 = H + 3$
23) $2x + 4 = x - 9$
24) $5x - 2 = 4x + 13$
25) $(5x - 2) - 2(x - 3) = 2x$
26) $3x + 5 - x = 6x - 3 - 5x$
27) $x + 5 - 6 = (2x - 7) + 11$
28) $11h - 5h + 9 = 5h - 2$

5.4 Equations Solved by Multiplication and Division

Each of the equations that were solved in section 5.3 were solved by applying axiom 4 after some preliminary simplification. There are many equations that could not be solved if axiom 4 were the only one we had at our disposal. $3x = 9$ or $x/5 = 4$ are but two examples of equations of this type. Both of these equations are solved by applying axiom 5 to each member as shown below.

Example 1. $3x = 9.$

$$\frac{\not{3}x}{\not{3}} = \frac{\not{9}}{\not{3}} \qquad \text{Divide both sides by 3.}$$
$$x = 3.$$

Example 2. $\dfrac{x}{5} = 4.$

$$\not{5} \cdot \frac{x}{\not{5}} = 4 \cdot 5 \qquad \text{Multiply both sides by 5.}$$
$$x = 20.$$

Notice that in each case we have reduced the coefficient of the variable to 1 to effect a solution. Where the given equation had multiplication associated with the variable, $3x$ meaning 3 times x, we solved the equation by use of division. Where the given equation had division associated with the variable, $x/5$ meaning x divided by 5, we solved the equation by use of multiplication.

Some equations require both multiplication and division to solve for the variable.

Example 3. $\dfrac{2x}{5} = 7.$

$$5 \cdot \frac{2x}{5} = 7 \cdot 5 \qquad \text{Multiply both sides by 5.}$$
$$2x = 35$$
$$\frac{2x}{2} = \frac{35}{2} \qquad \text{Divide both sides by 2.}$$
$$x = 35/2.$$

The same result would have been obtained, if we had multiplied both sides by 5/2.

Example 4. $\dfrac{2x}{5} = 7.$

$$\frac{\not{5}}{\not{2}} \cdot \frac{\not{2}x}{\not{5}} = \frac{5}{2} \cdot 7$$
$$x = \frac{35}{2}.$$

Example 5. $-\dfrac{2}{11}x = \dfrac{2}{3}.$

$$\left(-\frac{\not{11}}{\not{2}}\right)\left(-\frac{\not{2}}{\not{11}}x\right) = \frac{\not{2}}{3}\left(-\frac{11}{\not{2}}\right) \qquad \text{Multiply both sides by } -11/2.$$
$$x = -\frac{11}{3}.$$

Example 6. $\frac{6x}{5} = 4.$

$$\frac{\cancel{5}}{\cancel{6}} \cdot \frac{\cancel{6}x}{\cancel{5}} = 4 \cdot \frac{5}{6} \qquad \text{Multiply both sides by 5/6.}$$

$$x = \frac{20}{6} \quad \text{or} \quad \frac{10}{3}.$$

Exercise 27

Solve each of the following equations.

1) $3x = 9$
2) $2x = 14$
3) $4n = 28$
4) $\frac{1}{3}n = 6$
5) $\frac{n}{2} = 12$
6) $2x/3 = 6$
7) $5x = 13$
8) $-2x = 7$
9) $3x/5 = -9$
10) $-2x = -10$
11) $\frac{-2x}{3} = -6$
12) $\frac{6x}{5} = -9$
13) $4x = 20$
14) $9y = 36$
15) $4t = -24$
16) $\frac{x}{.27} = .63$
17) $-8z = -1.6$
18) $\frac{-x}{6} = \frac{-2}{9}$
19) $\frac{1}{.9}V = .6$
20) $2t = \frac{1}{8}$
21) $3V = \frac{3}{11}$
22) $.3z = .931$
23) $\frac{-x}{4} = \frac{-26}{3}$
24) $16V = 9t$ for V
25) $16V = 9t$ for t

5.5 Equations Involving Multiple Operations

The practicing technician will frequently encounter equations which involve more than one operation. Should this occur, the axioms are applied one at a time until the desired variable has been isolated and the equation solved.

Example 1. $2x + 1 = 3.$

$$2x + 1 - 1 = 3 - 1 \qquad \text{Subtract 1 from each side.}$$
$$2x = 2$$
$$\frac{2x}{2} = \frac{2}{2} \qquad \text{Divide each side by 2.}$$
$$x = 1.$$

Example 2. $\frac{1}{6}x - 3 = 9.$

$$\frac{1}{6}x - 3 + 3 = 9 + 3 \qquad \text{Add 3 to each side.}$$
$$\frac{1}{6}x = 12$$
$$6 \cdot \frac{1}{6}x = 12 \cdot 6 \qquad \text{Multiply both sides by 6.}$$
$$x = 72.$$

In many cases, the variable may appear on each side of the equation. If this should occur we apply the appropriate axiom to isolate the variable on one side of the equation.

Example 3. $6x + 9 = 2x - 4.$

$$6x - 2x + 9 = 2x - 2x - 4 \qquad \text{Subtract } 2x \text{ from each side.}$$
$$4x + 9 = -4$$
$$4x + 9 - 9 = -4 - 9 \qquad \text{Subtract 9 from each side.}$$
$$4x = -13$$
$$\frac{4x}{4} = -\frac{13}{4} \qquad \text{Divide each side by 4.}$$
$$x = -\frac{13}{4}.$$

As the complexity of equations increases, it becomes desirable to have a logical scheme for solving them. The following illustrates a convenient order to follow.

1) Remove symbols of grouping on each side of the equation.
2) Simplify each side of the equation by collecting like terms.
3) Isolate the variable on one side of the equation by adding or subtracting the proper terms.
4) Solve for the variable by multiplying or dividing by the proper factor.

Example 4. $6x + 2(x + 1) = 3.$

$6x + 2x + 2 = 3$	Distributive Law
$8x + 2 = 3$	Collect like terms
$8x = 1$	Axiom 4
$x = 1/8.$	Axiom 5

Example 5. $3(2x + 6) + 7x = 9(2x + 7) + 5(x + 1).$

$6x + 18 + 7x = 18x + 63 + 5x + 5$	Distributive Law
$13x + 18 = 23x + 68$	Collect like terms
$-10x + 18 = 68$	Axiom 4
$-10x = 50$	Axiom 4
$x = -5.$	Axiom 5

Finally we will solve equations in which fractions are present in several of the terms. When this occurs, we first eliminate the fractions by multiplying each side of the equation by the lowest common denominator. This is referred to as clearing the equation of fractions.

Example 6. $\frac{x}{3} + \frac{2x}{7} = \frac{5x + 1}{42} - \frac{2x}{21}.$

$$42\left(\frac{x}{3}\right) + 42\left(\frac{2x}{7}\right) = 42\left(\frac{5x + 1}{42}\right) - 42\left(\frac{2x}{21}\right)$$

Axiom 5, multiply by the L.C.D.

$14x + 12x = 5x + 1 - 4x$	
$26x = x + 1$	Collect like terms
$25x = 1$	Axiom 4
$x = \frac{1}{25}.$	Axiom 5

Check:

$$\frac{\frac{1}{25}}{3} + \frac{2 \cdot \frac{1}{25}}{7} = \frac{5 \cdot \frac{1}{25} + 1}{42} - \frac{2 \cdot \frac{1}{25}}{21}$$

$$\frac{1}{75} + \frac{2}{175} = \frac{6}{210} - \frac{2}{525}$$

$$\frac{7 + 6}{525} = \frac{30 - 4}{1050}$$

$$\frac{13}{525} = \frac{13}{525}.$$

Exercise 28

Solve each of the following equations and check the root.

1) $3y + 2 = 8$

2) $3x + 6 = 12$

3) $5a + 11 = 16$

4) $3a + 4 = 9$

5) $6 - 2x = 18$

6) $3x + 7 = 2x$

7) $6 - 5a = 11$

8) $5a - 7 = -19$

9) $9a - 17 = 13$

10) $.3 - .25x = 2.4$

11) $2x - 3 = x + 6$

12) $6 - 3x = x - 28$

13) $3x - 12 = 2x + 17$

14) $.29x + .31 = .27x + .01$

15) $8 - 2x = x + 12$

16) $5r - 18 = 44 + 4r$

17) $x + 2(x - 3) = x + 2$

18) $7 - 4x = -19$

19) $3(x + 5) = 5(2 - x)$

20) $x - 2(x - 1) = -9$

21) $4x - (20 + x) = 40$

22) $1/3(2x + 6) = 9$

23) $x - 6 = 6 - x$

24) $3(2x + 4) + 5 = 41$

25) $5x - 3(x - 4) = 4x - 2$

26) $2(x + 1) = 3(x + 2) - 6$

27) $9x - 11 = 7x - 11$

28) $5x + 7 - x = 2x + 15$

29) $\frac{2x - 1}{5} = \frac{x + 1}{2}$

30) $\frac{5x}{2} - 1 = x - \frac{1}{2}$

31) $\frac{7x}{6} - 4 = x - 42$

32) $\frac{x}{5} - \frac{x}{2} = 9$

33) $\frac{3x}{2} + \frac{x}{3} = \frac{7}{6}$

34) $\frac{5}{6}x + 3 = \frac{1}{6}x - 9$

35) $\frac{2x - 4}{2} = \frac{x + 3}{6}$

36) $\frac{x}{4} - \frac{2x - 5}{3} = \frac{x}{2} + 1$

37) $\frac{x}{2} + \frac{3x + 1}{5} = \frac{x + 3}{10}$

38) $\frac{x + 7}{4} = \frac{3x + 2}{5} - 1$

39) $\frac{8x + 3}{6} - \frac{7x - 1}{4} + \frac{1}{2} = 0$

40) $\frac{7x - 1}{9} = 1 + \frac{7x + 4}{18}$

41) $(x + 2)(x + 3) = (x + 1)(x - 7)$

42) $3 - s(s - 2) = 5 - s^2$

43) $3a^2 - 2a(a - 3) + 6 = a^2 + 4(a - 13)$

44) $y^2 - y(3y - 5) = y^2 - 2y^2 - 2y - y^2 - 14$

45) $3x^2 - (x + 4)(3x - 2) = 2(1 - 5x) + 6$

5.6 Ratio and Proportion

Many of the problems of technical mathematics are concerned with comparing quantities. The student of auto mechanics is familiar with the terms compression ratio, gear ratio, and final gear ratio, for example. The student of electronics has probably encountered problems of resistance and wire sizes stated as ratios. The machine shop student is familiar with the turning speed of the headstock and its relationship to gear ratios. It is impossible to become involved in any technical field without a working knowledge of ratios.

A *ratio* is simply the comparison of two quantities by division or, stated more formally, *the ratio of one quantity to another like quantity is the quotient of the first quantity divided by the second.* Ratios can be written in many forms. For example, the ratio of 3 to 7 can be written as 3/7, $3 \div 7$, or 3:7. No matter what the form, however, the two quantities being compared carry the *same units of measure.* We will not compare a quantity expressed in dollars and cents to one expressed in pounds and ounces. The ratio itself, however, carries no unit of measurement and is therefore simply a number.

A *proportion is a statement that two or more ratios are equal.* Since 5/10 and 1/2 are ratios with the same value, the statement of this fact,

$$\frac{5}{10} = \frac{1}{2},$$

is a simple proportion. If the ratio a/b is equal to the ratio c/d, then $a/b = c/d$ is a proportion. This is read *a is to b as c is to d.* In this proportion a and d are called the *extremes* and b and c are called the *means.* If the means of a proportion are equal, then that number is said to be the *mean proportional* between the extremes.

Example 1. $\frac{2}{4} = \frac{4}{8}$. 4 is the mean proportional between 2 and 8.

The student has probably observed by this time that a proportion is a fractional equation, and therefore that all the axioms which relate to equations and equalities also hold for proportions. Several properties of proportions are illustrated by examples below. No attempt at formally proving these properties is made.

1. *In a proportion, the product of the means is equal to the product of the extremes.*

Example 2. If $\frac{2}{5} = \frac{4}{10}$, then $(2)(10) = (4)(5)$, and $20 = 20$.

Example 3. If $\frac{6}{9} = \frac{2}{3}$, then $(6)(3) = (9)(2)$, and $18 = 18$.

Example 4. If $\frac{a}{b} = \frac{c}{d}$, then $ad = bc$.

2. *In any proportion the means or extremes may be interchanged without losing equality.*

Example 5. If $\frac{6}{9} = \frac{2}{3}$, then $\frac{3}{9} = \frac{2}{6}$ or $\frac{6}{2} = \frac{9}{3}$.

Notice that the ratios change, but the equality remains.

Example 6. If $\frac{a}{b} = \frac{c}{d}$, then $\frac{a}{c} = \frac{b}{d}$ or $\frac{d}{b} = \frac{c}{a}$.

3. *In a proportion, the ratios may be inverted without losing equality.*

Example 7. If $2{:}9 = 4{:}18$, then $9{:}2 = 18{:}4$.

Example 8. If $a{:}b = c{:}d$, then $b{:}a = d{:}c$.

Several other theorems about proportions exist, but will not be mentioned here. The student is encouraged to try to discover one or two from his knowledge of the basic axioms of equality.

Exercise 29

Express the following as ratios and reduce where possible.

1) 20 to 25
2) 15 to 130
3) 1/2 to 3/10
4) 5/6 to 10/9
5) 3 yards to 2 feet
6) 20 cm. to 35 mm.
7) 6 lb. 3 oz. to 5 lb. 2 oz.
8) 15 cm. to 7 ft.
9) 2812 hectometers to 21 kilometers
10) 2 ft. 7 in. to 7 ft. 9 in.

Find the solution set such that each of the following proportions is a true statement.

11) $3{:}x = 5{:}18$
12) $5{:}20 = x{:}100$
13) $6{:}M = 54{:}86$
14) 34.2 in.:2.7 in. = 14.8 in.:x in.
15) 2000 km.:1240 km. = x km.:240 km.
16) $x{:}3 = 5{:}8$
17) 36 in.:2 in. = 72 in.:x in.
18) 1725 ft.:3000 feet = x ft.:2000 feet

Modify each of the following proportions in three different ways by use of the theorems.

19) $\frac{x}{y} = \frac{3}{5}$.

20) $\frac{x}{3} = \frac{y}{7}$.

21) $\frac{x}{5} = \frac{y}{3}$.

22) $a:2 = b:3$

Find the ratio of x to y in each of the following.

23) $3x = 7y$

24) $\frac{1}{2}x = \frac{2}{5}y$

25) $x = y$

Find the mean proportional of each of the following.

26) 2 and 18

27) 3 and 12

28) 3 and 27

29) 9 and 16

30) 16 and 4

31) M and R

32) 2 and 32

33) 6 and 7

34) Bring to class at least two examples of problems from your particular field of study which use ratios and proportions.

35) A blueprint is drawn to scale, and the scale is given as 1/8 in. = 1 ft.
   a) What is the ratio of the size of the drawing to the final product?
   b) If one dimension measures $4\frac{3}{8}$ inches on the blueprint, what is the size of the corresponding part of the final product?

36) An automobile has a 16-quart cooling system. If the ratio of antifreeze to water is three to one, how much of each does the system contain?

37) Tapers are frequently used for plugs, gauges, spindles, and bearings. Compute the taper per foot in diagrams a, b, and c below.

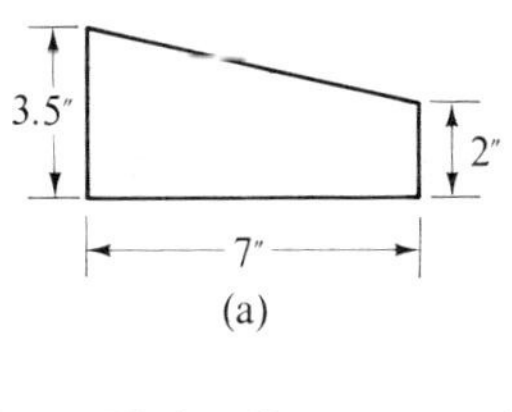

(a)

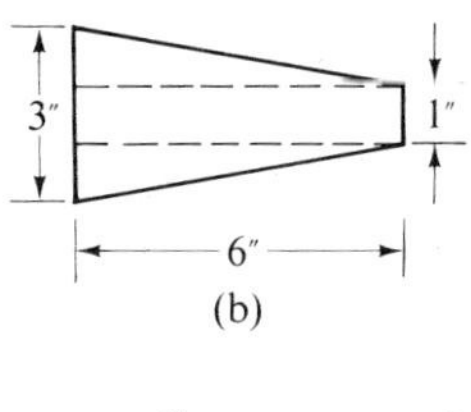

(b)

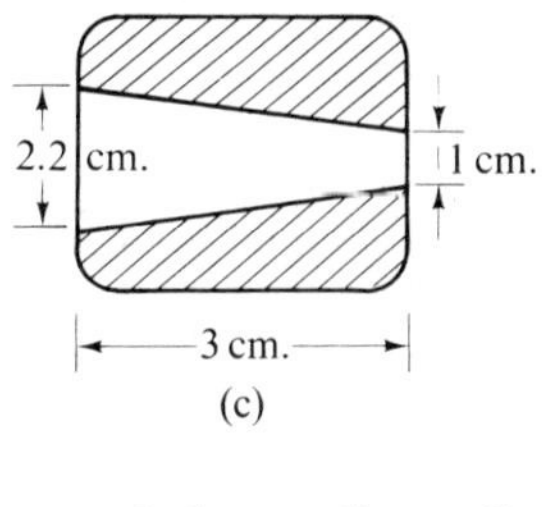

(c)

38) If the diameters of two pulleys are as shown and the smaller pulley turns 800 rpm, how fast does the larger pulley turn?

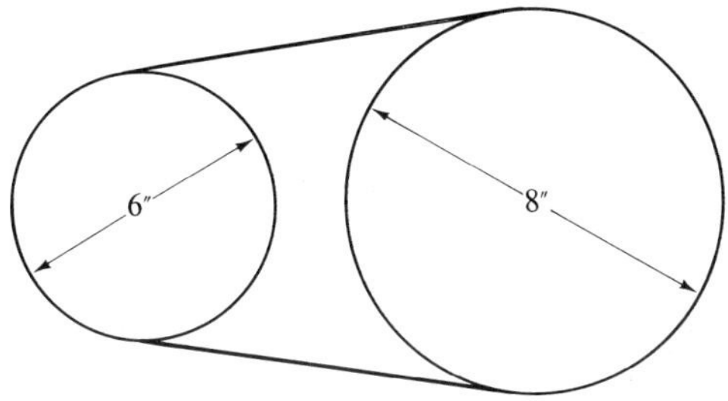

39) Pulley a has a radius of 2 inches, pulley b has a radius of $1\frac{1}{2}$ inches, while pulleys c and d have radii of 4 inches. If pulley a is driven by a motor making 1760 rpm, how fast will the shaft to which pulley d is attached turn?

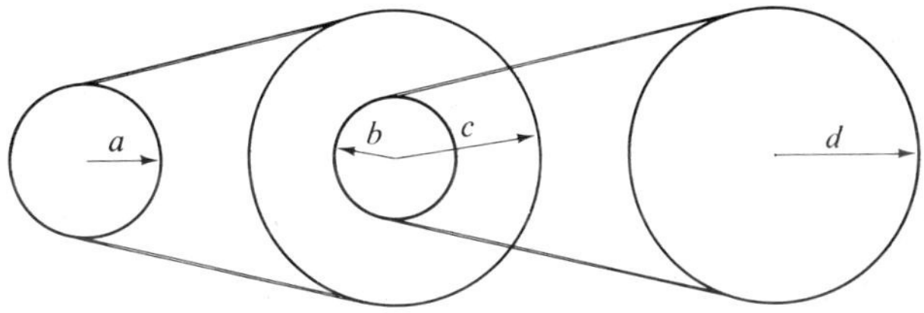

40) The relationship between two gears can be written as $TR = tr$, where T is the number of teeth on the large gear, t the number of teeth on the small gear, and R and r the number of revolutions per minute of the large and small gears, respectively. Fill in the missing values in the following table.

T	t	R	r
80	32		45
12		5	15
75		90	360
?	320	420	210

41) Power (in foot pounds per minute) is given by the ratio between work w and time t. Find the power exerted in pushing a load of 60 pounds 150 feet in 5 minutes. (Work = force × distance.)

42) The pitch of a rafter is the ratio between the rise and the run of the rafter. Find the pitch of a rafter with a rise of 30 inches and a run of 12 feet.

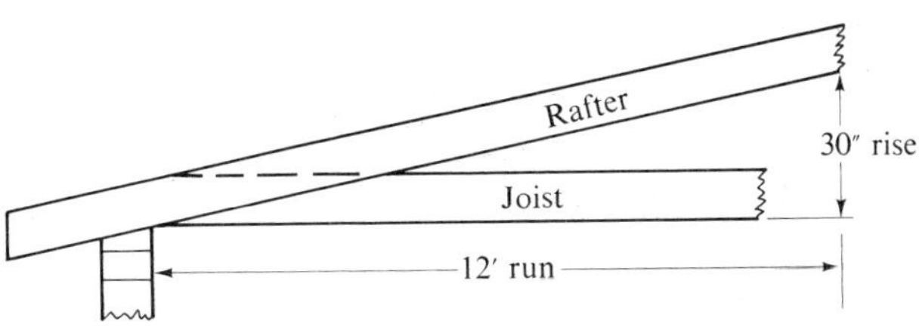

43) Find the run of a rafter having a pitch of 1/5 if the rise is 3 feet. (See problem 42.)

5.7 Formulas

In all fields of technology many of the basic facts are represented by equations. The area of a rectangular piece of sheet metal is found by multiplying the length by the width:

$$A = l \cdot w.$$

In electrical theory, the voltage E is found to be the product of the current I and the Resistance R:

$$E = I \cdot R.$$

In automotive technology, the car speed S is given by the quotient of the product of the crankshaft speed RPM and tire size W divided by the rear axle gear ratio R times 168:

$$S = \frac{RPM(W)}{R(168)}.$$

Such generalized expressions are called *formulas.* The student will encounter many others as he pursues his studies.

Most of the formulas can be found in tables. However, they may not be in the most convenient form for solving a particular problem. If this is the case, we may find it convenient to change the form.

Example 1. Find the resistance R, given the formula $E = I \cdot R$.

$$E = IR.$$

$$\frac{E}{I} = \frac{IR}{I}.$$

$$\frac{E}{I} = R.$$

Example 2. Solve the formula $S = \dfrac{RPM(W)}{R(168)}$ to find the revolutions per minute RPM.

$$S = \frac{RPM(W)}{R(168)}.$$

$$S \cdot R(168) = RPM \cdot W.$$

$$\frac{S \cdot R(168)}{W} = RPM.$$

$$RPM = \frac{S \cdot R(168)}{W}.$$

Example 3. Solve the formula $I = \frac{E}{R + r}$ encountered in electrical theory for r.

$$I = \frac{E}{R + r}$$
$$I(R + r) = E.$$
$$IR + Ir = E.$$
$$Ir = E - IR$$
$$r = \frac{E - IR}{I}.$$

Exercise 30

Solve each of the following formulas for the variable indicated:

1) $A - B + C = 180$ for B.
2) $VP = bT$ for P.
3) $C = 2\pi r$ for r.
4) $D = \frac{NP}{\pi}$ for P.
5) $D = rt$ for t.
6) $V = LWH$ for H.
7) $A = \frac{1}{2}(x + y)$ for x.
8) $\frac{1}{R_t} = \frac{1}{R_2}$ for R_2.
9) $\frac{1}{C} = \frac{1}{r} + \frac{1}{s} + \frac{1}{t}$ for r.
10) $V = \frac{1}{3}\pi r^2 h$ for h.
11) $M = \frac{Lt + g}{t}$ for g and L.
12) $S = \frac{n(a + 1)}{2}$ for n.
13) $I = prt$ for t.
14) $P = 2l + 2w$ for l.
15) $F = ma$ for m.
16) $I = \frac{E}{R}$ for E.
17) $E = Mc^2$ for M.
18) $S = \pi r^2 + \pi rL$ for L.
19) $S = \frac{1}{2}at^2$ for a.
20) $A = \frac{1}{2}bh$ for h.
21) $F = \frac{KmM}{d^2}$ for M.
22) $A = \frac{1}{2}h(b_1 + b_2)$ for b_2.
23) $S = \frac{a - r1}{1 - r}$ for a.
24) $S = \frac{a}{1 - r}$ for a.
25) $F = \frac{WV}{gR}$ for R.
26) Ratio $= \frac{\text{driven gear}}{\text{drive gear}}$ for drive gear.
27) Decimetral Pitch $= \frac{\text{number of teeth} + 2}{\text{outside diameter}}$ or $DP = \frac{N + 2}{od}$ for N.
28) Horsepower $= \frac{\text{torque} \cdot RPM}{5252}$ for torque.

5.8 Functions

A *function* is a *rule that assigns each element of one set to one and only one element of another set.* Such a rule is frequently expressed as an equation in which two variables are present. The equation $y = 6x + 1$ is such a rule since each different value that is assigned to x produces a different value for y. For example,

$$\text{if } x = 0, \text{ then } y = 1; \text{ if } x = 1, y = 7; \text{ if } x = -2, y = -11.$$

If we let the set D contain the elements 0, 1, and -2, and the set R contain 1, 7, 11, then the equation establishes a one-to-one correspondence between the members of the set D and the members of the set R. See diagam below.

D		R
0	$\rightarrow$	1
1	$\rightarrow$	7
-2	$\rightarrow$	-11

The set D is called the *domain* of the function while the set R is called the *range*. For the above equation, the domain was arbitrarily restricted to the set containing 0, 1, and -2. In reality, the domain for this function could have been any real number, since each would produce a distinct element in the range when substituted into the rule. Furthermore, there is no value of x that would make the equation meaningless.

Example 1. Does $y = \frac{1}{x}$ describe a function? If so, what is its domain?

Each distinct replacement for x in the domain produces a distinct value of y in the range. $x = 1/2$ corresponds to $y = 2$; $x = 3$ corresponds to $y = 1/3$, etc.

x cannot assume the value of 0 since division by zero is undefined. The domain is all real numbers except zero.

Functions frequently are defined by using a single letter such as f, g, m, t, which allows us to consider several functions simultaneously without danger of confusion. For example, the notation

$$f\colon x \rightarrow 6x + 1$$

describes a function and is read: f is the function that pairs x with $6x + 1$ or that maps x into $6x + 1$. This notation shows the pairing between any x from the domain and the corresponding element of the range. We could have used any other letter instead of f to describe this function, but the remaining parts of the notation would have been the same.

$$g\colon\ x \to 6x + 1$$
$$m\colon\ x \to 6x + 1$$
$$t\colon\ x \to 6x + 1$$

Example 2. If $s\colon x \to x^2 + x + 1$ and $D = \{1,2,3\}$ find R.
If $x = 1$, then $x^2 + x + 1 = 1 + 1 + 1 = 3$
If $x = 2$, then $x^2 + x + 1 = 4 + 2 + 1 = 7$
If $x = 3$, then $x^2 + x + 1 = 9 + 3 + 1 = 13$
Therefore, $R = \{3,7,13\}$.

Yet another type of notation for functions exists. Using this type, the original equation $y = 6x + 1$ would have been written $f(x) = 6x + 1$. This is read: a function of x is equal to $6x + 1$. One distinct advantage to this type notation is that it permits identification of the domain and range elements simultaneously.

Example 3. $f(x) = 6x + 1$.

$$f(0) = 1 \qquad f(-2) = -11$$
$$f(1) = 7 \qquad f(t) = 6t + 1$$

Here the domain elements are 0, 1, -2, and t; while the corresponding range elements are 1, 7, -11, and $6t + 1$.

Notice that the notation, $f(x)$, does not mean f times x. If the product were desired, we would have used $f \cdot x$ or just fx.

Example 4. $g(x) = x^3 - 1$ and $h(x) = 3x + 1$. Find $g[h(2)]$.
We first find $h(2) = 3 \cdot 2 + 1 = 7$. Then $g[h(2)] = g[7] = 7^3 - 1 = 343 - 1 = 342$.

Functions are not necessarily always defined by a single equation. Consider the function

$$t\colon\ t(x) = \begin{cases} 2x, & \text{if } x > 0 \\ 1, & \text{if } x = 0 \\ -3x, & \text{if } x < 0 \end{cases}$$

This notation states that t is the function that pairs x with $2x$ if $x > 0$, with 1 if $x = 0$ and with $-3x$ if $x < 0$. Thus we have

$$t(5) = 2 \cdot 5 = 10$$
$$t(0) = 1$$
$$t(-2) = -3(-2) = 6.$$

Exercise 31

State the range of each of the following functions.

1) $g: x \rightarrow 3x - 2 \qquad D = \{2,5,9\}$

2) $t: x \rightarrow 2x^2 - 1 \qquad D = \{-1,1,6\}$

3) $r: x \rightarrow 3x^2 - 2x \qquad D = \{0,1,2\}$

Given that each of the following describe a function, what is its domain?

4) $f: x \rightarrow \dfrac{1}{x - 1}$ 5) $g: x \rightarrow x^2 + 1$ 6) $s: x \rightarrow \dfrac{1}{(x + 1)(x - 3)}$

Given $f(x) = 3x^2 + x + 2$. Find:

7) $f(2)$ 8) $f(3)$

9) $f(-1)$ 10) $f(6)$

Given $g(x) = (x + 1)^3$. Find:

11) $g(1)$ 12) $g(-3)$

13) $g(2)$ 14) $g(-1)$

Given $s(x) = 2x - 1$ and $t(x) = x^2 - 3$. Find:

15) $s(4) - t(1)$ 16) $s(-2) + t(3)$

17) $[s(t)]^2$ 18) $s(t + h)$

19) $s(t)$ 20) $\dfrac{s(t + h) - s(t)}{h}$

21) $s[t(1)]$ 22) $t[s(1)]$

Given the domain of each of the following functions, $D = \{-2,0,2\}$. Find the range.

23) $s: s(x) = \begin{cases} x, \text{ if } x < 0 \\ -1, \text{ if } x = 0 \\ 2x + 1, \text{ if } x > 0 \end{cases}$

24) $t: t(x) = \begin{cases} 0, \text{ if } x \geq 0 \\ 1, \text{ if } x < 0 \end{cases}$

25) $R: R(x) = \begin{cases} -9x + 6, \text{ if } x > 0 \\ (3 - 2x)^2, \text{ if } x \leq 0 \end{cases}$

5.9 Graphs

At one time or another, you have encountered various figures that have been used to visually interpret statistical data. Such figures are called graphs, and they occur in numerous ways. Pie graphs, (Figure 5.1) and bar graphs, (Figure 5.2) are but two of the many forms in existence.

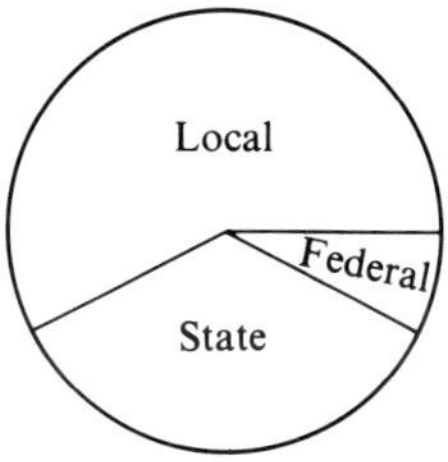

Division of the support of education from local, state, and federal sources.

Figure 5.1

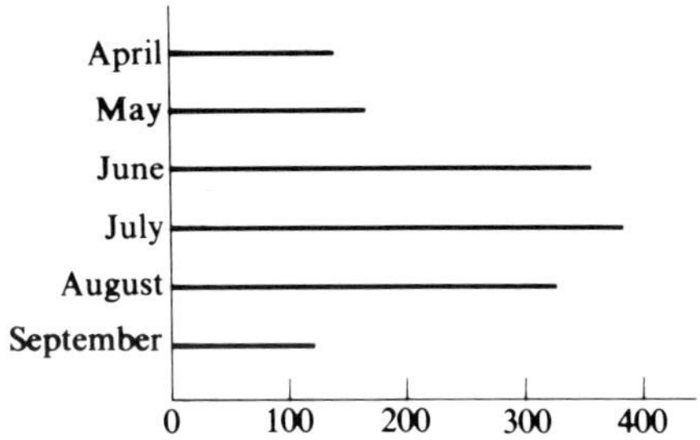

Typical earnings of a college student during six successive months. (In hundreds of dollars.)

Figure 5.2

In this section we will study a method of geometrically interpreting algebraic relationships between two variables. In other words, we will study a means by which we can draw the graphs of algebraic equations. Such a means is called *the rectangular coordinate system.*

5.10 The Rectangular Coordinate System

In an earlier section of the book we showed how numbers could be represented on a horizontal line. A point is chosen at which to begin, and the number zero is assigned to it. Numbers to the right increase in magnitude and are considered positive; while numbers to the left decrease in magnitude and are considered negative. Since an equation may relate two variables, we are going to need a means of considering both at the same time, or rather how a change in one will affect the other. To accomplish this, we construct a second number line at right angles to the first one and passing through the point labeled zero. We can now locate a number corresponding to the second variable along this line. In much the same manner as above, we consider the numbers to increase in magnitude as we move up and call the numbers above zero positive, while the numbers below are negative.

The point at which the two lines intersect is called the *origin*, the horizontal line is called the *x axis* and the vertical line the *y axis*. The four sections of the plane formed by these two lines are called *quadrants* and are numbered counterclockwise with Roman numerals as in Figure 5.3. Such a system is known as *the rectangular coordinate system.*

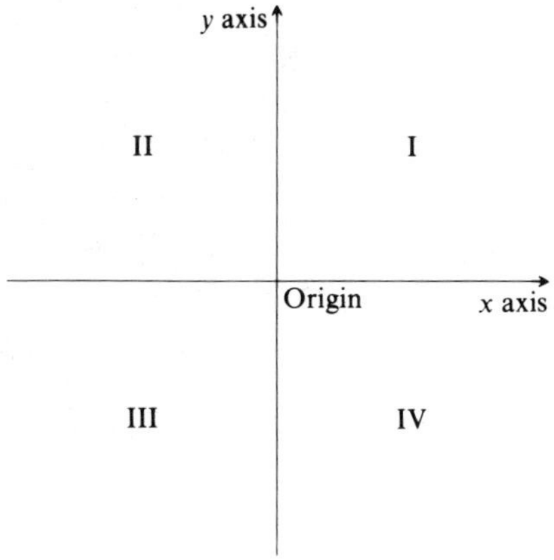

Figure 5.3

Every point P in the plane can then be located as an ordered pair (x,y). In an ordered pair the x value is always listed first and is called the *abscissa*, while the y value is listed second and is called the *ordinate*. Using this type of designation, it can be seen that the x value, or abscissa, is the perpendicular distance of the point from the y axis while the y value, or ordinate, is the perpendicular distance of the point from the x axis. See Figure 5.4.

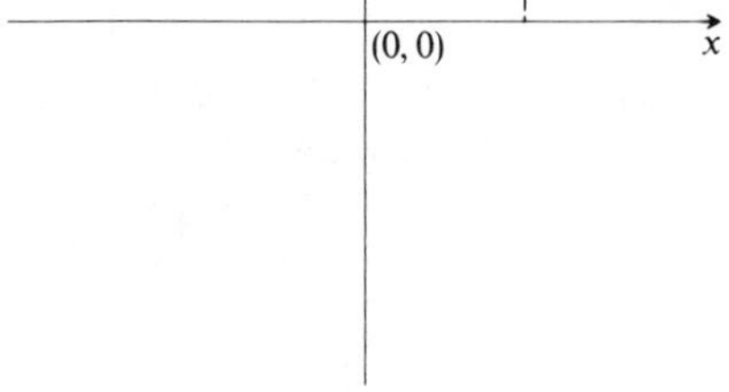

Figure 5.4

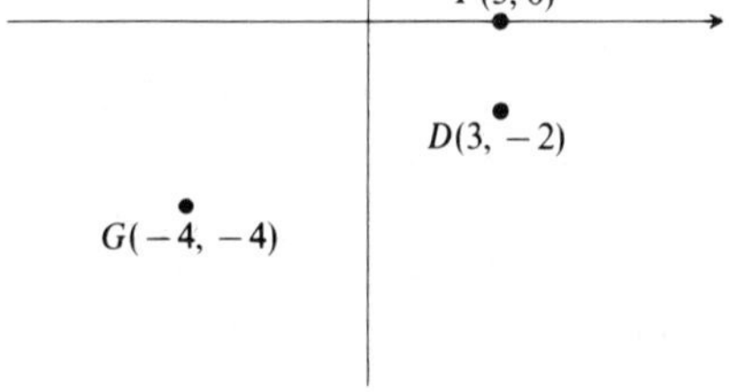

Figure 5.5

When spoken of collectively, the two elements that form the ordered pair are called the coordinates of the point P.

The position of the points $A(6,2)$, $B(-1,2)$, $C(-2,6)$, $D(3,-2)$, $E(0,3)$, $F(3,0)$, and $G(-4,-4)$ are shown in Figure 5.5.

Notice that in the first quadrant, both the abscissa and ordinate are positive, (Point A). In the second quadrant the abscissa is negative, while the ordinate is positive, (Points B and C). In the third quadrant the abscissa and ordinate are both negative, (Point G), and in the fourth quadrant the abscissa is positive and the ordinate negative, (Point D). Further, if a point lies on one of the axis, at least one of its coordinates must be zero, (Points E and F).

5.11 Linear Equations

The word *linear* means of or relating to a line or lines, and it is in this sense that we first consider algebraic equations whose geometric representations are straight lines. To illustrate what we mean, consider the equation $y = x$. If we arbitrarily allow x to assume the integral values -3 through 3, y assumes these same values also, but in such a way that $y = -3$ when $x = -3$, $y = 0$ when $x = 0$, $y = 2$ when $x = 2$ and so on. This relationship can be illustrated with the ordered pairs $(-3,-3)$, $(-2,-2)$, $(-1,-1)$, $(0,0)$, $(1,1)$, $(2,2)$, $(3,3)$.

If we locate each of these points on a rectangular coordinate system, it is easy to see that they all lie on the same straight line. (See Figure 5.6.) If we join these points by a straight line and extend the line within the confines of the graph, all other points that lie on the line will satisfy the equation, and conversely, all points that satisfy the equation will lie on the line. Notice that $(4,4)$, $(1/2,1/2)$, $(-3/2,-3/2)$ are coordinates of points on the line. The arrowheads indicate the line extends to infinity in both directions.

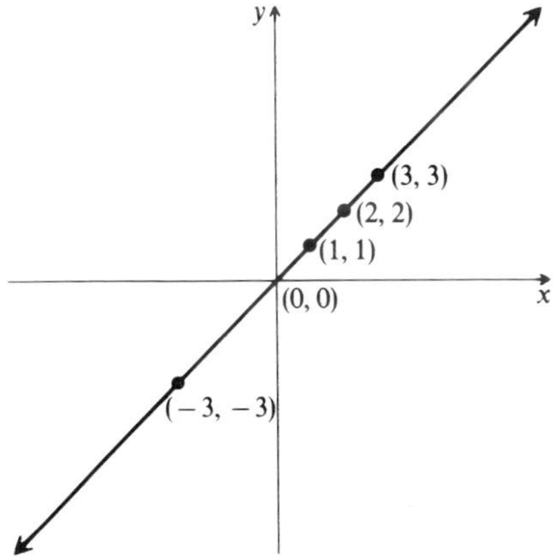

Figure 5.6

Now consider the equation $y = 3x + 6$. By arbitrarily assigning the values of $-3,0,1,2$ to x from its domain, we find the range elements, $-3,6,9,12$. If we write this one-to-one correspondence as ordered pairs and sketch them on a coordinate system, we will again find that they lie on a straight line as shown below.

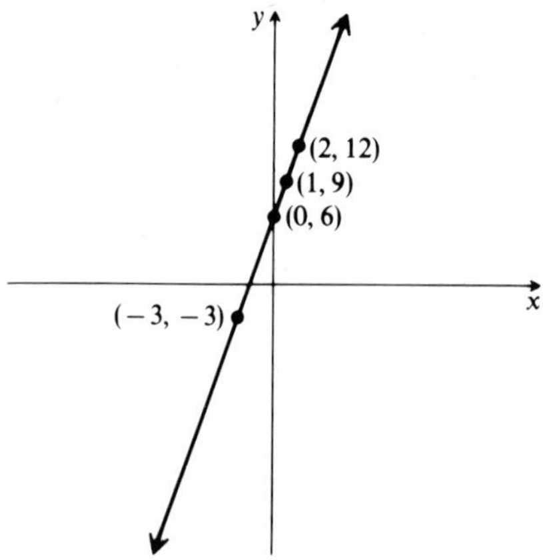

Figure 5.6b

Example 1. Sketch the graph of $2y = 3x + 4$.
If we let x assume the values 0, 2, and 4, then we find the ordered pairs (0,2), (2,5), and (4,8) satisfy the equation. The graph is shown below.

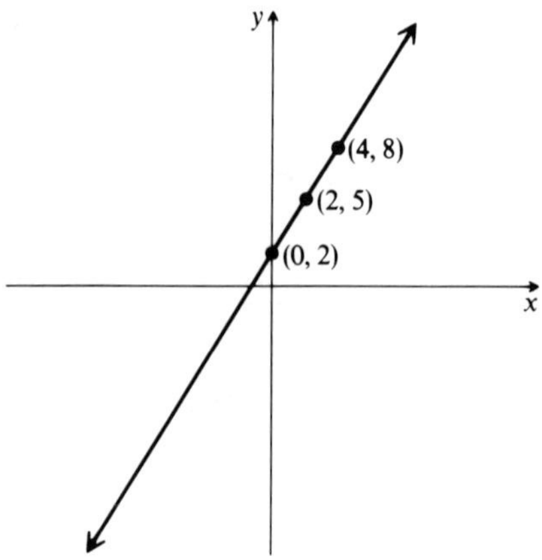

Example 2. Sketch the graph of $F = \frac{9}{5}C + 32$.

In this case let F assume the part that y has played and C the part that x has played in the previous discussion. If we let C assume the values 0, 50 and 100 and substitute these values into the given equation, then F will assume the values 32, 122 and 212 respectively. This information produces the ordered pairs, (0,32), (50,122) and (100,212), which we use to sketch the graph.

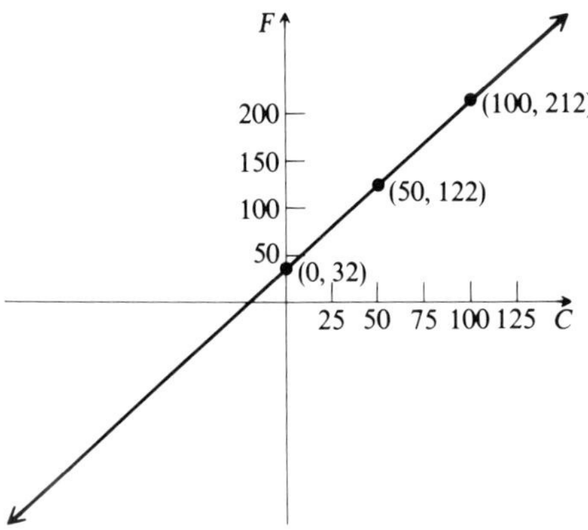

(The above equation is used for converting from Fahrenheit to Centigrade temperatures.)

Two special cases of linear equations are often overlooked because of their simplicity. These cases are $x = c$ and $y = c$, where c is any constant. The first is a vertical line through $x = c$, while the second is a horizontal line through $y = c$. The graphs are shown below in Figure 5.7.

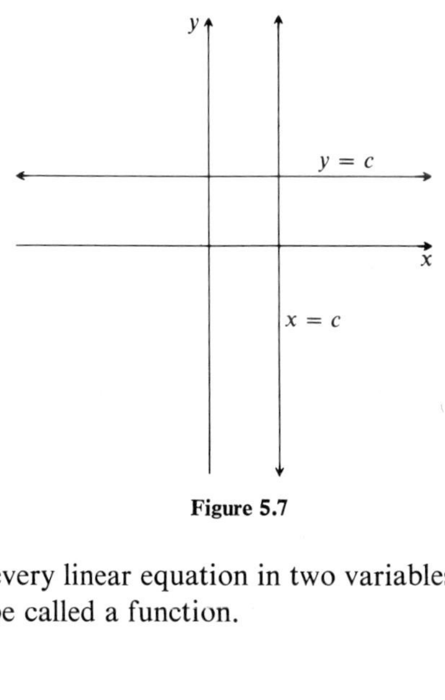

Figure 5.7

Notice that every linear equation in two variables, with the exception of $x = c$, could be called a function.

How could we have recognized that the graph of the two equations in the previous examples would have been straight lines without actually plotting the points? Does the form of the linear equation suggest any information about its graph? These and other questions will be discussed in the following sections.

Exercise 32

Locate the quadrant or axis where each of the following points lies.

1) $(2,6)$
2) $(-2,3)$
3) $(3,-7)$
4) $(-2,-5)$
5) $(7,0)$
6) $(0,0)$
7) $(-6,7)$
8) $(1/2,2/3)$
9) $(1/4,-1.3)$
10) $(-3,0)$
11) $(0,-9)$
12) $(0,\sqrt{3})$

Plot each of the following points on a separate coordinate system. Which points seem to lie on the same straight line?

13) $(0,0)$, $(1,1)$, $(2,2)$
14) $(-1,3)$, $(6,2)$, $(13,1)$
15) $(-3,4)$, $(-1,2)$, $(1,-1)$
16) $(-2,-3)$, $(0,0)$, $(2,3)$

Sketch the graph of each of the following. Functional notation has been used in some cases.

17) $2y = 3x - 5$
18) $4y - x = 3$
19) $4x - 5y = 6$
20) $2x - 3y = 2$
21) $f(x) = 2x + 1$
22) $f\colon x \to 2x - 4$
23) $g\colon x \to \frac{1}{2}x + 5$
24) $y = 9$
25) $x = 9$

5.12 The Slope of a Line

One of the basic concepts associated with graphing straight lines is that of *slope*. The steepness of a roadway, the grade of a lot, and the glide path of an airplane are all illustrations of slope. Each of these depends on the ratio of the rise (the vertical change) to the run (the horizontal change) from one point to a second point. If this ratio is constant between any two points on a line, the line is a straight line. With this in mind, we form the following definition. *The slope of a line is the ratio of the rise to the run.* The slope is commonly designated m. Thus we have

$$m = \text{slope} = \frac{\text{rise}}{\text{run}} = \frac{\text{change in vertical coordinates}}{\text{change in horizontal coordinates}}$$

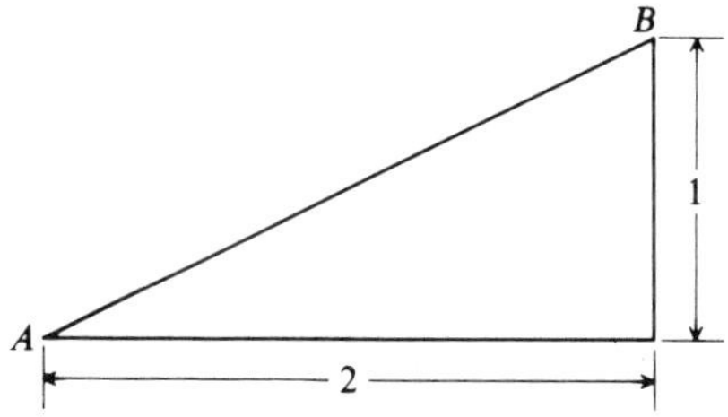

Figure 5.8

In the diagram above, the slope of the line is 1/2, since there is a vertical change of one unit and a horizontal change of two units as we move along the line from A to B.

Now suppose that $P_1(x_1,y_1)$ and $P_2(x_2, y_2)$ are any two points on a line as shown in Figure 5.9.

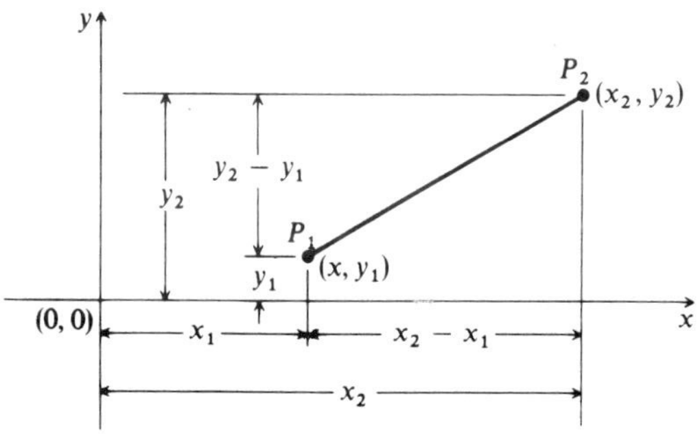

Figure 5.9

As we move from P_1 to P_2 we traverse a horizontal distance of $x_2 - x_1$ units and a vertical distance of $y_2 - y_1$ units. Thus the slope of the line joining P_1 to P_2 is

$$m = \frac{y_2 - y_1}{x_2 - x_1}.$$

Example 1. Determine the slope of the line joining (3,6) to (5,7).
We arbitrarily let $(x_1,y_1) = (3,6)$ and $(x_2,y_2) = (5,7)$, and we have

$$m = \frac{y_2 - y_1}{x_2 - x_1} = \frac{7 - 6}{5 - 3} = \frac{1}{2}.$$

If we had interchanged the choices for (x_1,y_1) and (x_2,y_2) we would have had

$$m = \frac{y_2 - y_1}{x_2 - x_1} = \frac{6 - 7}{3 - 5} = \frac{-1}{-2} = \frac{1}{2}.$$

Example 2. Find the slope of the line joining $(-1,7)$ to $(-3,-4)$.

$$m = \frac{7-(-4)}{-1-(-3)} = \frac{7+4}{-1+3} = \frac{11}{2}.$$

Notice that if we choose (x_2,y_2) to be the righthand most point of the two points (x_1,y_1) and (x_2,y_2), then $x_2 - x_1$ will always be positive. $y_2 - y_1$ will be positive if the line moves upward from left to right and negative if it moves downward. Thus the slope is always positive if a line moves upward from left to right and negative if it moves downward. Also, notice that if a line is vertical then $x_2 - x_1 = 0$, and the slope fails to exist. For this reason a vertical line is said to have no slope. If a line is horizontal, $y_2 - y_1 = 0$. Therefore, the slope of a horizontal line is 0.

5.13 The Slope Intercept Form of the Equation of a Line

Consider the specific point $A(0,b)$ and the general point $B(x,y)$. (Figure 5.10.)

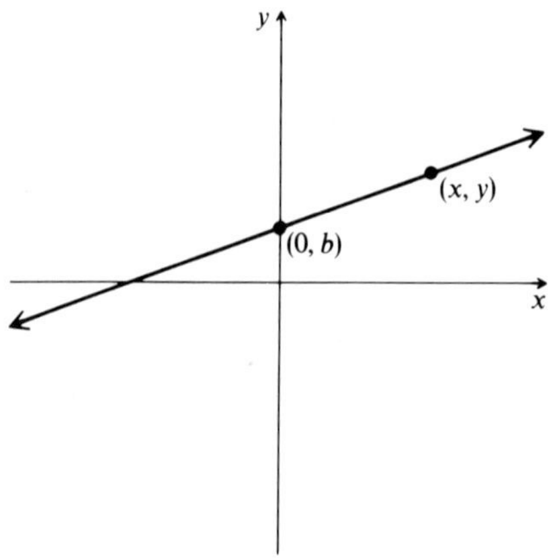

Figure 5.10

If $\overleftrightarrow{AB}$ is a line joining the two points then the slope of the line is

$$m = \frac{y-b}{x-0}.$$

Solving this equation for y, we have $y = mx + b$. This equation is known as the *slope intercept form* for the equation of a line since m is the slope and b is the y coordinate of the point where the line crosses or intercepts the y axis.

Example 1. Find the slope and the y intercept of the line whose equation is $y = \frac{2}{3}x + 6$.
In this equation $m = 2/3$ and $b = 6$. Therefore, the slope is $2/3$ and the y intercept is 6.

Example 2. Find the slope and y intercept for the equation $2x + 6y = 8$.
First we solve the equation for y obtaining

$$6y = -2x + 8 \quad y = \frac{-1}{3}x + \frac{4}{3};$$

thus

$$m = \frac{-1}{3} \text{ and } b = \frac{4}{3}.$$

Example 3. Sketch the graph of the line whose equation is

$$y = \frac{-2}{3}x + 4.$$

We first locate the point at which the graph crosses the y axis. Since $b = 4$, this point is (0,4). We now locate a second point (3,2) by starting at (0,4) and counting off the slope. (Down 2 and 3 to the right). If we had interpreted $-2/3 = 2/-3$, we could have found another point $(-3,6)$ which would have been on the line also. The graph is shown in the following figure.

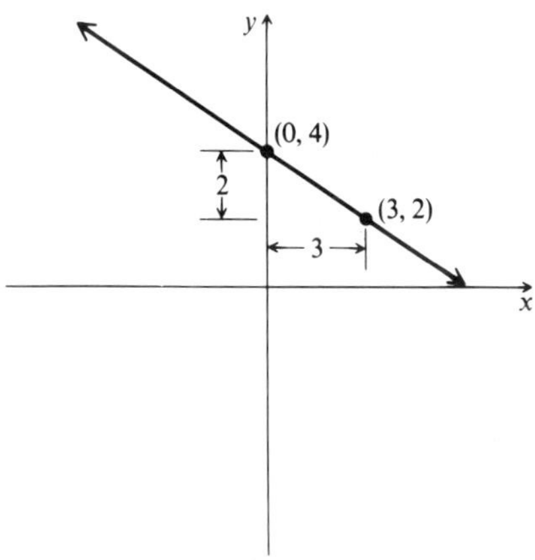

Figure 5.11

5.14 The Intercept Form of the Equation of a Line

Another useful form for the equation of a line is the *intercept form*. It is generally written as

$$\frac{x}{a} + \frac{y}{b} = 1.$$

If we let $y = 0$ and solve the resulting equation, we find $x = a$ and that $(a,0)$ is a point on the line. Similarly, if we let $x = 0$, we find $y = b$ Thus $(0,b)$ is a point on the line. Since these two points are the x and y intercepts, we can see that they are identifiable from the basic form.

Example 1. Find the x and y intercepts for the line whose equation is

$$\frac{x}{3} + \frac{y}{2} = 1.$$

Since $a = 3$ and $b = 2$, the x intercept is the point $(3,0)$ and the y intercept is the point $(0,2)$.

Example 2. Find the x and y intercept for the line whose equation is

$$\frac{x}{2} - \frac{y}{5} = 1.$$

We first rewrite the line in the basic form.

$$\frac{x}{2} + \frac{y}{-5} = 1.$$

We now identify the x intercept as $(2,0)$ and the y intercept as $(0,-5)$.

Example 3. Find the x and y intercepts for the line

$$2x + 7y = 4.$$

Dividing both sides of the equation by 4 yields

$$\frac{x}{2} + \frac{7y}{4} = 1.$$

We now rewrite this to put it in the intercept form

$$\frac{x}{2} + \frac{y}{4/7} = 1.$$

The x intercept is 2 and the y intercept is $4/7$. The graph is shown in the figure below.

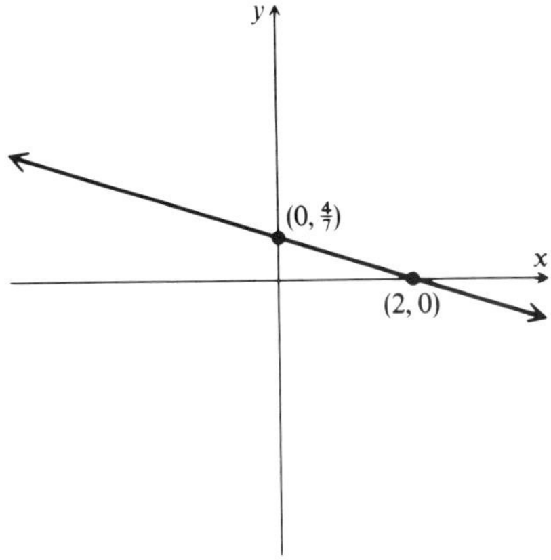

Exercise 33

Find the slope of the line joining the following pairs of points.

1) (2,3), (7,1)
2) (5,9), (−2,6)
3) (−3,−4), (7,2)
4) (−2,−7), (7,3)
5) (1/2,1/3), (−1/3,1/2)
6) (7,−2), (11,−4)
7) $(3,-5), (\frac{1}{2},2)$
8) (0,5), (9,2)
9) (0,0), (−3,−2)
10) (5,0), (0,3)

Write the equation of each of the following lines whose slope and y intercept are given.

11) $m = 6, b = 3$
12) $m = \frac{-1}{9}, b = 5$
13) $m = -4, b = 2$
14) $m = 5/11, b = -3$

Find the slope and y intercept for each of the following lines and sketch their graphs.

15) $y = 3x + 9$
16) $y = \frac{-2}{3}x + 5$
17) $2y = 4x - 8$
18) $3y = 7x + 9$
19) $-2y = -6x - 3$
20) $x + 3y = 2$
21) $2x - 5y = 11$
22) $7x - y = 14$
23) $3x + 2y - 6 = 0$
24) $4y - 8x = 38$

Place each of the following lines in the intercept form. Identify the x and y intercepts and sketch the graphs.

25) $3x + 8y = 24$
26) $6y = -4x - 12$
27) $5x - 2y + 10 = 0$
28) $2x - 7y = 9$
29) $5x - 9y = 11$

30) A linear resistor is one which possesses a constant E/I ratio. Sketch the graph of $E = 6I$. What is the E/I ratio?

31) Ohms law states that $E = IR$. Sketch the graph for $R = 1$, $R = 2$ and $R = 5$ on the same set of axes. What do you notice about the results?

5.15 Other Forms for the Equation of a Line

Two additional forms that will allow you to write the equation of a line with the proper data are the *two point form* and the *one point slope form.*

If two points $P_1 = (x_1,y_1)$ and $P_2 = (x_2,y_2)$ are known, we can find the equation of the line passing through them. Let P_1 and P_2 be any two points on a line and $P_3 = (x,y)$ be any other point on the line (Figure 5.12).

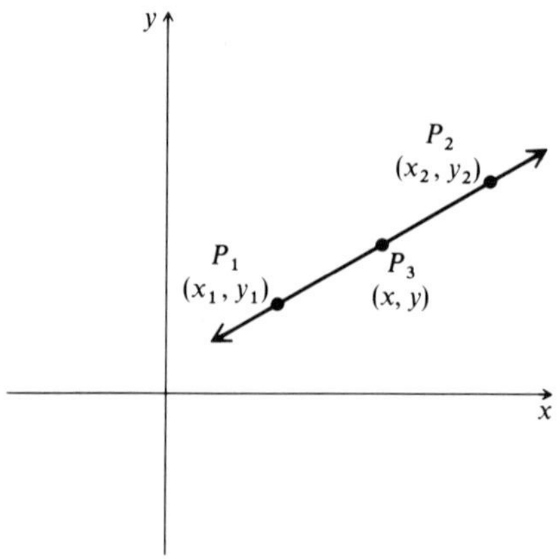

Figure 5.12

Since all three of the points lie on the same line, the slope between P_1 and P_3 must be equal to the slope between P_1 and P_2. Computing these slopes we have

$$\frac{y - y_1}{x - x_1} = \frac{y_2 - y_1}{x_2 - x_1}.$$

Multiplying both sides by $x - x_1$ yields

$$y - y_1 = \frac{y_2 - y_1}{x_2 - x_1}(x - x_1).$$

This latter equation is known as the *two point form of the equation of a line.*

Notice that this form is invalid whenever $x_1 = x_2$ since the line would be vertical. (Refer to section 5.12.)

Example 1. Find the equation of the line through (3,5) and (7,2). Letting $(3,5) = (x_1,y_1)$ and $(7,2) = (x_2,y_2)$, we have

$$y - 5 = \frac{2-5}{7-3}(x-3) \text{ or } y - 5 = \frac{-3}{4}(x-3).$$

Clearing of fractions and collecting all terms on the left hand side of the equation we have

$$4y + 3x - 29 = 0.$$

Here, as in the computation of slope, the choice of (x_1,y_1) and (x_2,y_2) is arbitrary. The reader is encouraged to interchange (x_1,y_1) and (x_2,y_2) and verify that the same equation is obtained.

When an equation is written in the form

$$ax + by + c = 0,$$

it is said to be in *general form.*

Example 2. Find the equation of the line through (6,3) and (−2,4). Find the point where the line crosses the y axis. Letting $(x_1,y_1) = (6,3)$ and $(x_2,y_2) = (-2,4)$ we have

$$y - 3 = \frac{4-3}{-2-6}(x-6)$$

$$y - 3 = -\frac{1}{8}(x-6).$$

Solving for y we have

$$y = -\frac{1}{8}x + \frac{15}{4}.$$

From this we see the y intercept is the point (0,15/4).

Perhaps you have noticed that the coefficient of the right hand member of the equation

$$y - y_1 = \frac{y_2 - y_1}{x_2 - x_1}(x - x_1)$$

is the slope of the line joining the point (x_1,y_1) to (x_2,y_2). For this reason we can replace this quantity by m, obtaining

$$y - y_1 = m(x - x_1).$$

This latter form is known as the *one point slope form* of the equation of a line. This form can be used to write the equation of a nonvertical line whenever the slope m and one point (x_1,y_1) are known.

Example 1. Find the equation of the line through $(-1,7)$ with slope 2. Substituting in the one point slope form, we have $y - 7 = 2(x + 1)$. If we put this in the general form, then $2x - y + 9 = 0$ is the desired equation.

Example 2. Find the equation of the line through (6,7) with zero slope. Substituting in the one point slope form yields

$$y - 7 = 0(x - 6),$$

or

$$y = 7.$$

Exercise 34

Find the equation of the line through each of the following pairs of points. Write the equation in general form.

1) $(3,4), (-5,7)$
2) $(1,-2), (2,3)$
3) $(0,0), (5,6)$
4) $(2,-3), (6,-4)$
5) $(2,7), (-1,7)$
6) $(3,9), (3,0)$ (Why does the two point form fail to apply?)
7) $(0,7), (7,0)$
8) $(5,2), (13,4)$

Find the equation of the line through the indicated point with the indicated slope. Write the equations in the slope intercept form.

9) $(6,2),\ m = 4$
10) $(0,0),\ m = \frac{-1}{2}$
11) $(3,-4),\ m = \frac{2}{3}$
12) $(-6,2),\ m = 0$
13) $(5,9)$, No slope.
14) Use the two point form for the equation of a line to find the equation of the line through $(a,0)$ and $(0,b)$. Put the equation in the intercept form.
15) In the section on measurement we learned that 1 foot is equal to approximately 30.48 centimeters. If we express this relationship as a linear equation, we have $c = 30.48\,f$, where c and f represent centimeters and feet respectively. Sketch the graph of this relationship from $f = 0$ to $f = 5$.

Chapter 5 Review Exercise

Solve the following equations

1) $x - 9 = 15$
2) $3x + 4 = 16$
3) $\frac{1}{5} = \frac{2}{3}x$
4) $(3 - m) + 2 = 4 + 2m$

5) $-2(2y - 5) = 3(y + 2)$ 6) $\frac{2x + 1}{4} + 3 = \frac{3x - 5}{2}$

Express as a ratio and simplify

7) 5 yds. to 6 ft. 8) 6 qts. to 2 gal.

Solve for x

9) $A + B + x = 520$ 10) $\frac{x + y}{2} = 4$

11) $F = xa$ 12) $M = \frac{1}{2}xt^2$

State the range of the following functions if the domain is $\{4,5,6\}$.

13) $f\colon x \to x + 7$ 14) $h\colon x \to x^2 - 3$

Given $g(x) = 3x^2$, find

15) $g(4)$ 16) $g(y + 1)$ 17) $g(-2)$

Sketch the graph of the following:

18) $y + x = 6$ 19) $3x = y - 5$ 20) $x - 2y = 4$

21–23. Determine the slope, x intercept and y intercept of exercises 18–20.

Find the equation of lines having the following properties.

24) Passing through the points $(-2,3)$ and $(4,-1)$.

25) With a slope of $-(1/3)$ and containing the point $(6,-1)$.

6 Logarithms

6.1 The General Exponential Equation

If we consider the equation $y = b^x$ for a moment, we notice that if $0 < b < 1$ and $x > 0$, the value of y decreases as x increases. Also, if $b > 1$ and $x > 0$, then y increases as x increases. This equation for some constant base b is called the general *exponential equation*. The value of y is determined from successive powers of a constant base b. A graph of $y = 2^x$ is shown in Figure 6.1, and a graph of $y = (1/2)^x$ is shown in Figure 6.2.

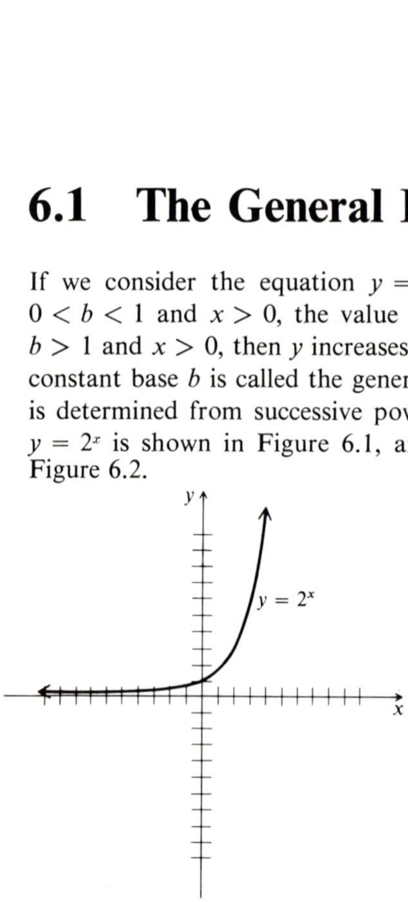

Figure 6.1

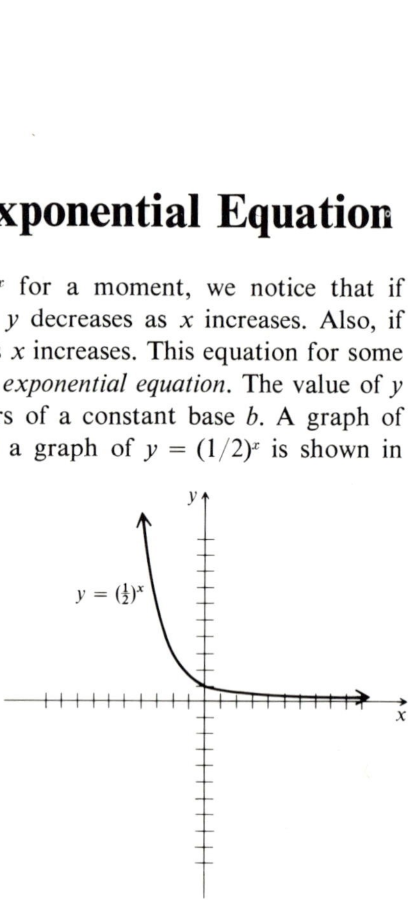

Figure 6.2

By use of the graphs we could obtain approximations for various values of y, such as

$$y = 2^{1.5},\ 2^{\sqrt{2}} \text{ or } (1/2)^{1/3}.$$

6.2 The Logarithmic Equation

When the values of x and y are interchanged in an equation, the resulting equation is called the *inverse* of the given equation. The inverse of the exponential equation is therefore $x = b^y$ for some positive value of b. Since equations are normally solved for y in terms of x, this leads to the definition of a logarithm: *A logarithm of a number x is the power of some positive base b necessary to obtain x.* This relation is written as

$$y = \log_b x$$

and by definition is equivalent to the expression

$$b^y = x,$$

the inverse of the exponential equation.

The definition of the logarithm lets us solve the inverse of the exponential equation for y in terms of x. If $b > 1$, the graph of $y = \log_b x$ appears as in Figure 6.3. If $0 < b < 1$, the graph of $y = \log_b x$ appears as in Figure 6.4.

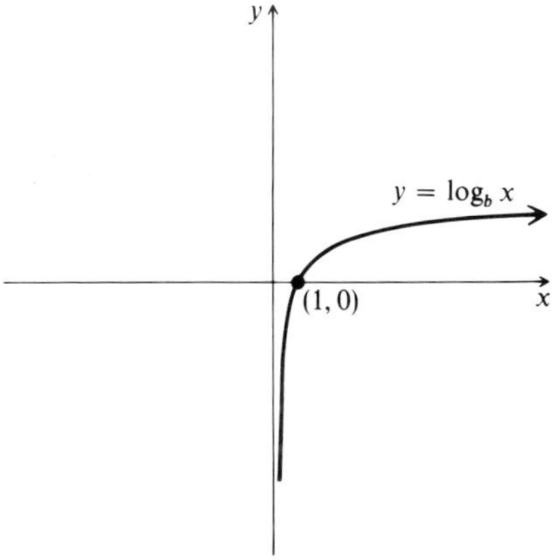

Figure 6.3

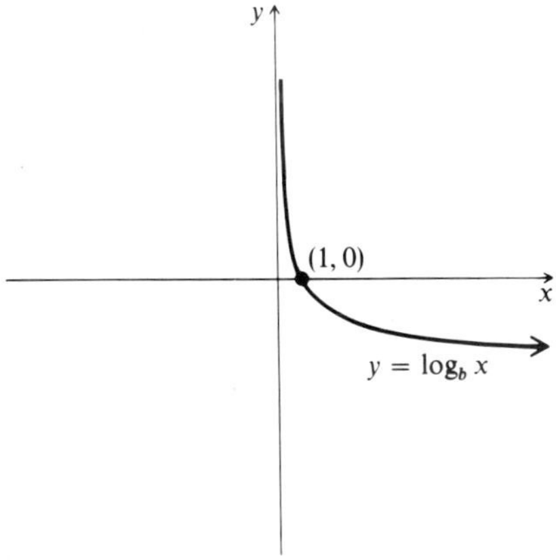

Figure 6.4

Example 1. $\log_2 8 = 3$ or $2^3 = 8$.

Example 2. $\log_{10} 100 = 2$ or $10^2 = 100$.

Example 3. $\log_3 1/9 = -2$ or $3^{-2} = 1/9$.

Example 4. If $\log_2 x = 5$, find x.
By definition $2^5 = x$. Since $2^5 = 32$, $x = 32$.

Example 5. If $\log_9 3 = x$, find x.

$$9^x = 3 \text{ or } 3^{2x} = 3^1.$$

Since the bases are equal, the exponents must be equal and $2x = 1$ or $x = 1/2$. Therefore,

$$\log_9 3 = 1/2 \text{ or } 9^{1/2} = 3.$$

Since $b^p \cdot b^q = b^{p+q}$, we can consider some interesting properties of logarithms. Consider $\log_b (p \cdot q)$. If $m = \log_b p$, $n = \log_b q$, then $b^m = p$ and $b^n = q$, and $p \cdot q = b^{m+n}$, from which we obtain $(m + n) = \log_b p \cdot q$. Substituting the values for m and n, we obtain

$$\log_b p \cdot q = \log_b p + \log_b q.$$

By the same method using exponents we can arrive at an expression for the logarithm of a quotient.

$$\log_b \frac{p}{q} = \log_b p - \log_b q \text{ if } p, q > 0.$$

Also, since $p^q = p \cdot p \cdot p \cdots p$ (q times) an extension of the above expressions allows us to write $\log_b p^q = q \log_b p$ and $\log_b \sqrt{p} = \frac{1}{2} \log_b p$.

Example 6. $\log_2 (8 \cdot 2) = \log_2 8 + \log_2 2 = 3 + 1 = 4$. Multiplying first, we obtain $\log_2 16 = 4$.

Example 7. $\log_3 (27/9) = \log_3 27 - \log_3 9 = 3 - 2 = 1$ and $\log_3 (27/9) = \log_3 3 = 1$.

Example 8. $\log_5 5^3 = 3 \log_5 5 = 3 \cdot 1 = 3$, or $\log_5 5^3 = \log_5 (125) = 3$.

Notice that for any permissible base b, $\log_b b = 1$.

Example 9. $\log_2 \sqrt[5]{32} = \log_2 32^{1/5} = 1/5 \log_2 32 = 1/5(5) = 1$, or $\log_2 \sqrt[5]{32} = \log_2 2 = 1$.

In the next section we will utilize these properties in performing calculations with logarithms.

Exercise 35

Graph the following.

1) $y = 3^x$ 2) $y = 10^x$ 3) $y = \left(\frac{1}{5}\right)^x$

4) $y = \log_3 x$ 5) $y = \log_{10} x$ 6) $y = \log_{12} x$

Write the following in logarithmic form.

7) $5 = 25^{1/2}$ 8) $\left(\frac{1}{2}\right) = (16)^{-1/4}$ 9) $9 = 3^2$

10) $27 = 9^{3/2}$ 11) $64 = 2^6$ 12) $2 = 64^{1/6}$

Write the following in exponential form.

13) $\log_{10} 100 = 2$ 14) $\log_{10} 1000 = 3$ 15) $\log_{10} .00001 = -5$

16) $\log_{16} 2 = 1/4$ 17) $\log_2 16 = 4$ 18) $\log_9 27 = 3/2$

19) $\log_5 25 = 2$ 20) $\log_{25} 5 = 1/2$

Solve the following for the unknown value using the properties of logarithms and exponents.

21) $\log_x 625 = 2$ 22) $\log_{10} x = 5$ 23) $\log_{10} x = -4$

24) $\log_3 9 = x$ 25) $\log_7 49 = x$ 26) $4^x = 16$

27) $x^3 = \frac{1}{27}$ 28) $5^4 = x^{12}$ 29) $27^{2/3} = x$

30) $x^{1/5} = 2$

Write the following as the sum or difference of logarithms.

31) $\frac{5\sqrt{75}}{16}$ 32) $\frac{\sqrt[5]{625}}{\sqrt{32}}$ 33) $\frac{(57.5)(8.35)}{\sqrt[3]{425}}$

Write the following as the logarithm of a single quantity.

34) $\frac{1}{5}\log_{10} 75 - \frac{1}{3}\log_{10} 15 + 2\log_{10} 25$

35) $\frac{1}{3}[\log_{10} 45 + \log_{10} 5] - \frac{1}{2}\log_{10} 27$

6.3 Common Logarithms

As we have discussed previously, we may have logarithms with any base greater than zero. Since our number system utilizes a base of ten, it will be useful to pay special attention to logarithms having a base of 10. These logarithms are called *common logarithms*.

We already know some common logarithms. $\text{Log}_{10} 1 = 0$, $\log_{10} 10 = 1$, $\log_{10} 100 = 2$, $\log_{10} 1000 = 3$ and in general, $\log_{10} 10^n = n$. Also, since $\sqrt{10} \simeq 3.162$, $\log_{10} 3.162 \simeq .5$. Notice that we can discuss 10 raised to any real number power. Since 10 to any real number power would be positive, we can consider the logarithm to the base 10 of any positive number. Note also that 10 to any positive power less than 1 lies between 1 and 10. Therefore, the common logarithm of any number between 1 and

10 must lie between 0 and 1. If we recall scientific notation for a moment, we can write the common logarithm of any positive number as an integer plus the logarithm of a number between 1 and 10.

Example 1. $\log_{10} 250 = \log_{10} (2.50 \times 10^2) = \log_{10} 2.50 + \log_{10} 10^2 = \log_{10} 2.50 + 2 = 2 + \log_{10} 2.50.$

Example 2. $\log_{10} .00035 = \log_{10} (3.5 \times 10^{-4}) = \log_{10} 3.5 + \log_{10} 10^{-4} = \log_{10} 3.5 + (-4) = -4 + \log_{10} 3.5.$

When we consider the common logarithm of a number, the integral part of the number, that is, the power of ten when the number is written in scientific notation, is called the *characteristic* of the logarithm. The power of 10 necessary to obtain the digits of the number, that is, the number between 1 and 10 when written in scientific notation, is called the *mantissa* of the logarithm. Table II in the back of the text gives the mantissas of common logarithms to four decimal places.

If we look at the logarithm of 8.75 in the table (this is done by finding 8.7 in the column headed n and move to the right under the column headed 5), we find $\log_{10} 8.75$ to be .9420. This means that $10^{.9420} = 8.75$. Also, $\log_{10} 87.5 = 1.9420$, $\log_{10} 875 = 2.9420$ and so on. The characteristics were determined by the method discussed above.

Example 3. Find the common logarithm of 450,000.
Since $450{,}000 = 4.50 \times 10^5$, the characteristic of the logarithm is 5. From Table III we find the mantissa to be .6532. Therefore, $\log_{10} 450{,}000 = 5.6532$. That is, $10^{5.6532} = 450{,}000$.

Example 4. Find $\log_{10} 35.2$.
$\log_{10} 35.2 = 1.5465.$

Example 5. $\log_{10} 772 = 2.8876.$

Since all the logarithms in this section are to the base 10, the base is not written and the base 10 is understood.

6.4 Negative Characteristics

As we saw in a previous section, $\log .00035 = -4 + \log 3.5$. We can now write this logarithm as $-4 + .5441$. Since the mantissa of a logarithm is always a positive number, this is not generally written as $-4.000 + .5441 = -3.4559$. Since this number is negative, we could not find it in the table of mantissas. In order to keep the logarithm written with a positive mantissa, we notice that $-4 = 6 - 10$ so that $-4 + .5441$ can be written as $6.5441 - 10$. In this manner, we can utilize the table of mantissas to perform calculations with numbers smaller than one.

A common error in using negative characteristics is to write the characteristics as the integral part of a decimal number such as -4.5441 instead of $-4 + .5441$. This would assume that the mantissa is negative which is, of course, false.

6.5 Antilogarithms

Since $\log 2 = .3010$, then the number whose logarithm is .3010 (that is, 2) is called the *antilogarithm* of the number .3010. In other words, the antilogarithm of a number is the result of 10 being raised to that power or the number whose logarithm is the given number.

Example 1. Find the antilogarithm 2.4346.
Since $\log 272 = 2.4346$, antilogarithm $2.4346 = 272$.

Example 2. Antilogarithm $1.7520 = 56.5$.

6.6 Linear Interpolation

Many times in reading tables of approximations of values on a curve, such as the table of mantissas of common logarithms, it becomes necessary to read the table to more digits than are available in the table. To approximate this value between tabular values, we use a process called *linear interpolation*. Linear interpolation is a process by which the values of a curve are approximated by use of a straight line. For instance, suppose a curve is as shown in Figure 6.5.

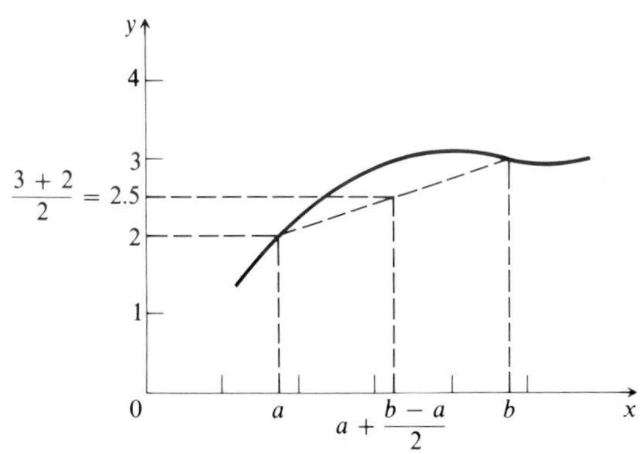

Figure 6.5

If the value corresponding to a is 2 and the value corresponding to b is 3, we can then use interpolation to approximate the value for the point midway between a and b, that is,

$$a + \frac{b - a}{2}.$$

This is accomplished by taking the difference between the tabular values, that is $3 - 2 = 1$, and dividing it by 2. This value is then added to the smaller value. Therefore, the approximate value corresponding to $\frac{b - a}{2} + a$ is $2 + 1/2 = 2.5$. Note that 2.5 is found on the line, not on the curve, and hence only approximates the value midway between a and b.

Using this method, let us approximate the logarithm of 7.885 using Table II as follows.

$$\begin{aligned}\log 7.890 &= .8971 \text{ (from Table II)}\\ \log 7.895 &= .8965 + .0003 = .8968\\ \log 7.880 &= .8965 \text{ (from Table II)}\end{aligned}$$

Since we are halfway between the log 7.890 and log 7.880, we take 1/2 the difference in the tabular value (.0006). Then $(.5)(.0006) = .0003$ which is added to the smaller tabular value so that the log 7.885 is approximately $.8965 + .0003 = .8968$.

Example 1. Find log 85.76.

log 85.80 = 1.9335

.10 —— log 85.80 = 1.9335 —— .0005

.06 —— log 85.76

log 85.70 = 1.9330

$$\frac{.06}{.10} = \frac{6}{10} = .6$$

$.6(.0005) = .0003.$

Therefore, $\log 85.76 \simeq 1.9330 + .0003 = 1.9333$.

Example 2. Find log .3214.

log .3220 = 9.5079 − 10

log .3214 —— .0014

log .3210 = 9.5065 − 10

$.4(.0014) = .00056 \simeq .0006.$

Therefore, $\log .3214 \simeq 9.5065 - 10 + .0006 = 9.5071 - 10$.

Example 3. Find log .002758.

$$\left.\begin{array}{l}\log .002760 = 7.4409 - 10 \\ \log .002758 \\ \log .002750 = 7.4393 - 10\end{array}\right] .0016$$

$(.8)(.0016) = .00128 \simeq 0.0013$

$\log .002758 \simeq 7.4393 - 10 + .0013 = 7.4406 - 10.$

This process can be reversed to find the antilogarithm of a number, if the exact value of the mantissas is not found in the tables.

For instance, suppose log N = .7799, and we desire to find N. We would look in the body of the table to find mantissa values on either side of our value and the numbers to which they correspond.

$$.01\left[\begin{array}{l}\log 6.03 = .7803 \\ \log N \quad = .7799 \\ \log 6.02 = .7796\end{array}\right. \qquad \rightarrow .0007, \quad \rightarrow .0003$$

We next find the difference in the tabular values; first between the two on either side of our value and then between our value and the lower tabular value. These differences are written as a proportion and multiplied by the difference in the numbers.

$$\frac{.0003}{.0007} = 3/7(.01) \simeq .0043, \; N \simeq 6.02 + .004 = 6.024.$$

Example 4. If log N = 3.3886, find N.

log 2450 = 3.3892

log N = 3.3886

log 2440 = 3.3874

$\frac{12}{18}(10) \simeq 6.67, \; N \simeq 2440 + 7 = 2447$

Example 5. Find antilog of 8.4978 − 10.

log .0315 = 8.4983 − 10

log N = 8.4978 − 10

log .0314 = 8.4969 − 10

$\frac{9}{14}(.0001) = .000064. \; N = .03146.$

6.6 Calculations with Logarithms

Since $\log (a \cdot b) = \log a + \log b$, then $\log (45 \times 8.7) = \log 45 + \log 8.7 = 1.6532 + .9395 = 2.5927$ and antilogarithm $2.5927 = 391.5$. Therefore, $45 \times 8.7 = 391.5$. Also, since $\log \frac{a}{b} = \log a - \log b$, $\log \frac{45}{8.7} = \log 45 - \log 8.7 = 1.6532 - .9395 = .7137$. Since antilogarithm $.7137 = 5.172$, $\frac{45}{8.7} = 5.172$.

Example 1. Using logarithms, find (523)(43.7).
If $N = (523)(43.7)$, then $\log N = \log 523 + \log 43.7 = 2.7185 + 1.6405 = 4.3590$.
Since antilog $4.3590 = 22{,}850$, $N = 22{,}850$.

Example 2. Find $\sqrt[4]{426}$.
If $N = \sqrt[4]{426} = (426)^{1/4}$, $\log N = 1/4 \log 426 = 1/4(2.6294) = .6574$.
Antilogarithm $.6574 = 4.543$. Therefore, $\sqrt[4]{426} = 4.543$.

Example 3. $\sqrt[3]{\dfrac{483 \times .0852}{.725}}$.

$$\begin{aligned}
\log N &= \tfrac{1}{3}[\log 483 + \log .0852] - \log .725 \\
&= \tfrac{1}{3}[2.6839 + 8.9304 - 10] - (9.8603 - 10) \\
&= \tfrac{1}{3}[11.6144 - 10] - (9.8603 - 10) \\
&= \tfrac{1}{3}[31.6144 - 30] - (9.8603 - 10) \\
&= 10.5384 - 10 - (9.8603 - 10) \\
&= .6780
\end{aligned}$$

Antilog $.6780 = 4.765$
Therefore, $N = 4.765$.

Example 4. $\dfrac{3.55 \times 75.2 \times .0392}{.00575 \times 89.3}$.

$$\log N = \log 3.55 + \log 75.2 + \log .0392 - \log .00575 - \log 89.3.$$

$$\log N = .5502 + 1.8762 + (8.5933 \quad 10) - (7.7597 - 10) - (1.9509).$$

$$\begin{array}{rl}
 & .5502 \\
 & 1.8762 \\
 & \underline{8.5933 - 10} \\
= & 11.0197 - 10 - (9.7109 - 10)
\end{array}
\qquad\qquad
\begin{array}{r}
7.7597 - 10 \\
\underline{1.9509} \\
9.7106 - 10
\end{array}$$

$$= 1.3091$$

$$N = 20.37$$

Example 5. $83.2 \div .0375$.
$\log (83.2 \div .0375) = \log 83.2 - \log .0375 =$
$1.9201 - [8.5740 - 10] = 1.9201 - 8.5740 = 10.$
We need to write 1.9201 as $11.9201 - 10$ since subtraction would have yielded a negative mantissa.
Thus we have $(11.9201 - 10) - (8.5740 - 10) = 3.3461$.
Antilog $3.3461 = 2219$. Therefore, $\dfrac{83.2}{.0375} = 2220.$

Exercise 36

Find the common logarithms of the following using Table II.

1) 4.56 2) 45.6 3) 456
4) 752,000 5) .000693 6) 93.2
7) 186,000 8) 6.02×10^{-23} 9) 2.723
10) .0375

Find the antilogarithms of the following.

11) 1.8129 12) .9969 13) 4.4487
14) $6.1206 - 10$ 15) $9.3304 - 10$ 16) 2.5024
17) $8.6522 - 10$ 18) $3.8267 - 10$ 19) $6.8727 - 10$
20) 5.9763

Perform the following calculations using logarithms. Interpolate where necessary.

21) $(75.3)(.579)$ 22) $(83.9)(146)$ 23) $(975)(45.7)$
24) $(.0347)(.00592)$ 25) $\dfrac{15000}{754}$ 26) $\dfrac{97.8}{.0578}$
27) $596\sqrt{787}$ 28) $\dfrac{\sqrt[5]{475{,}000}}{\sqrt{265}}$ 29) $\dfrac{97.5\sqrt{4.32}}{\sqrt[3]{48.8}}$
30) $\dfrac{\sqrt[3]{975}}{.392\sqrt{598}}$ 31) $(.5)^6$ 32) $(3)^5\sqrt[3]{78}$
33) $(9.75)^4(.0475)^2$ 34) $\dfrac{(55.7)^3\sqrt[3]{9.487}}{(8.65)^4}$

Chapter 6 Review Exercise

Sketch the graph of the following.

1) $y = 5^x$ 2) $y = \left(\dfrac{1}{5}\right)^x$ 3) $x = 5^y$ 4) $x = \left(\dfrac{1}{5}\right)^y$

Write in logarithmic form.

5) $9 = 3^2$ 6) $1000 = 10^3$ 7) $\frac{1}{1000} = 10^{-3}$

Write in exponential form.

8) $\log_3 9 = 2$ 9) $\log_{10} 1000 = 3$ 10) $\log_2 64 = 6$

Solve for the variable.

11) $\log_x 27 = 3$ 12) $\log_5 125 = x$ 13) $\log_4 x = 3$

Find the common logarithm of the following.

14) 45.7 15) .00753 16) .6324 17) 8613.0

Find the antilogarithm of the following.

18) 3.2414 19) .4935 20) $8.3456 - 10$ 21) $.8735 + (-3)$

Calculate using logarithms.

22) $\sqrt[3]{756}$ 23) $\frac{4.973}{16.35}$ 24) $\frac{\sqrt[5]{(419)(3.65)}}{(18.7)^2}$

25) $\frac{(86.3)(\sqrt[3]{9.31})}{\sqrt{75.7}}$

7
Geometry

7.1 Geometry

The word **geometry** comes from the Greek roots *geo* and *metria*, meaning *measurement of the earth*. The subject of geometry deals with the measurement, size, and relationships to one another of different figures.

In this section we will discuss some basic principles of **plane geometry** or the geometry of flat surfaces. Emphasis is on applications rather than formal proofs.

7.2 Undefined Concepts

In the study of geometry, in order to discuss properties we must start with certain concepts which are not defined. These concepts are those of point, line, and plane. We will accept the meaning of these words as undefined and represent a point as a dot, a line as a straight mark, and a plane as a flat surface. A point is labeled by a capital letter at the dot, a line is labeled by indicating two points on the line or by a single small letter on the line, and a plane is labeled by three or more points on the plane. Keep in mind that these representations are not definitions but ideas to help in our understanding. These concepts are then used to define words necessary for a discussion of geometric ideas.

7.3 Rays and Line Segments

We use the concept of line to define other related ideas. A **line segment** is the set of points on a line between two given points of the line. Line segments are labeled by using a capital letter at each endpoint of the segment. Therefore segment AB indicates all the points on a line between point A and point B and is written as $\overline{AB}$ or simply AB. Also, because geometry deals with figures, we must know how to represent figures with a picture. Segment AB would be represented by

A **ray** differs from a line segment in that a ray is the set of all points in one direction from a given point on a line. In other words, a ray is bounded at one end and extends indefinitely in the other direction. A ray is labeled by its endpoint, called its origin, and one other point on the ray. Therefore ray AB, written $\overrightarrow{AB}$, would be all the points in a line on the same side of A as point B. A picture of $\overrightarrow{AB}$ is

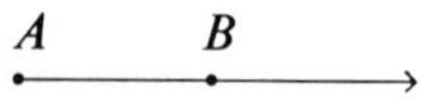

Lines intersect if they have one point in common (Figure 7.1). The common point is the **point of intersection.** If two lines in the same plane do not intersect no matter how far they are extended, they are called **parallel lines** (Figure 7.2).

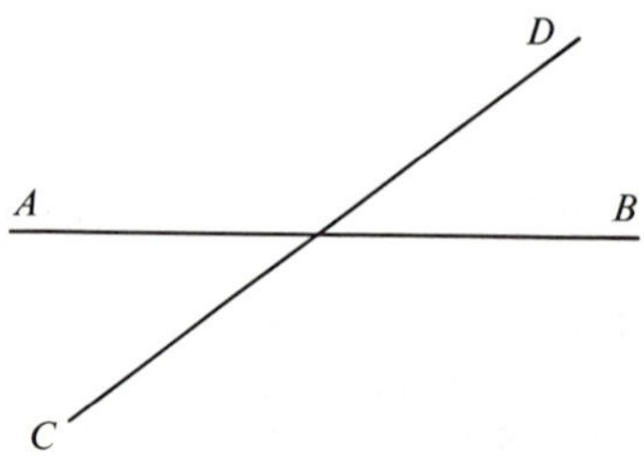

Figure 7.1

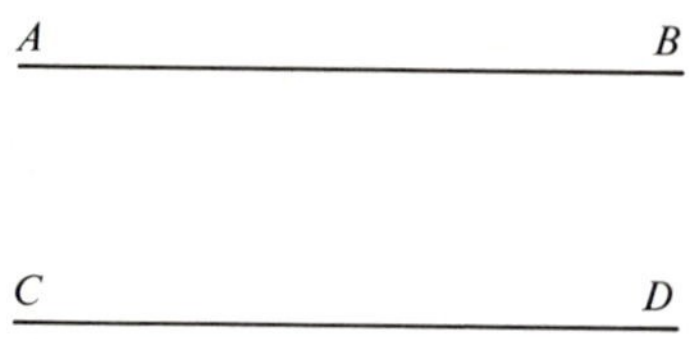

Figure 7.2

7.4 Angles

The union of two rays is called an **angle.** That is, when two rays have the same endpoint, they form an angle. The common point is the **vertex** of the angle. Angles are measured by the rotation between the rays, in degrees. A degree is $1/360$ of the total rotation about a point. A tool for measuring angles is called a protractor. Angles are labeled by a capital letter at the vertex as in Figure 7.3, by a small letter or numeral between the rays as in Figure 7.4, or by three capital letters as in Figure 7.5. The symbol for angle is $\angle$.

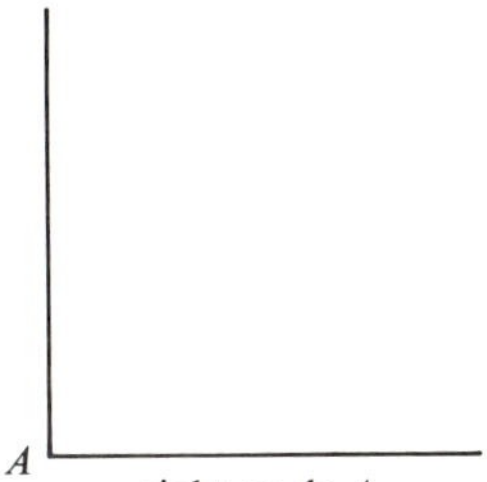

right angle A

Figure 7.3

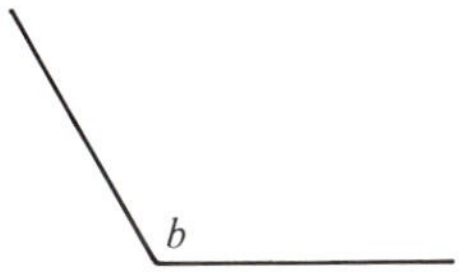

obtuse angle b

Figure 7.4

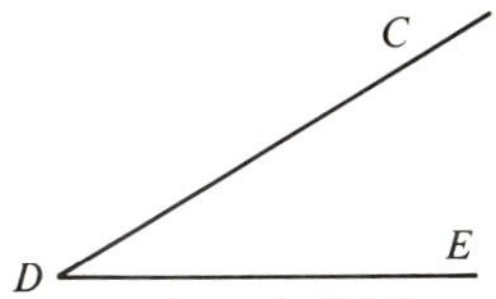

acute angle CDE

Figure 7.5

When two lines intersect, four angles are formed. Those angles having a common endpoint and a common ray between them are called **adjacent angles** and those which have only a common vertex are called **vertical angles** (Figure 7.6). Observe that when two lines intersect, the vertical angles formed have the same number of degrees and are equal. No such statement can be made for adjacent angles.

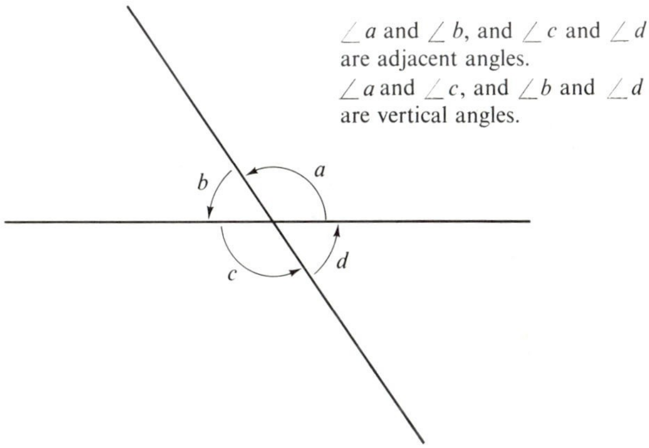

Figure 7.6

Angles of 90° are called **right angles** (Figure 7.3). When two lines intersect in such a way that they form right angles, the lines are said to be **perpendicular.** Since all right angles have 90°, all right angles are equal. An angle which consists of two adjacent right angles is a **straight angle,** because the noncommon sides lie in a straight line. Angles containing less than 180° and more than 90° are called **obtuse angles** (Figure 7.4), and angles containing less than 90° and more than zero degrees are called **acute angles** (Figure 7.5).

7.5 Triangles

A **triangle** is a closed three-sided figure whose sides are straight lines. A triangle has three angles, and may be classified according to the type of angle it contains. A triangle containing a right angle is called a **right triangle** (Figure 7.7), one containing an obtuse angle is called an **obtuse triangle** (Figure 7.8), and one containing three acute angles is called an **acute triangle** (Figure 7.9). Triangles are labeled by a letter at each vertex, as indicated in the figures.

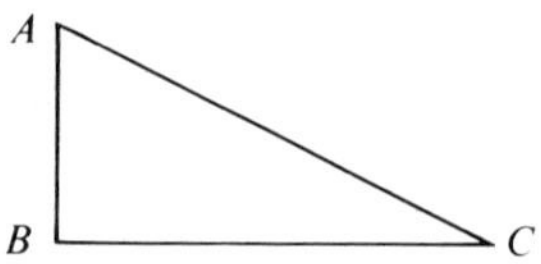

Figure 7.7

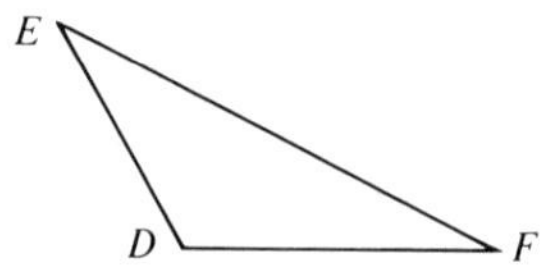

Figure 7.8

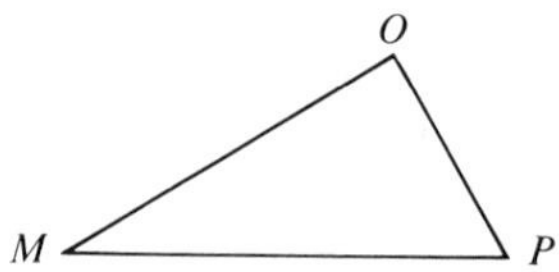

Figure 7.9

The total measure of all the angles of a triangle is equal to 180°. For this reason, a triangle can have at most one right angle or one obtuse angle. The other two angles must be acute.

Triangles are also classified according to their sides. A triangle with three equal sides is referred to as an **equilateral triangle.** An equilateral triangle also has three equal angles, and for this reason may also be called an **equiangular triangle.** Each of these angles measures 60° (Figure 7.10).

A triangle with two equal sides is an **isosceles triangle** (Figure 7.11). The unequal side is the **base** and the angle opposite the base is the **vertex angle** of the triangle.

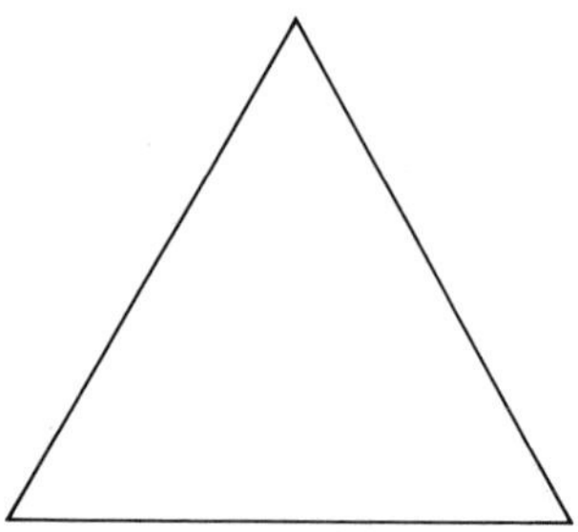

Figure 7.10

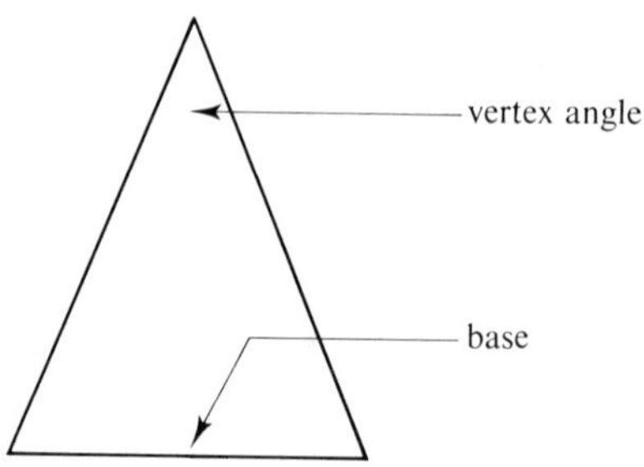

Figure 7.11

A triangle with no two sides equal is called a **scalene triangle.** A scalene triangle must also have no two angles equal.

7.6 The Pythagorean Theorem

Right triangles have an important property which other triangles do not have. This property is described by the **Pythagorean theorem:** *in a right triangle, the sum of the squares of the two legs* (the sides of the triangle adjacent to the right angle) *is equal to the square of the hypotenuse* (the side of the triangle opposite the right angle). If a right triangle is labeled ABC, with $\angle C = 90°$, the theorem would be written

$$(AB)^2 = (AC)^2 + (BC)^2.$$

Figure 7.12 shows a representation of this theorem.

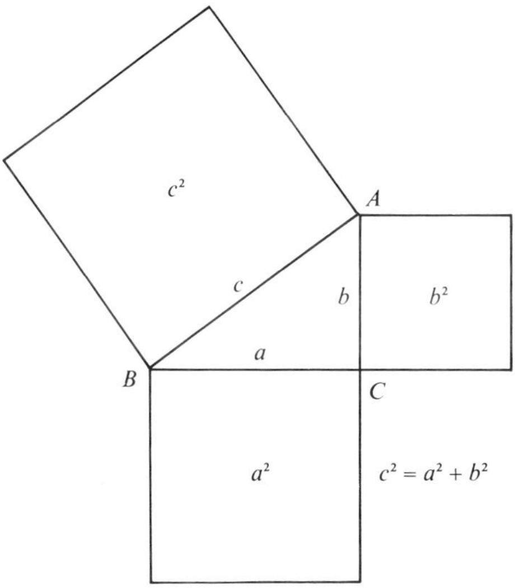

Figure 7.12

Example 1. If the legs of a triangle are 4 inches and 3 inches, find the hypotenuse.

$(4)^2 + (3)^2 = 16 + 9 = 25.$

Therefore the hypotenuse equals $\sqrt{25} = 5$ in.

Exercise 37

Answer the following:

1) If two angles of a triangle measure 45° and 65° respectively, how large is the third angle?

2) Is it possible to have a triangle with angles measuring 65°, 75°, and 45°? Why?

3) Is it possible for a right triangle to be constructed with sides of lengths 3 inches, 5 inches, and $\sqrt{27}$ inches?

4) If two angles of a triangle each contain 60° and if one side is 5 inches, how long are the other two sides?

5) An isosceles triangle has a vertex angle of 38°. How large are each of the two base angles?

6) If a right triangle has one acute angle measuring 65°, how large is the other acute angle?

Classify the triangles having the following parts:

7) Angles of 85°, 90°, and 5°

8) Sides of 15 in., 14 in., and 12 in.

9) Sides of 7 mm., 12 mm., and 12 mm.

10) Two angles of 45° each

11) Three angles of 60° each

12) Two sides of 21 in. and one side of 7 in.

13) Three unequal sides

14) Three sides of 15 in. each

15) Legs of 5 in. and 4 in.

16) Hypotenuse of 13 in. and one leg of 5 in.

17) Legs of 9 in. and 12 in.

18) One leg of 15 in. and one acute angle of 45°

Answer the following:

19) Name the type of angle shown as *a* through *f* in the diagrams below.

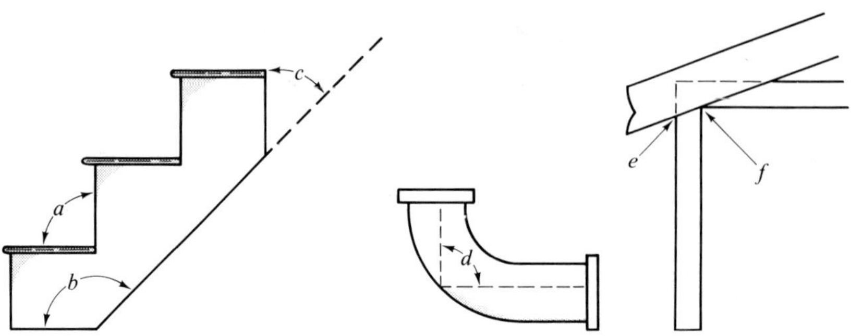

20) What kind of angle would you expect a drive belt to make with the radius of a pulley at the point of contact?

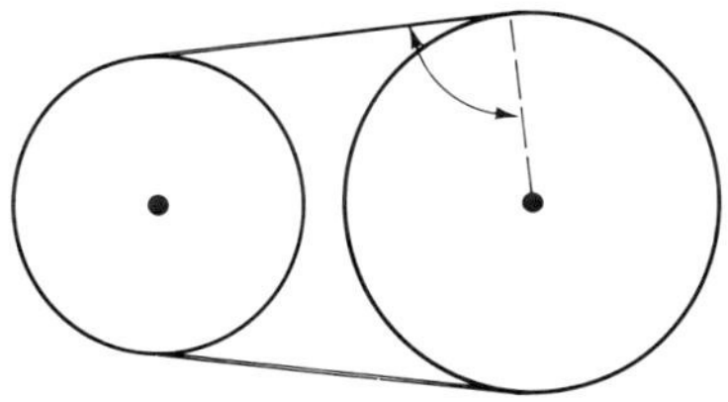

Use the Pythagorean theorem to determine the missing parts of the right triangles in the table below. If desired, a table of square roots or a slide rule may be used to determine the answers to the nearest tenth.

	x	y	z
21)	5	12	?
22)	?	15	17
23)	2.1	?	4.7
24)	3.2	4.1	?

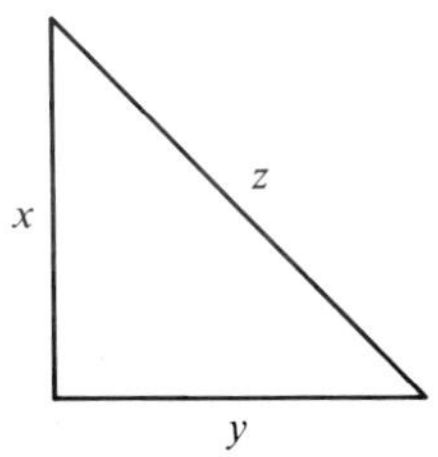

25) A surveyor runs a traverse having three courses that form a right triangle. If the two courses forming the legs are 225 and 318 feet, determine the distance along the hypotenuse.

26) In the figure shown, if $a = 11$ in., $b = 7$ in., and $c = 5$ in., find the value of v. Use the formula given.

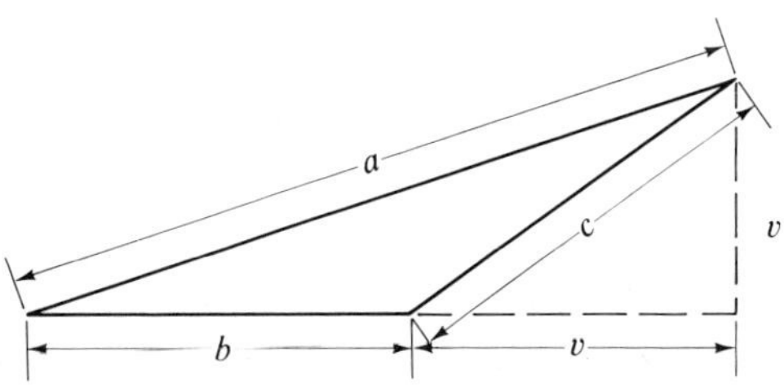

Find x in the figures below. Use the formula given in Problem 26 where applicable.

27)

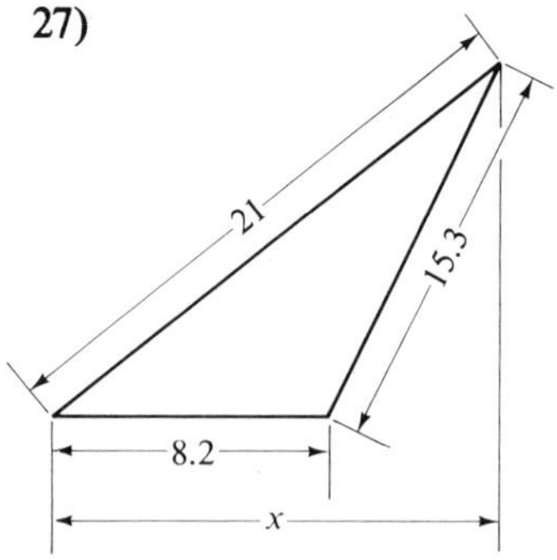

28)

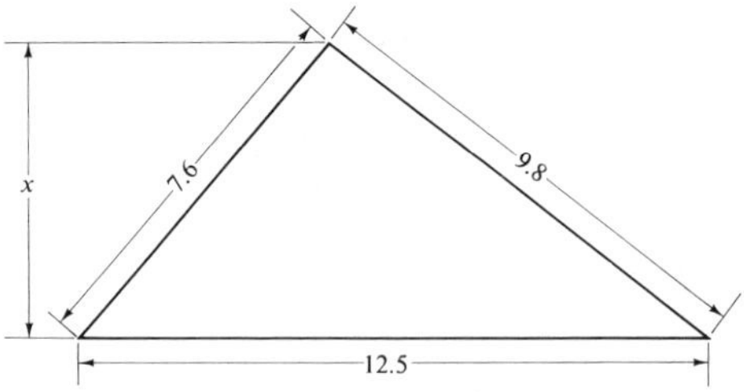

29)

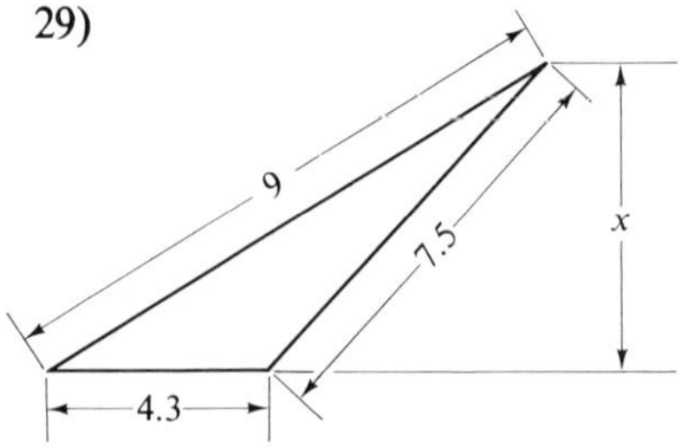

30)

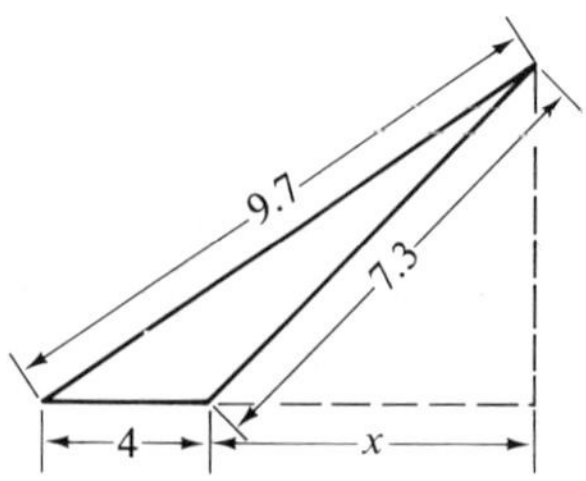

31)

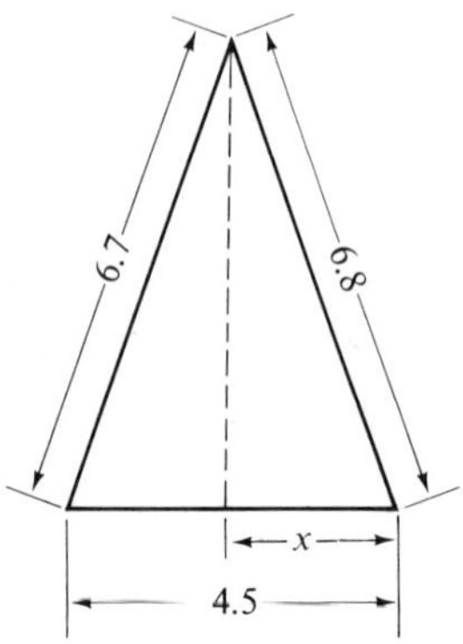

7.7 Similarity and Congruence

When two triangles have the same size and shape, the triangles are **congruent.** Two triangles must be congruent if two sides and the included angle of one triangle equal two sides and the included angle of the other;

if two angles and the included side of one triangle equal two angles and the included side of the other; or if the three sides of one triangle equal the three respective sides of the other. An included angle is the angle between the two sides being considered; an included side is the side used as a ray for two angles and hence "included" between them.

When two triangles have the same shape but not the same size, they are **similar** (Figure 7.13). In similar triangles each angle of one triangle is equal to the corresponding angle of the other triangle. Each side of one triangle is proportional respectively to the corresponding side of the other triangle.

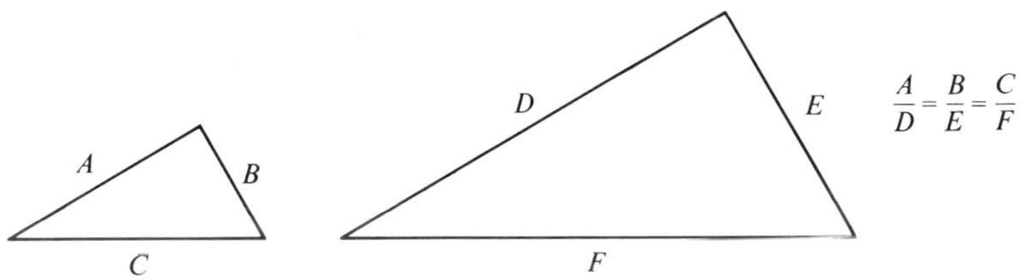

Figure 7.13

This property of proportional sides can be used to find the lengths of quantities when they cannot be measured directly. Consider a flagpole of height *h* feet. If one measures the shadow of the pole and finds it to be 80 feet long, and then forms a similar triangle with a 5-foot stake whose shadow is 8 feet long as shown in Figure 7.14, one can form a proportion with the sides of the similar triangles,

$$\frac{h}{5} = \frac{80}{8},$$

and solve for h:

$$h = \frac{5 \cdot 80}{8} = 50.$$

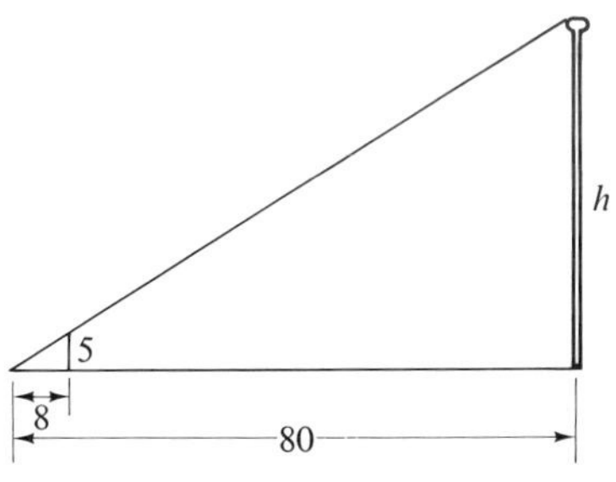

Figure 7.14

Many other applications can be found for similar triangles. In Chapter 10 these facts will be used to develop the trigonometry of right triangles.

Exercise 38

State whether the following triangles are congruent, similar, or neither.

1) Triangles ABC and DEF with $\angle A = \angle D$, $\angle F = \angle C$, and $AB = DE$
2) Triangles ABC and DEF with $\angle B = \angle E$ and $\angle A = \angle F$
3) Triangles ABC and DEF with $\angle C = \angle E$, $AB = DE$, and $AC = DF$
4) Triangles ABC and DEF with $AB = DE$, $BC = EF$, and $AC = DF$
5) Triangles ABC and DEF with $\angle A = \angle D$, $AB = DE$, and $BC = EF$

Find x in each of the following problems by using similar triangles:

6)

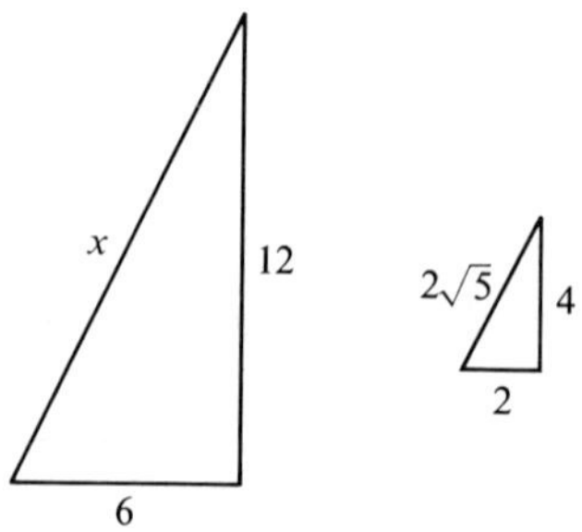

7)

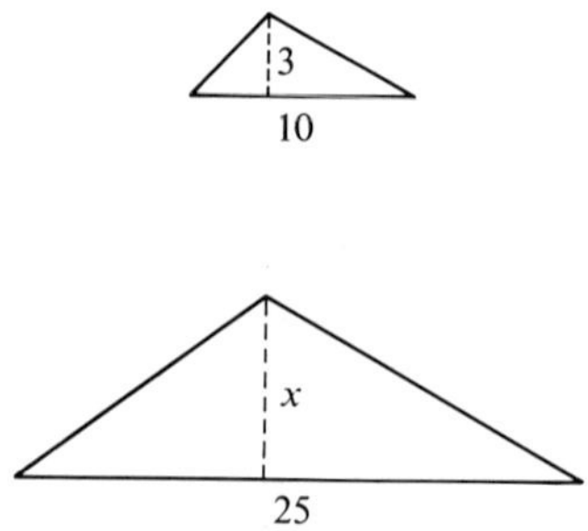

8)

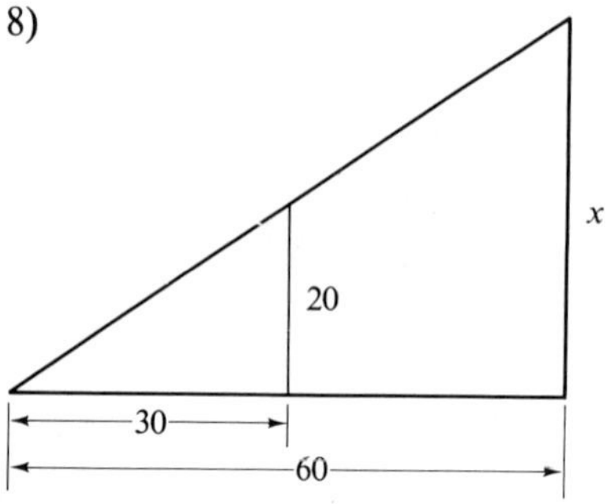

9)

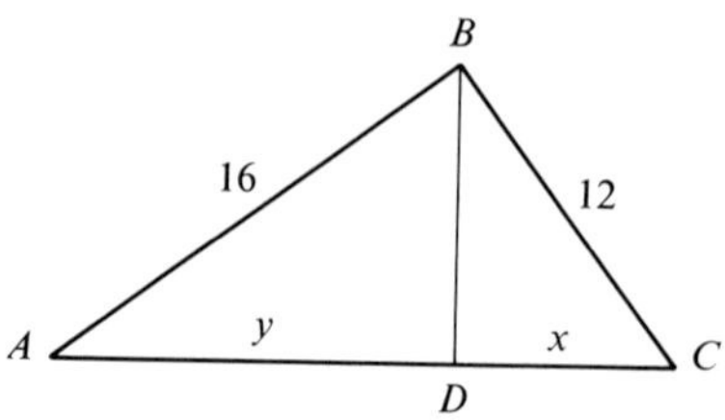

If $\angle BAD = \angle DBC$, find x.

10) In Problem 9, find y.
11) In the following drawing, if DC represents the distance across a canal and triangle AEB is similar to triangle DCB, find DC if $AE = 20$ ft., $AB = 20$ ft., and $BC = 100$ ft.

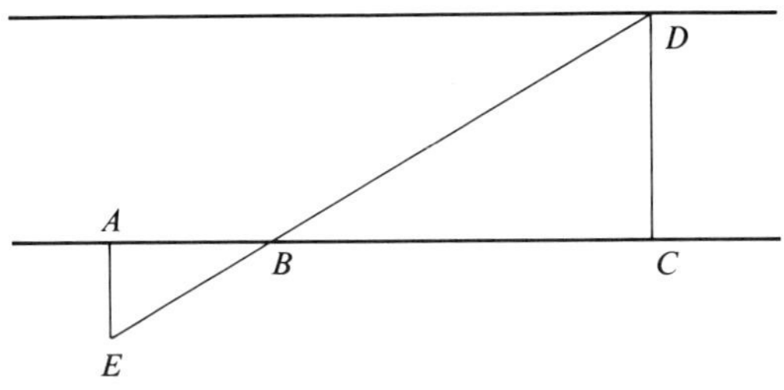

12) If the sun is shining in such a way that a 9-foot pole casts a shadow 15 feet long, how tall is a pole casting a shadow of 85 feet?

13) Two braces are placed across a corner as shown below. Find the length of the brace marked x.

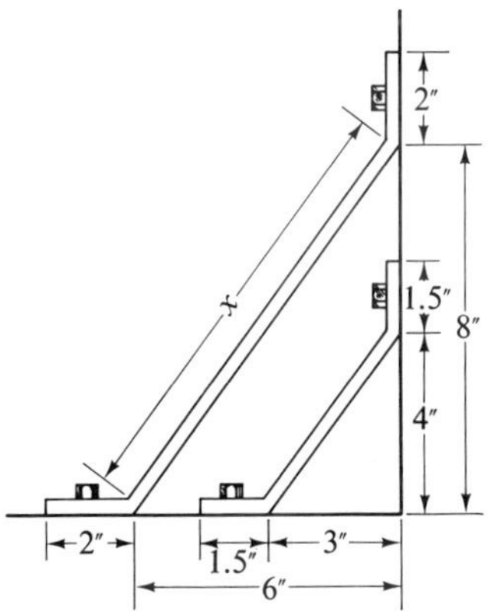

7.8 Polygons

Many figures in geometry have more than three sides. These figures are called **polygons.** Polygons are named by the number of sides they have. Four-sided figures are called **quadrilaterals,** five-sided figures are called **pentagons,** six-sided figures are **hexagons,** eight-sided figures are **octagons,** ten-sided figures are **decagons,** and so on. If polygons have all sides equal they are said to be **equilateral,** and if they have all angles equal they are called **equiangular.** If polygons are both equilateral and equiangular they are said to be **regular** polygons. The line segment connecting two non-consecutive vertices of a polygon is called a **diagonal** of the polygon. In Figure 7.15, each broken line is a diagonal.

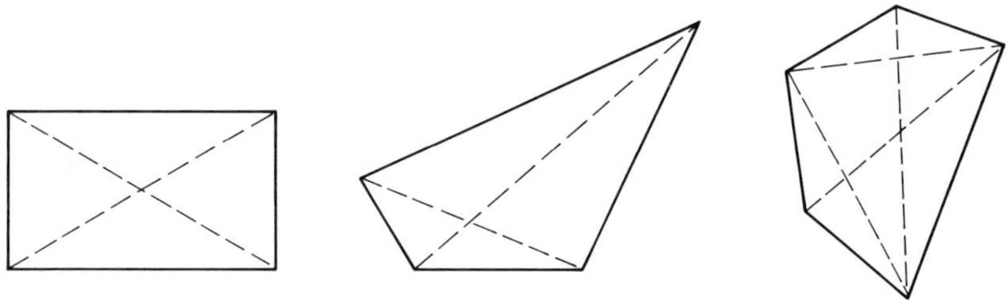

Figure 7.15

7.9 Quadrilaterals

We will discuss two major classifications of four-sided polygons or quadrilaterals. The first is the **parallelogram,** which is any quadrilateral with two pairs of parallel sides. Parallelograms, as well as having two pairs of parallel opposite sides, have equal opposite sides and equal opposite angles. In addition, a diagonal of a parallelogram divides it into two congruent triangles. This follows directly from the two previous properties. An **altitude** of a parallelogram is the distance between a pair of parallel sides, measured along a line perpendicular to both parallel lines.

Rectangles are parallelograms having all right angles (Figure 7.16), and **squares** (Figure 7.17) are equilateral rectangles. **Rhombuses** (Figure 7.18) are equilateral parallelograms without right angles. These figures have all the properties of parallelograms as well as their own individual properties.

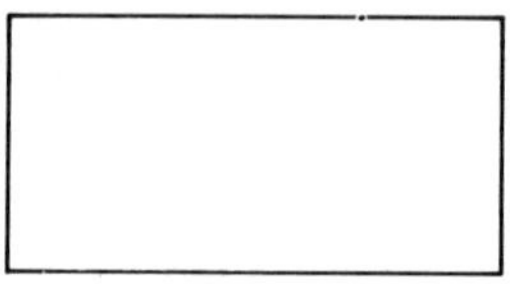

Figure 7.16

Figure 7.17

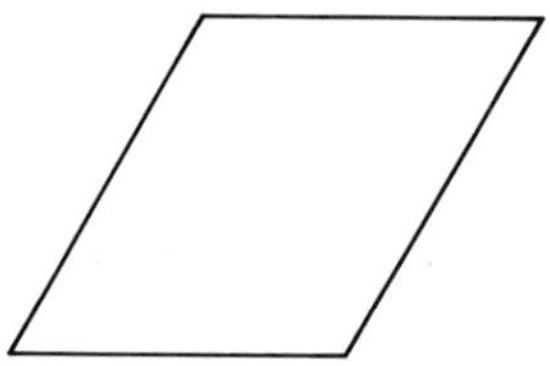

Figure 7.18

The second major type of quadrilateral is the **trapezoid,** which has only one pair of parallel sides (Figure 7.19). The altitude of a trapezoid is the distance between the parallel sides. These parallel sides are called the **bases** of the trapezoid. The line connecting the midpoints of the two nonparallel sides of a trapezoid is the **median** of the trapezoid. The median of a trapezoid is equal in length to one-half the sum of the two parallel sides of the trapezoid. An **isosceles trapezoid** is a trapezoid in which the nonparallel sides are equal.

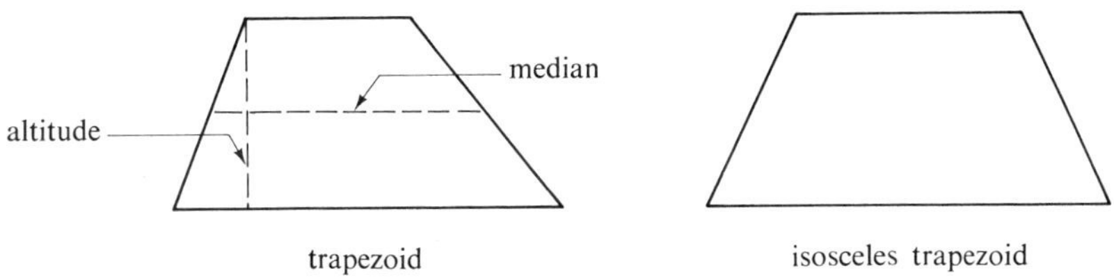

Figure 7.19

Properties of other polygons will not be discussed here. Complete proofs of these and other properties can be found in most plane geometry textbooks.

Exercise 39

1) Find the perimeter of a regular decagon having a 10-inch side.
2) How many sides does a pentagon have?
3) How many sides does an octagon have?
4) Find the perimeter of a regular hexagon of side k.
5) Is every square a rectangle?
6) Is every rectangle a square?
7) What is the altitude of a parallelogram?
8) Name two geometric figures that could be called equilateral quadrilaterals.
9) What are the parallel sides of a trapezoid called?
10) Geometric figures of more than three sides are identified by a general classification. What is it?
11) If the diagonal is drawn in a parallelogram as shown, what do we know about $\angle BCA$ and $\angle CAD$? Why?

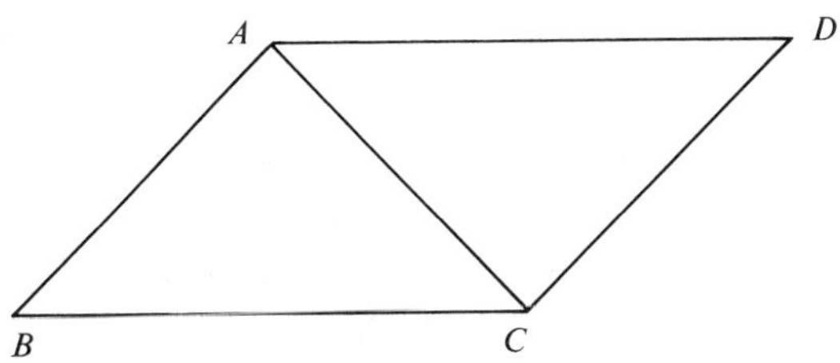

12) Is every square a parallelogram?
13) Is every parallelogram a square?
14) Can a trapezoid be equilateral? Why?
15) If the bases of an isosceles trapezoid are 9 inches and 15 inches and the altitude is 4 inches, how long is one of the nonparallel sides?
16) If the bases of an isosceles trapezoid are 10 inches and 16 inches and the altitude is 4 inches, what is the perimeter of the trapezoid?
17) Find the length of the median of the trapezoid with parallel sides of 2.5 and 7.3 centimeters.
18) What do squares and rhombuses have in common?
19) If the sides of a rectangle are 9 inches and 6 inches, how long is the diagonal of the rectangle?
20) If the altitude of an isosceles trapezoid is 15 inches, one of the nonparallel sides is 17 inches, and the longer base is 26 inches, what is the length of the median?
21) If two right triangles are similar and the legs of one are 9 inches and 12 inches, and if the short leg of the smaller triangle is 3 inches, how long is the hypotenuse of the smaller triangle?
22) Give the perimeter of each of the following figures if they are regular polygons and one side of each is 12 feet.

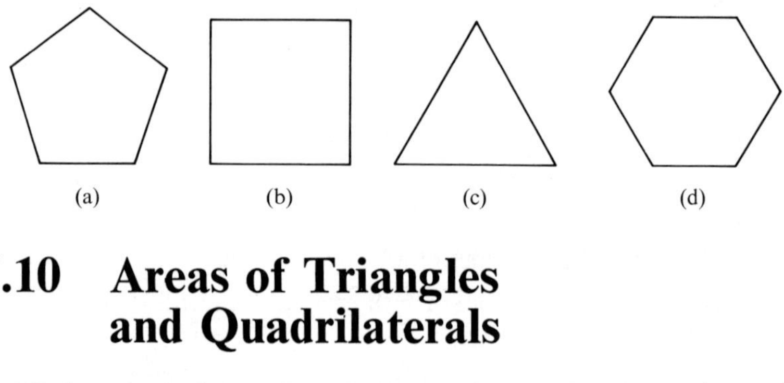

(a) (b) (c) (d)

7.10 Areas of Triangles and Quadrilaterals

Recall that area is the number of square units contained in a certain figure or space. In this section we will discuss formulas for finding areas of geometric figures.

The **altitude** of a triangle is the perpendicular distance from any vertex to the opposite side or to the extension of the opposite side. This opposite side is the **base** of the triangle. A triangle has three altitudes and therefore three bases. The area of a triangle is found by multiplying one-half the altitude times the base. Figure 7.20 shows a triangle in three positions indicating the altitude h and the base b of each. The formula for finding the area of a triangle is $A = \frac{1}{2}bh$.

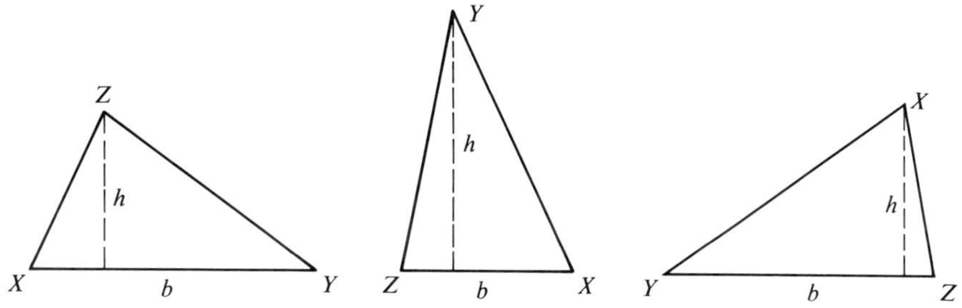

Figure 7.20

Example 1. Find the area of a triangle with an altitude of 6 feet, 5 inches and a base of 1 yard.

Before the area can be found, all dimensions must be converted to the same units.

6 ft. 5 in. = 77 in.

1 yd. = 36 in.

$$A = \frac{1}{2}(77 \text{ in.})(36 \text{ in.})$$

$$= 1386 \text{ sq. in.}$$

The area of a parallelogram is equal to the altitude times the side to which it is drawn. This side is called the base. The formula for this area then is

$$A = bh.$$

Since rectangles and squares are parallelograms, their areas can be found by the same formula. However, if one side of a rectangle, say the length, is considered the base, the other side, the width, would be the altitude and the formula for the area of a rectangle could be given as

$$A = lw.$$

This is the formula most commonly used. The altitude of a square is equal to the base. Each of these parts is a side s of the square and the most common formula for the area of a square is

$$A = s^2.$$

Keep in mind the properties of these figures discussed here. They become very useful in problems involving area. In most cases, a sketch of the problem helps.

Example 2. Find the area of the triangles formed by the diagonal of a parallelogram having an altitude of 8 inches and a base of 16 inches.

Since the diagonal of a parallelogram divides the parallelogram into two congruent triangles, we find the area of the parallelogram to be 8 in. $\times$ 16 in. = 128 sq. in. and take $\frac{1}{2}$ of it. The area of each triangle is 64 square inches.

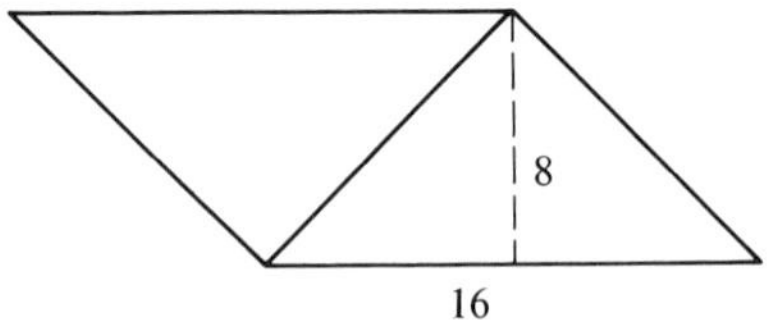

Example 3. If the area of a rectangle is 256 square inches and the width is 8 inches, find the length.

$$A = lw,$$

$$256 = l \cdot 8.$$

$$l = \frac{256}{8} = 32 \text{ in.}$$

To find the area of a trapezoid, perpendiculars can be drawn from the endpoints of its short base to the other base, forming two triangles and a rectangle (Figure 7.21). The area of each of the figures can be found independently to find the area of the trapezoid. The area of the rectangle is $(h)(y)$, the area of one triangle is $\frac{1}{2}(h)(x)$, and the area of the other triangle is $\frac{1}{2}(h)(z)$. The total area is

$$(h)(y) + \frac{1}{2}(h)(x) + \frac{1}{2}(h)(z) = (h)\left(\frac{1}{2}x + \frac{1}{2}z + y\right)$$

$$= \frac{1}{2}(h)(x + z + 2y)$$

$$= \frac{1}{2}(h)[(x + y + z) + y]$$

$$= \frac{1}{2}(h)(B + y) \qquad (\text{where } B = x + y + z)$$

$$= \frac{1}{2}(h)(B + b) \qquad (\text{where } b = y).$$

Therefore, a general formula for the area of any trapezoid is

$$A = \frac{1}{2}(h)(B + b),$$

where B is the long base and b is the short base.

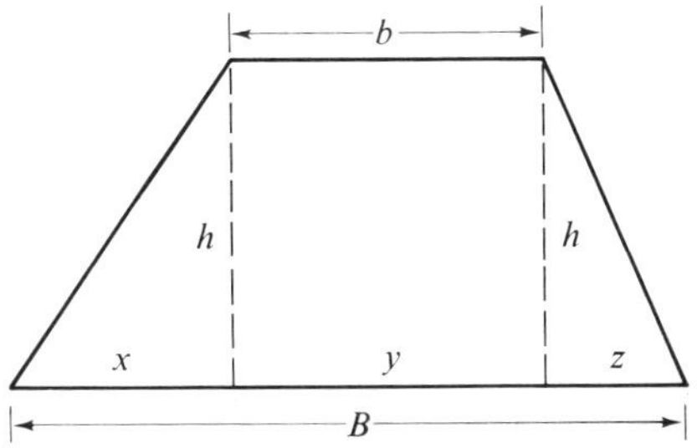

Figure 7.21

Example 4. Find the area of a trapezoid having bases of 13 inches and 15 inches and an altitude of 8 inches.

$$\begin{aligned} A &= \frac{1}{2}(8)(15 + 13) \\ &= \frac{1}{2}(8)(28) \\ &= 4(28) = 112 \text{ sq. in.} \end{aligned}$$

Example 5. Find the long base of a trapezoid which has an area of 289 square inches if the short base is 15 inches and the altitude is 17 inches.

$$289 = \frac{1}{2}(17)(B + 15)$$

$$289 = \frac{17}{2}(B) + \frac{17}{2}(15),$$

$$289 - \frac{17}{2}(15) = \frac{17}{2}(B),$$

$$\frac{578 - 255}{2} = \frac{17}{2}(B),$$

$$B = \frac{323}{2} \cdot \frac{2}{17} = \frac{323}{17} = 19 \text{ in.}$$

Exercise 40

Find the area of each of the following figures:

1)

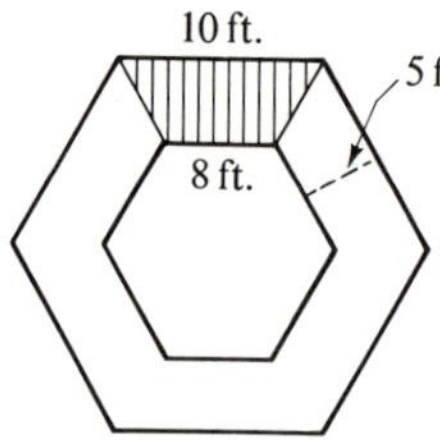

2)

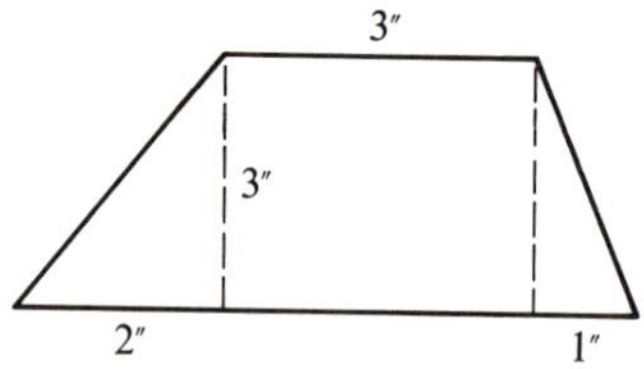

3)

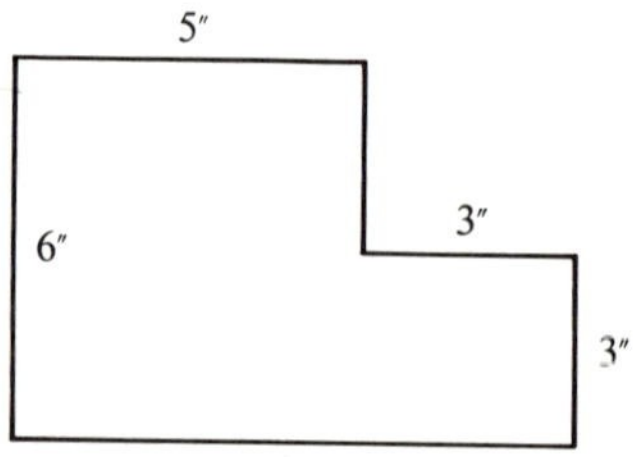

4)

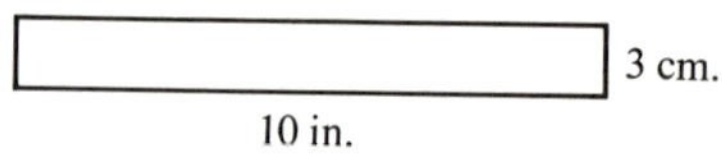

5)

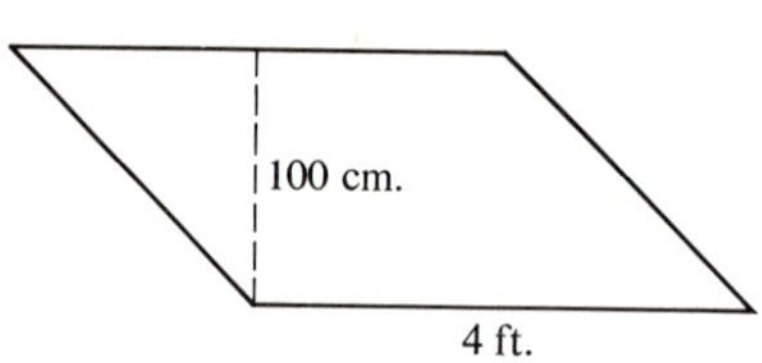

6)

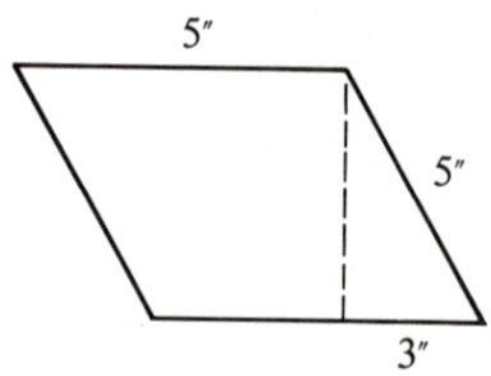

7)

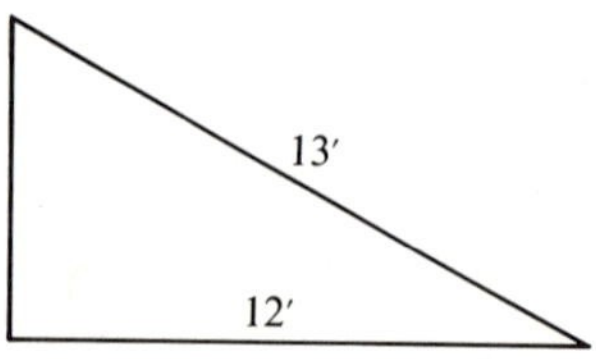

8)

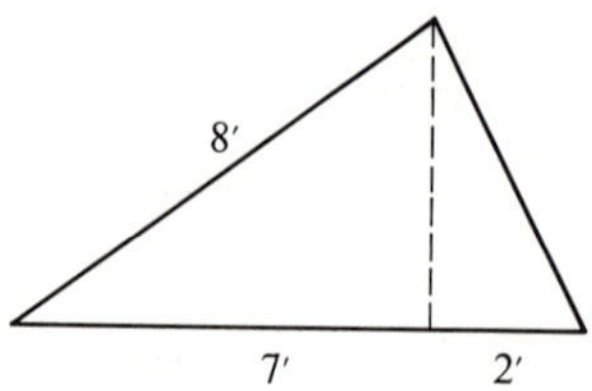

9)

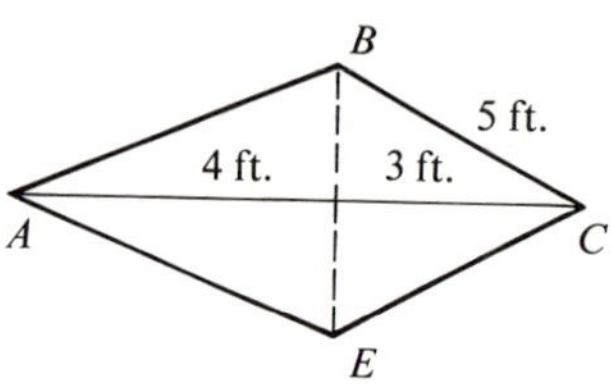

Triangle *ABC* is congruent to triangle *AEC*.

10)

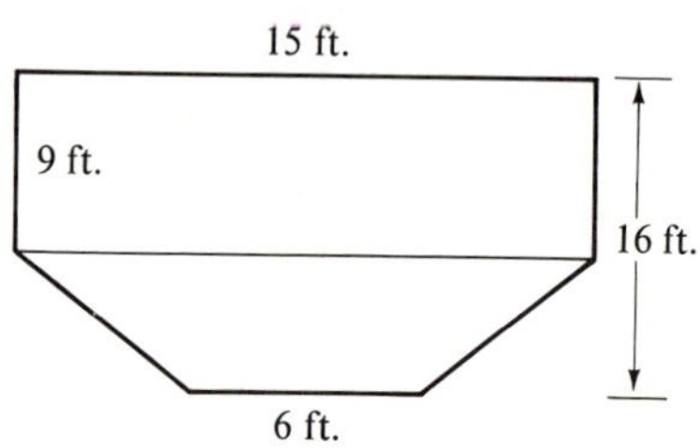

11) What is the altitude of a triangle?

12) How many bases does a triangle have; how many altitudes?

13) What property of the diagonal of a parallelogram allows one to determine the area as twice the area of either triangle with a diagonal as one side?

14) Determine the altitude of a parallelogram whose area is given as 320 square centimeters and whose base is 40 centimeters.

15) Determine the altitude of a trapezoid whose area is 256 square inches and whose bases are 7 inches and 11 inches.

16) The front and side views of a house are shown below. Assuming the back and other sides have the same shape, how much paint would be required to cover the stucco if one gallon covers 420 square feet. Allow 90 square feet for doors and windows.

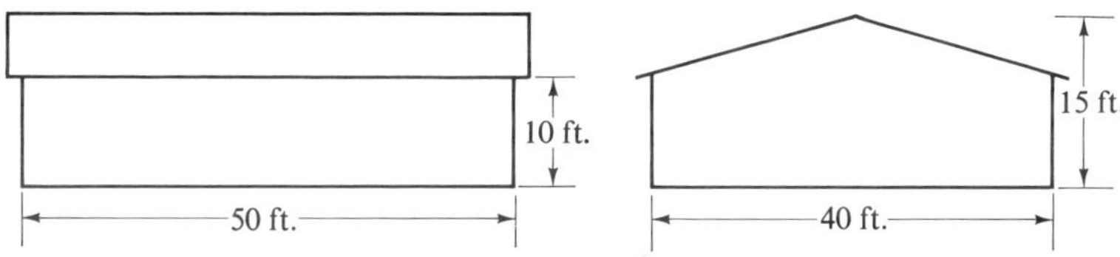

17) Determine the surface area of the stamping. All dimensions are in inches. Consider only one side of the stamping.

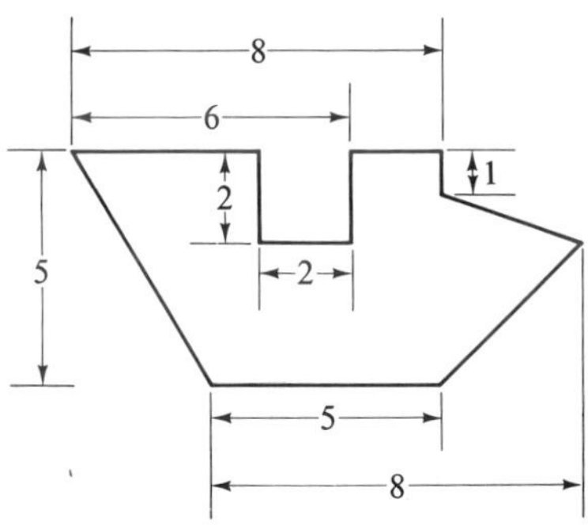

18) The cross section of a piece of metal has the shape and dimensions given. Find the area of the figure. All dimensions are in inches.

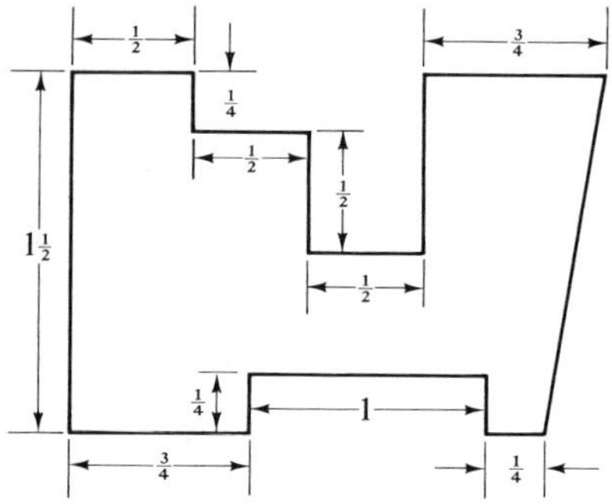

19) In the figure shown, find the area of the shaded portion if the areas of similar triangles are proportional to the squares of their respective bases.

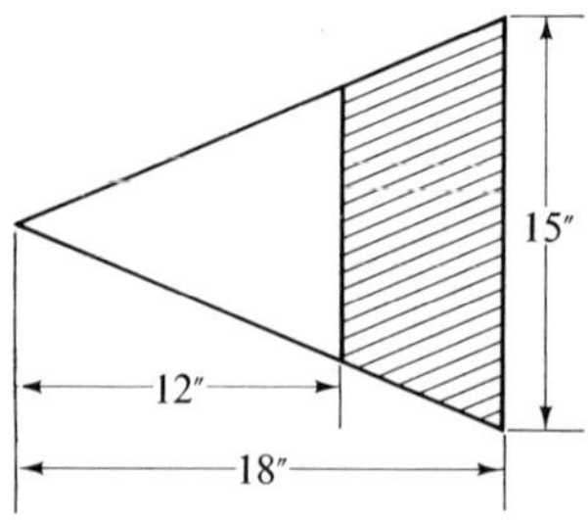

7.11 Circles

A **circle** is defined to be the set of all points in a plane at a given distance from some fixed point. This fixed point is called the **center.** One-half of a circle is called a **semicircle** (Figure 7.22) and one-fourth of a circle is a **quadrant** (Figure 7.23).

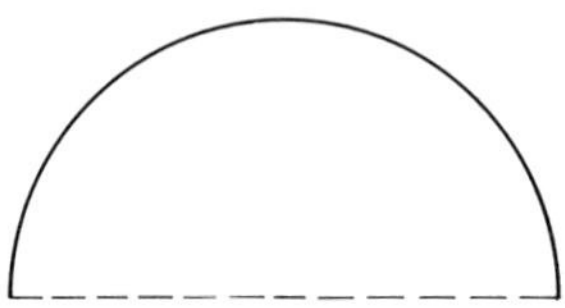

Figure 7.22

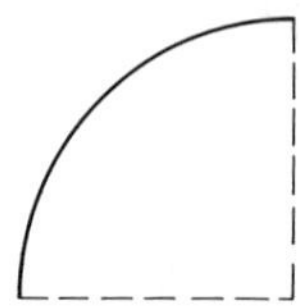

Figure 7.23

The distance around a circle is called its **circumference,** the distance from the center to a point on the circumference is called the **radius,** and a line segment from one point on the circumference through the center to another point on the circumference is the **diameter.** The diameter of any circle is twice as long as the radius.

The set of points on the circle between any two points on the circumference is an **arc.** Arc AB is written as $\widehat{AB}$. The straight line segment between two points on a circle is a **chord** of the circle. A line segment passing through the circle and intersecting it twice is a **secant** of the circle. A line segment touching the circle only once is a **tangent** to the circle. Figure 7.24 indicates these relationships.

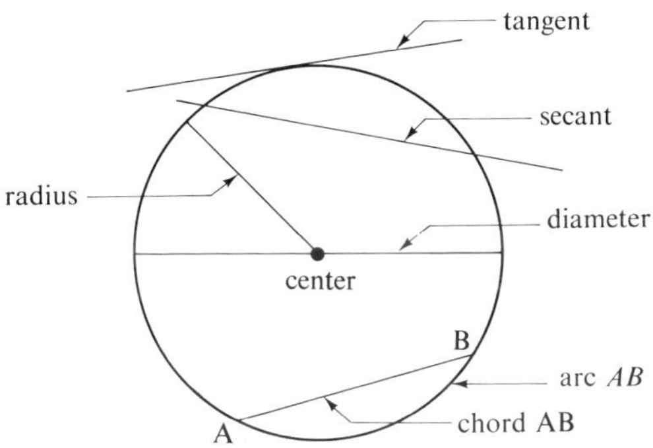

Figure 7.24

The ratio between the circumference of any circle and its diameter is constant. This ratio is given the Greek letter π, or *pi.* π has approximately the measure of 3.14159, but cannot be written in exact decimal form since it is an irrational number. (See Section 4.9.) Since $\pi = C/d$, where $C =$ circumference and $d =$ diameter, the formula for finding the circumference of a circle is

$$C = \pi d \qquad \text{or} \qquad C = 2\pi r.$$

The formula for the area of a circle is

$$A = \pi r^2.$$

Example 1. Find the circumference and area of a circle whose radius is 6 inches. Approximate π by 3.14.

$$\begin{aligned} C &= 2\pi r \\ &= (2)(3.14)(6) \\ &= (12)(3.14) \\ &= 37.68 \text{ in.} \end{aligned}$$

$$A = \pi r^2$$
$$= (3.14)(6)(6)$$
$$= (3.14)(36)$$
$$= 113.04 \text{ sq. in.}$$

7.12 Volume

In Chapter 6 volume was discussed as a measure of cubic units. We found that the volume of a four-sided, right-angled object can be found by taking the product of its three dimensions. Many figures have different shapes, however, and their volumes cannot be found in this way. In this section formulas for some of the more common geometric solids will be discussed. These formulas will be given without proof and will be used in the problem set.

The volume of a cube is found by cubing a side, and is written by the formula

$$V = s^3.$$

In general, the volume of a rectangular solid is found by taking the product of the three dimensions:

$$V = lwh.$$

(See Figure 7.25.)

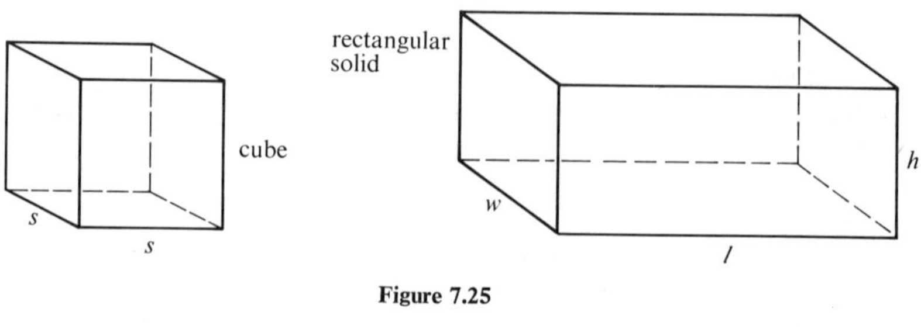

Figure 7.25

A **pyramid** is a solid with a polygon for a base and triangular faces which meet at one point called the **vertex.** The altitude of the pyramid is the perpendicular distance from the vertex to the base. The volume of a pyramid is found by taking one-third the altitude times the area of the base or

$$V = \frac{1}{3}Bh,$$

where B is the area of the base and h is the altitude (Figure 7.26).

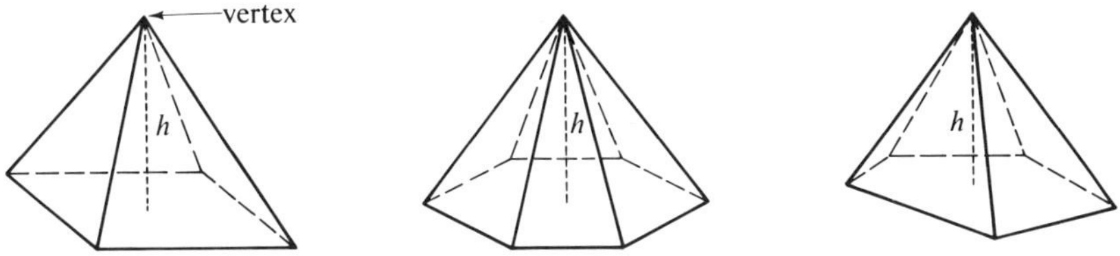

Square Pyramid Hexagonal Pyramid Pentagonal Pyramid

Figure 7.26

A **cone** is much like a pyramid, except that the base is a circle instead of a polygon (Figure 7.27). The volume of a cone can be found by $V = \frac{1}{3}Bh$, where B is the area of the base and h is the altitude. Since the base is a circle, the formula for finding B is πr^2, where r is the radius of the circle. The volume of a cone is then found by the formula

$$V = \frac{1}{3}\pi r^2 h.$$

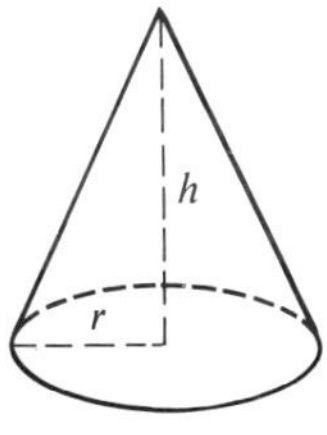

Figure 7.27

A **right circular cylinder** is a solid whose two bases are circles and whose lateral surface is perpendicular to these bases (Figure 7.28). The volume of a right circular cylinder is given by the formula $V = Bh$ or

$$V = \pi r^2 h,$$

where h is the distance between the bases.

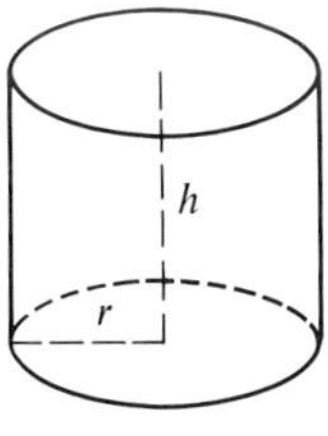

Figure 7.28

Exercise 41

If the letters in the table below represent parts of a circle, find the missing quantities. Use 3.14 for π.

	d	r	A	C
1)	14 in.	_____	_____	_____
2)	4 in.	_____	_____	_____
3)	_____	6 rd.	_____	_____
4)	_____	$3\frac{1}{2}$ in.	_____	_____
5)	15 cm.	_____	_____	_____

Solve the following:

6) Find the area of a circular walkway if the inside diameter is 72 feet and the outside diameter is 80 feet. Find the volume of cement required to pour the walkway if the cement is poured to a depth of 4 inches.

7) Find the volume of a cone with an altitude of 10 feet and a base with a radius of 3 feet.

8) Find the volume of a pyramid if the base is a square having a 6-inch side and the altitude is 15 inches.

9) If the volume of a cone is 72π cubic feet and the altitude is 8 feet, what is the radius of the base?

10) In Problem 6 above, if the cost of concrete is \$27.00 per cubic yard what is the total cost of the walkway?

11) Find the volume in gallons of a cylindrical tank having an altitude of 7 feet and a base diameter of 4 feet. (1 cubic foot contains approximately 7.5 gallons.)

12) If gasoline costs 27.9 cents per gallon and each gallon weighs 6.2 pounds, find the cost and weight of the tank of Problem 11.

13) If the diameter of a piston is 3.13 inches and the stroke is 4.75 inches, what is the cubic inch displacement of one cylinder of the engine?

14) If a cattle feeding trough is in the shape of a rectangular solid, find the volume of the trough if it is 12 feet long, 4 feet wide, and 8 inches deep.

15) Find the volume of a tetrahedron with an edge of 6 inches. (A tetrahedron is a triangular pyramid whose faces are congruent equilateral triangles.)

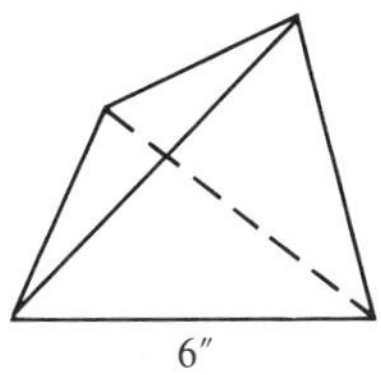

16) A water storage tank is in the shape of a sphere having a radius of 16 feet. How many gallons will it hold? The formula for the volume of a sphere is $V = \frac{4}{3}\pi r^3$.

17) A square pyramid is a pyramid whose base is a square. A gravel hopper is in the shape of an inverted square pyramid. If one side of the square is 15 feet, and the hopper is 12 feet deep, how many cubic yards of gravel does it hold?

18) A squelch basin is made by cutting a barrel in half lengthwise so that it forms a trough with semicircular ends. Find the volume of the trough if the barrel was 4 feet long and had a base radius of 14 inches.

19) Find the volume of steel used in making the cylinder shown in the diagram below. Convert the measures to decimals and find the volume to the nearest hundredth.

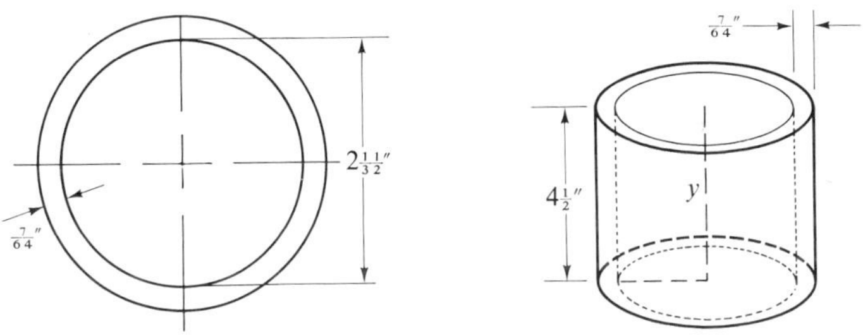

20) A gasoline tank is in the shape of a solid whose bases are isosceles trapezoids and whose sides are perpendicular to the bases. How many gallons of gasoline will the tank hold?

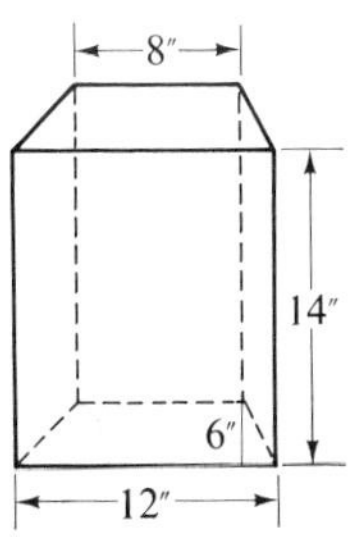

21) Sand is deposited by a conveyor belt so that it forms a pile the shape of a right circular cone. If it is 345 feet around the base of the pile and the pile is 15 feet high, how many cubic yards of sand are in the pile?

7.13 Geometric Constructions

In many areas of industry, particularly those in which design and structure become a part, one must be able to construct certain geometric figures. This construction needs to be done with a minimum of equipment. Historically, most geometric constructions were done by using only a **straight edge** (some device for drawing a straight line) and a **compass** (a tool for drawing a circle). In this section certain basic geometric constructions using only a compass and straight edge will be discussed. Although all the constructions discussed in this section can be proven by geometric methods, no attempt will be made to do this. Instead, they will be developed from an intuitive point of view.

Most constructions require duplicating a segment or marking off a segment of a given length on a given line. To do this, we choose a point of the line as one endpoint of the segment, and open the compass so that each point of it is at an endpoint of the segment being duplicated. We then move the compass so that one point of it is on the designated point on the line. The point at which the pencil end of the compass touches the line will be the desired duplicate of the given segment. (See Figure 7.29.) Distances are very often transferred from a scale to a drawing in this manner.

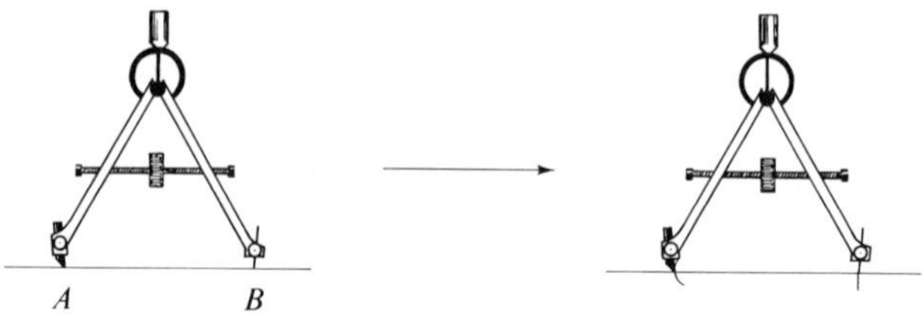

Figure 7.29

Example 1. If one inch on a blueprint represents 10 feet, using a ruler marked in tenths of an inch, mark off a segment representing 27 feet.

Since 1 inch represents 10 feet, 1/10 inch represents 1 foot and 2.7 inches represents 27 feet. Therefore, we set one point of the compass at zero on a ruler and open the compass until the other point is at 2.7 inches. Then we place the point of the compass at a designated point on a line. The point at which the pencil of the compass touches the line will be the end of the desired segment representing 27 feet.

Example 2. Lay off a line segment four times the length of given line segment AB.

Follow the procedure outlined above four times, each time beginning with the endpoint of the previous segment.

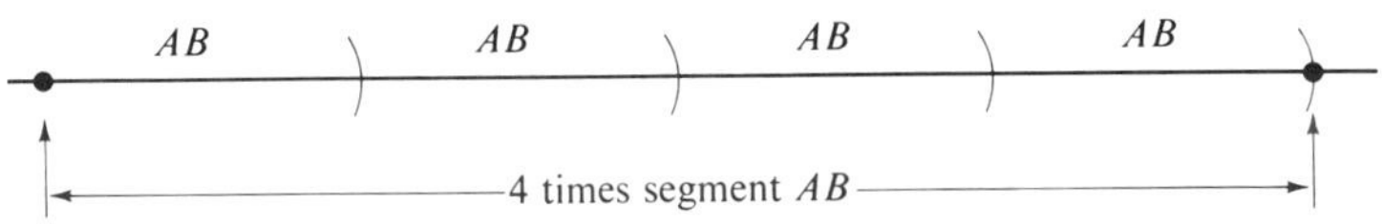

It is often necessary to **bisect** a line segment, that is, to divide the segment into two equal parts. To do this, we construct the **perpendicular bisector** of the line segment. This is a line perpendicular to the segment being bisected and passing through its midpoint. With the compass open a distance slightly greater than one-half the length of the segment, cast an arc above and below the segment from each of its endpoints so that the arcs intersect at points above and below the segment. A line through these two points is the perpendicular bisector of the line segment (Figure 7.30).

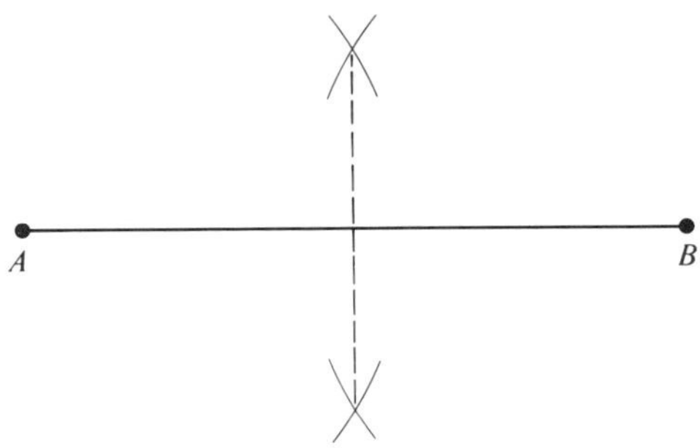

Figure 7.30

Notice from the last construction that every point on the perpendicular bisector of a line segment is equidistant from the endpoints of the segment, and that by connecting any point on the perpendicular bisector of a segment with the endpoints of the segment an isosceles triangle is formed with the segment as its base.

Example 3. Divide the given line segment into four equal parts by construction methods.

First bisect the given segment by the perpendicular bisector method, then use the same method to bisect each half, dividing the segment into four equal parts. This procedure can be used to divide a segment into 4, 8, 16, or more equal parts.

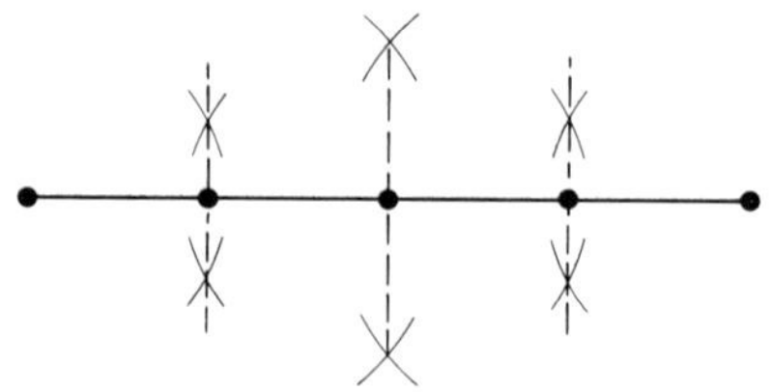

Another perpendicular construction is the construction of a perpendicular to a line at a point on the line, for instance, point P of given line AB. To perform this construction, place the point of the compass at point P on the line and with a convenient radius cast an arc intersecting the line at a point on each side of point P. This is shown by arcs at points X and Y in Figure 7.31. Point P is then the midpoint of segment XY. The construction can be completed by setting the compass length slightly greater than one-half XY and casting an arc from X and Y so that the arcs intersect either above or below the line. Connect the point of intersection of these arcs (point Q) to point P. Line QP is perpendicular to line AB at P, or $QP \perp AB$. The symbol $\perp$ means "is perpendicular to."

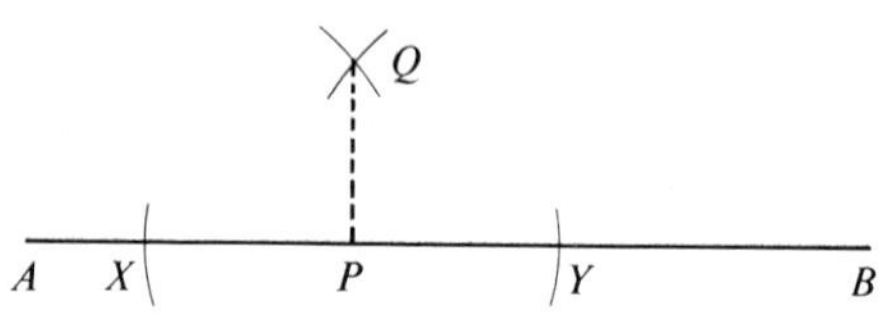

Figure 7.31

A third useful perpendicular construction is that of constructing a perpendicular to a line from a point not on the line. That is, construct a perpendicular to line AB from point P, where P is not on line AB (Figure 7.32). To perform this construction, set the compass so that the radius is a little greater than the shortest distance from the point to the line. Cast an arc intersecting the line twice, as shown by points M and N. Using points M and N as centers, cast equal arcs on the side of AB opposite from P so that these arcs intersect. This intersection is shown at point Q of Figure 7.32. The line PQ is the desired line perpendicular to AB from point P, or $PQ \perp AB$.

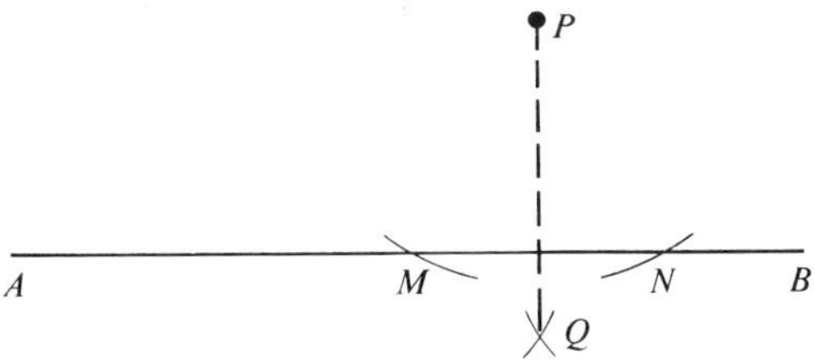

Figure 7.32

These constructions seem more useful when we consider that the distance from a point to a line is measured along the perpendicular. This gives us a method for finding the distance from a point to a line. We first construct the perpendicular and then measure the distance along this perpendicular.

Example 4. Find the distance from point P to line AB.

Construct the perpendicular from P to AB as before. Measure the distance from P to AB along this perpendicular.

Exercise 42

1) Construct a line segment equal to three times the length of segment AB.

   A $\qquad\qquad$ B

2) If on a drawing 1/16 in. = 1 ft., how many inches represent 45 feet? Construct a segment representing a length of 45 feet using this scale.

3) Draw a line segment 3 inches long. By construction divide the segment into 8 equal parts. How long is each part?

4) Construct an isosceles triangle having a base and an altitude equal to segment AB.

   A $\qquad\qquad$ B

5) Construct a square with sides $1\frac{1}{2}$ inches long. (Hint: the sides of a square are perpendicular.)

6) Construct a rhombus with one diagonal 3 inches long and the other diagonal 2 inches long. (Hint: the diagonals of a rhombus are perpendicular bisectors of each other.)

7) Construct the perpendicular to a line segment XY at a point 1/4 the length of XY from point X.

8) Find a point whose distance from a line segment XY is 1/2 the length of the segment.

9) Construct a perpendicular from P to a line MN, where P does not lie on the line.

10) Construct a rectangle whose length is twice its width.

11) A given rectangle is 2 feet by 4 feet. Divide it by construction into equal squares 1 foot on a side.

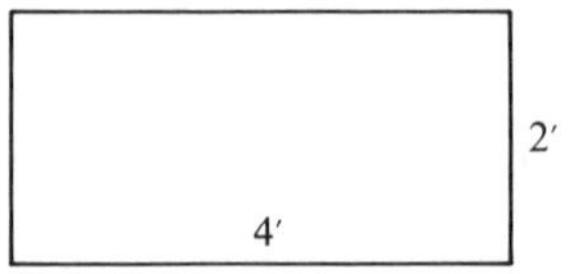

12) How many points are exactly ——x—— units away from a given line L? Locate at least three points satisfying this property.

13) If ———— represents 5 miles on a map, use a compass to find how far it is from point A to point B by following the road. Using a ruler, find approximately how much distance could be saved if it were possible to go directly from point A to point B by a straight line.

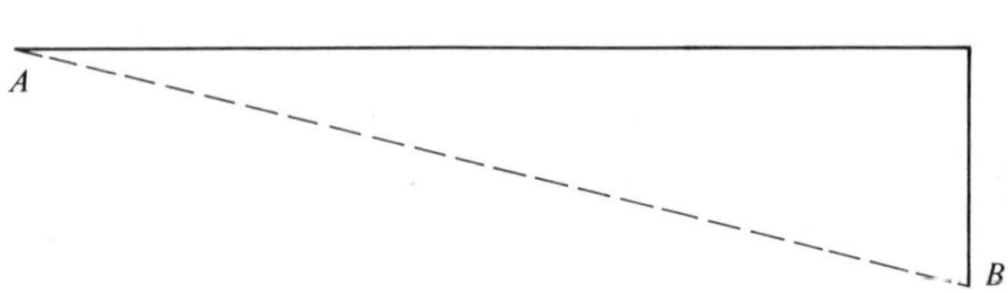

14) Construct a right triangle having legs of 2 inches and 3 inches.

15) Find the perpendicular bisector of a 2-inch segment AB. Locate point P 1 inch above the segment and point Q 1 inch below the segment on this bisector. Connect points P and Q to the endpoints

A and *B* of the segment. What kind of figure is *APBQ*? Give reasons for your answer. How long is *AP*? *BQ*?

7.14 Angle Construction

Recall that angles of a given size can be drawn by using a protractor. If, however, we do not need to know the size of a given angle but only to duplicate the angle, it can be done by construction.

Suppose we wish to duplicate the angle *CAB* on line *XY* with vertex at *Z* (Figure 7.33). Using a convenient radius of the compass and with the vertex of the given angle as center we cast an arc intersecting both rays of the angle. Without changing the radius and using point *Z* on line *XY* as center, we cast an arc intersecting line *XY* at *S*. We then put one point of the compass at one of the points of intersection of the arc with the rays of the given angle, and the other point of the compass at the point of intersection of the arc with the other ray of the angle. We then mark off the same distance on the other arc with one point of the compass at point *S*. The point where this new arc intersects the first arc, and the beginning point *Z*, determine an angle congruent to the given angle *CAB* with vertex at *Z*.

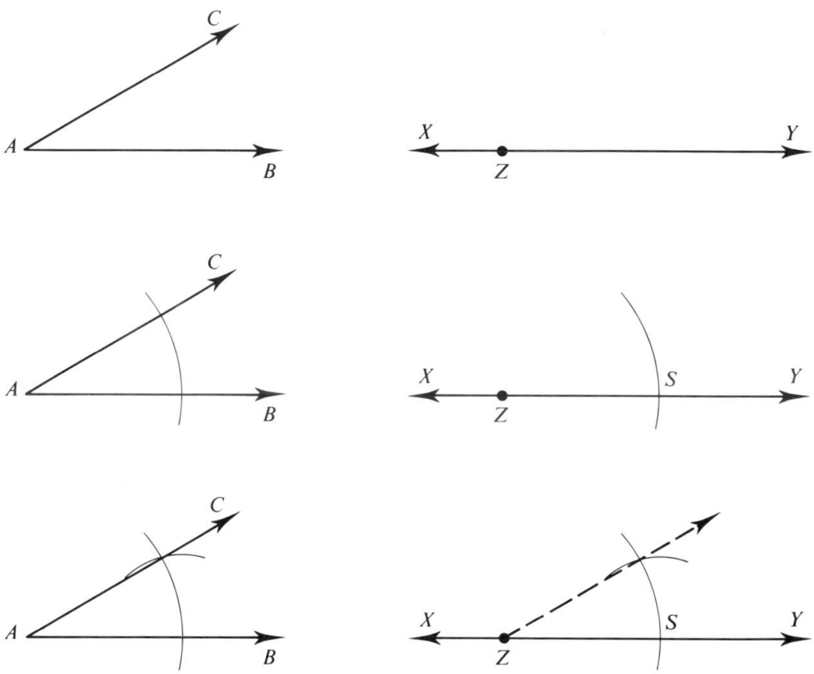

Figure 7.33

If instead of constructing a congruent angle we want an angle exactly half as large as the given angle, we can bisect the angle. To construct an angle bisector we begin as if we were duplicating an angle, that is, on a given angle ($\angle PQR$ of Figure 7.34) we cast an arc intersecting both rays of the angle. From each point of intersection of the arc with the angle rays we cast an arc opposite the vertex but between the rays so that the arcs intersect between the rays. These two arcs need not be the same radius as the first arc but they must have equal radii. The point of intersection of the latter two arcs and the vertex of the given angle determine the angle bisector of the given angle.

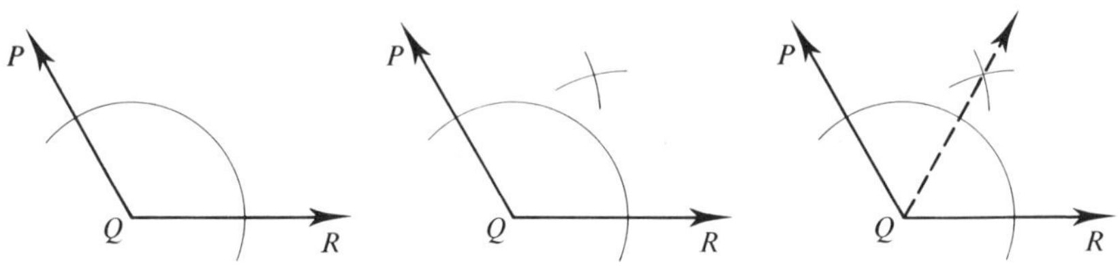

Figure 7.34

An angle can be divided into 4, 8, 16, or more equal parts by continued use of this same process with each half of the bisected angle. The method works as well with acute angles as it does for obtuse angles.

Example 1. Construct an angle equal to one-half angle ABC on line MN with vertex at P.

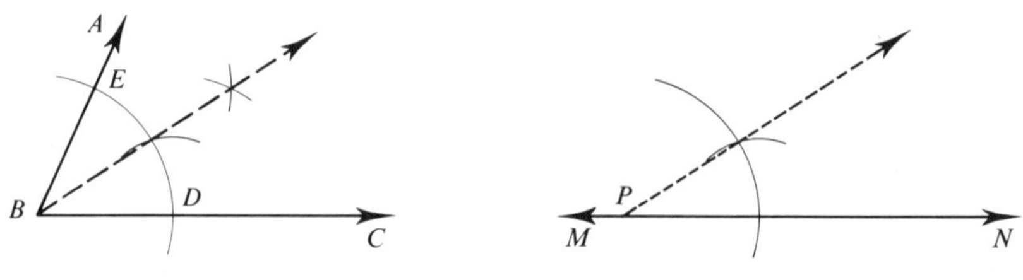

To solve this problem, angle ABC must be bisected and then one-half of it duplicated on the line MN. An arc is cast from B intersecting both rays of angle ABC at D and E, and the same radius is used to cast an arc with P as a center intersecting line MN. Now we complete the angle bisector of angle ABC as discussed previously. If we then measure the distance between the point of intersection of the angle bisector with our first arc and with point D and measure the same distance along the arc intersecting line *MN*, we immediately have an angle equal to one-half angle *ABC* with vertex at *P*.

Example 2. Construct an angle equal to $\frac{1}{4}$ $\angle XYZ$.

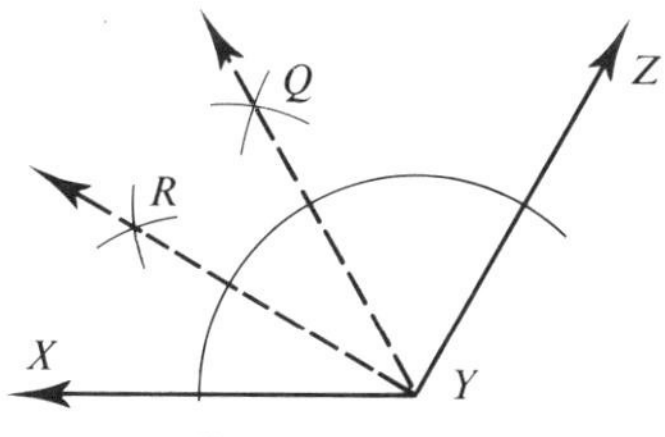

First bisect angle *XYZ* forming angle bisector *YQ*. Then bisect angle *QYX* forming angle bisector *YR*. Angle *RYX* is an angle equal to $\frac{1}{4}$ angle *XYZ*. Notice here again that it is not necessary to cast a new arc to bisect angle *QYX*. The same arc used to bisect angle *XYZ* is sufficient.

By using combinations of the constructions discussed thus far, certain angles of a specified size can be constructed without the use of a protractor. For instance, the construction of a perpendicular forms an angle of 90°. If this angle is bisected, an angle of 45° is formed. Further bisection can be done to form smaller angles of $22\frac{1}{2}$°, and so on.

Another frequently encountered angle is an angle of 60°. Recall that each angle of an equilateral triangle measures 60°. We can construct an angle of 60° (and so an equilateral triangle) as follows:

On a line choose any point *P* and with a convenient radius and this point as center, cast two arcs, one intersecting the line at *Q* and the other either above or below the line approximately midway between points *P* and *Q* (Figure 7.35). Using point *Q* and without changing the radius of the compass, cast a third arc intersecting the second arc. This point of intersection can be connected to either of the points *P* or *Q* to form an angle of 60° or to both of them to form an equilateral triangle. By continued bisection angles of 30°, 15°, $7\frac{1}{2}$°, or smaller can be constructed.

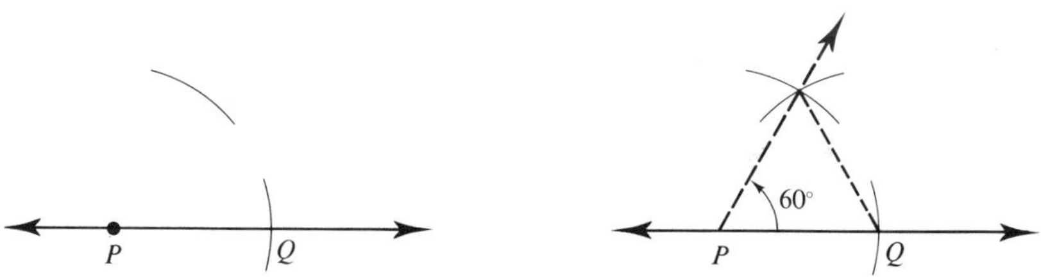

Figure 7.35

Example 3 illustrates the process of combining these constructions to form angles of various other sizes. The reader is asked to use this process further in the exercise set.

Example 3. Construct an angle of 75°.

75° is 60° + 15°, so we begin by constructing an angle of 60° on a given line (Step 1). Using the vertex of the 60° angle we construct a perpendicular to the given line (Step 2). Since 60° + 30° = 90°, the angle between the perpendicular and the terminal side of the 60° angle measures 30°. We now bisect this 30° angle (Step 3). This angle bisector forms an angle of 75° with the given line.

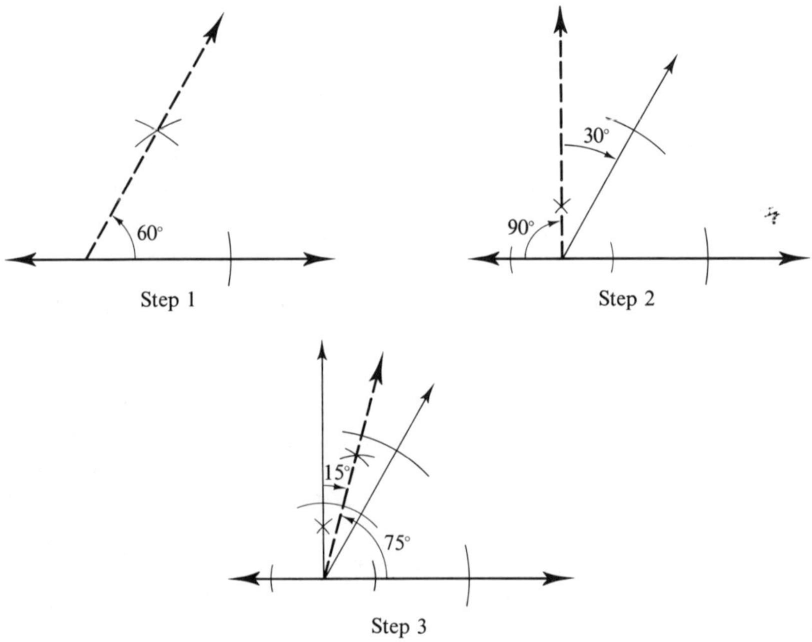

Example 4. We can use another method to construct the same angle.

75° is 30° + 45°. We first construct a perpendicular and bisect the right angle associated with it, forming an angle of 45° (Step 1). Next on another line we construct a 60° angle and bisect it, forming an angle of 30° (Step 2). Now on a new line we duplicate the 45° angle (Step 3). Using the terminal side of this angle and the same vertex, we duplicate the 30° angle (Step 4). The angle formed by the terminal side of the 30° angle and the beginning line is an angle of 75°.

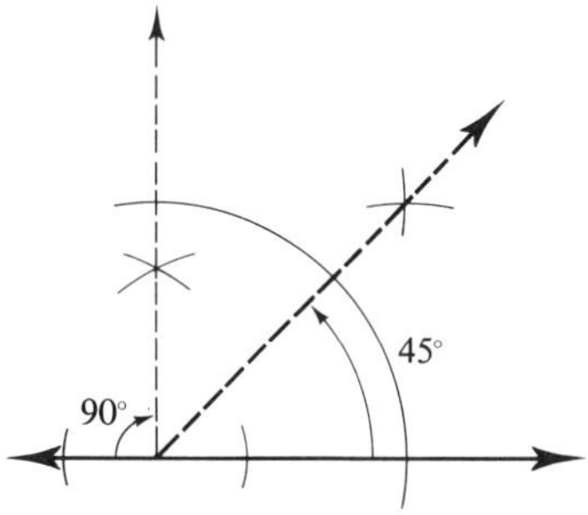

Step 1

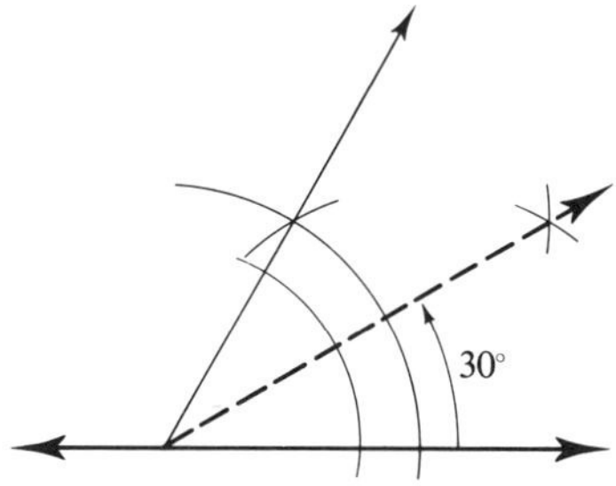

Step 2

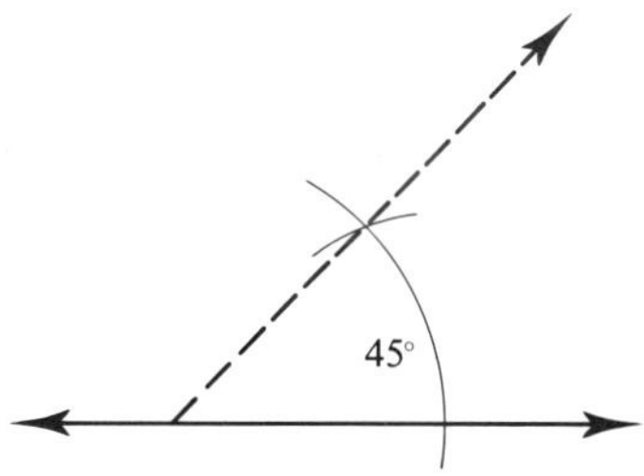

Step 3

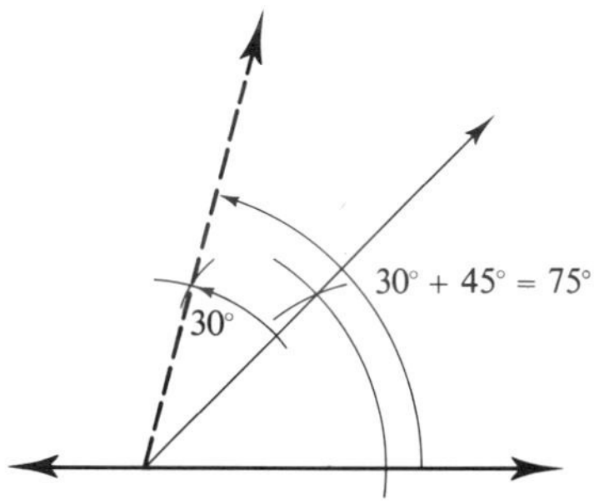

Step 4

Examples 3 and 4 are only two of many ways to construct an angle of 75°. The reader is encouraged to discover other methods.

Exercise 43

1) Construct an angle congruent to angle *ABC*.

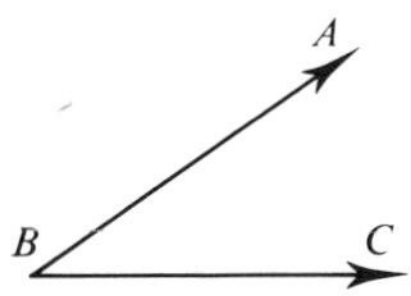

2) Bisect angle XYZ.

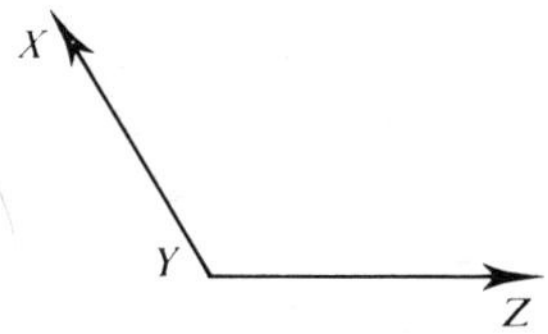

3) Construct an angle equal to twice the measure of the angle given in Problem 1.

4) Construct an angle equal to $1\frac{1}{2}$ times the measure of angle MNO.

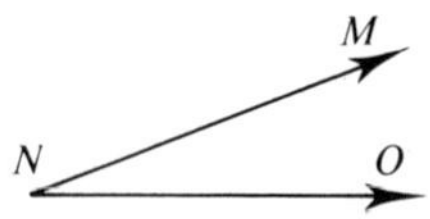

Construct angles having the following measures. Do not use a protractor.

5) 90° 6) 45° 7) 60° 8) 30°

9) 15° 10) $22\frac{1}{2}$° 11) 75° 12) 105°

13) 120° 14) 165° 15) 135° 16) $67\frac{1}{2}$°

17) $52\frac{1}{2}$° 18) $82\frac{1}{2}$°

19) Construct an isosceles triangle with base angles of 30°.

20) Construct an isosceles right triangle. How large must each base angle be?

The following constructions utilize material covered thus far. They will become more useful and meaningful as you progress in mathematics.

21) Draw a triangle. By construction, locate the midpoint of each side. Connect each midpoint to the opposite vertex. These lines are called the **medians** of the triangle. Notice that they have a common point of intersection. This point of intersection is called the **centroid** of the triangle.

22) Cut out a triangle from a piece of heavy paper or cardboard. Locate the centroid as discussed previously. Pass a knotted thread through the centroid of the triangle and suspend it. If the construction was done accurately, the triangle will hang horizontally. In an object of uniform density, the centroid is called the **center of gravity** of the object.

23) Draw a triangle. Construct the angle bisector of each angle of the triangle. All should intersect in a common point. Find the perpendicular distance from the point of intersection to a side and using this distance as a radius and the point as center, construct a circle. With an accurate construction, each side of the triangle is tangent to the circle. The circle is said to be **inscribed** in the triangle and the point is called the **incenter** of the triangle.

24) Draw a triangle. Construct the perpendicular bisector of each side. Extend the three perpendicular bisectors until they intersect. Again, they should intersect at a common point. With this point as center and the distance from this point to a vertex as radius, construct a circle. The circle should pass through all three vertices of the triangle, that is, the circle should **circumscribe** the triangle. This point is called the **circumcenter** of the triangle.

25) Draw a triangle. Construct the three altitudes of the triangle (the lines from the vertices perpendicular to the opposite sides or their extensions). Notice that these three altitudes intersect in a common point. This point is called the **orthocenter** of the triangle.

7.15 Construction of Regular Polygons

In Exercise 48 the student was asked to construct a square by using the construction of perpendicular sides. A square, as well as many other regular polygons, can also be constructed by inscribing the polygon in a circle, that is, by constructing the polygon so that the vertices of the polygon lie on the circle.

To construct an inscribed square, first draw a circle with a given point as center and draw a diameter of the circle. Through the center construct a line perpendicular to the diameter, forming a perpendicular diameter. Connect the end points of the diameters in order to form the desired inscribed square (Figure 7.36).

A regular inscribed octagon can be constructed by bisecting each central right angle* formed in the construction of the square and extend-

* A central angle is the angle formed by two radii of a circle with the vertex at the center.

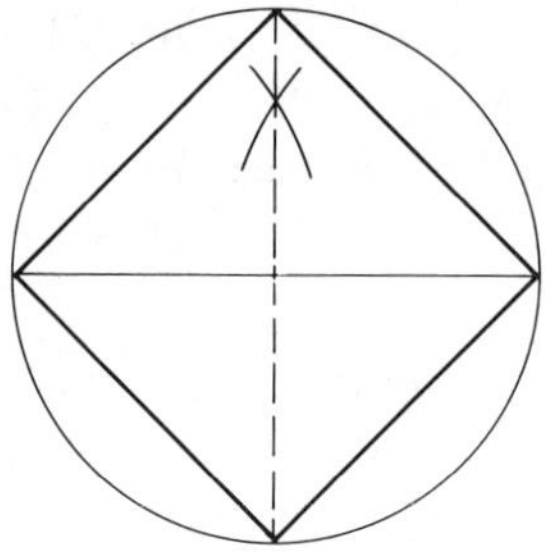

Figure 7.36

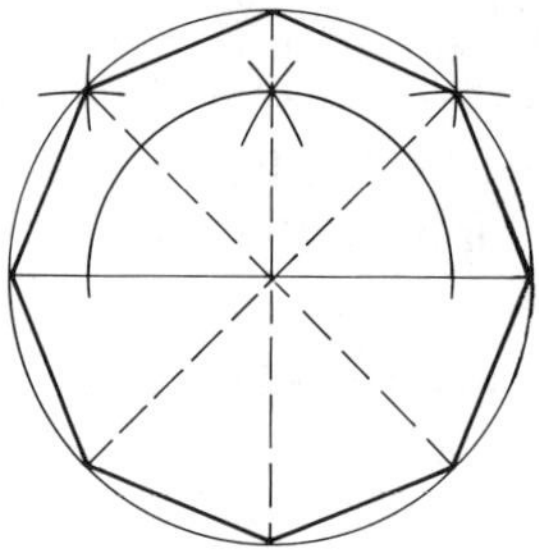

Figure 7.37

ing these angle bisectors forming diameters (Figure 7.37). If the process of bisecting the central angles is continued, regular polygons of 16, 32, 64, or more sides can be constructed.

A regular hexagon can be inscribed in a circle by considering the size of the central angle associated with each side. Each central angle must be 1/6 of 360° or 60°. If at the center of the circle an angle of 60° is constructed, one of the central angles is formed. By extending the rays forming diameters, another angle is formed (Figure 7.38). The remaining larger angles must both be 120°, so if they are bisected, six 60° central angles will have been constructed. The construction is completed by connecting the endpoints of the diameters in order.

If we remember, however, that all radii of a circle are equal, and that equilateral triangles have angles of 60°, we find that each side of a regular inscribed hexagon is exactly the same length as the radius of the circle. Thus a regular hexagon can be inscribed in a circle by setting the compass equal to the radius and casting successive arcs on the circumference of the circle using this radius (Figure 7.39). This divides the circumference into six equal arcs, and the chords of these arcs form the sides of a regular inscribed hexagon.

One further regular polygon will be discussed here, the regular inscribed pentagon. We begin the construction with a circle and construct perpendicular diameters in it (Figure 7.40). We then bisect one of the radii of the circle and set the compass on the length of the segment from the

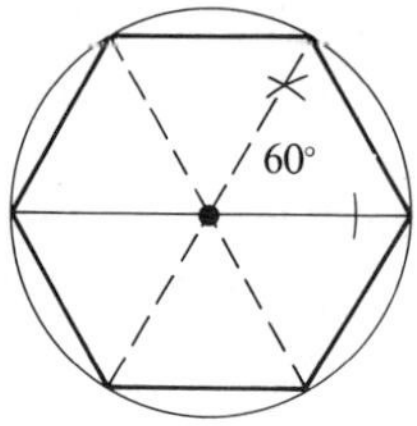

Figure 7.38

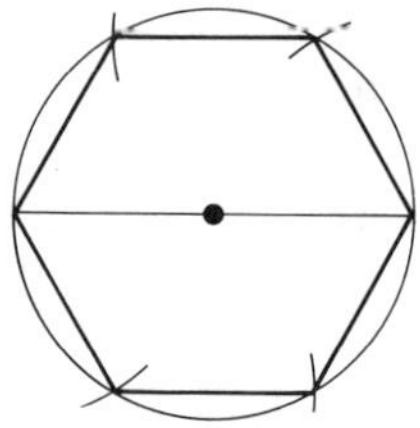

Figure 7.39

midpoint of this radius to the end of the other diameter (distance AB, Figure 7.41). Using this distance as a radius and A as the center, we cast an arc intersecting the diameter on the opposite side of the center (point C). We then measure the distance from this point to the endpoint of the other diameter (distance BC, Figure 7.42). This distance is the length of a side of the desired inscribed pentagon (Figure 7.43).

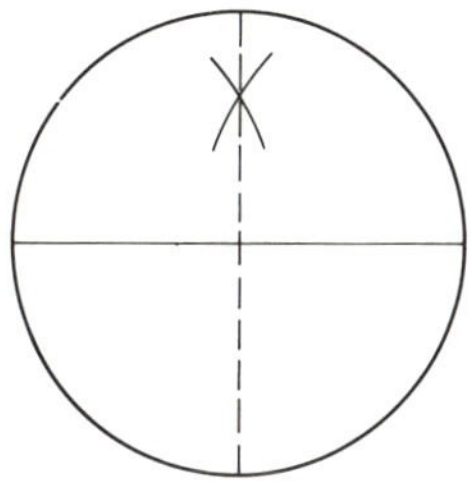

Figure 7.40

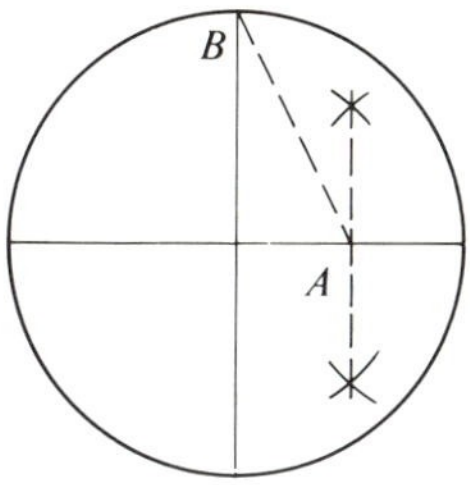

Figure 7.41

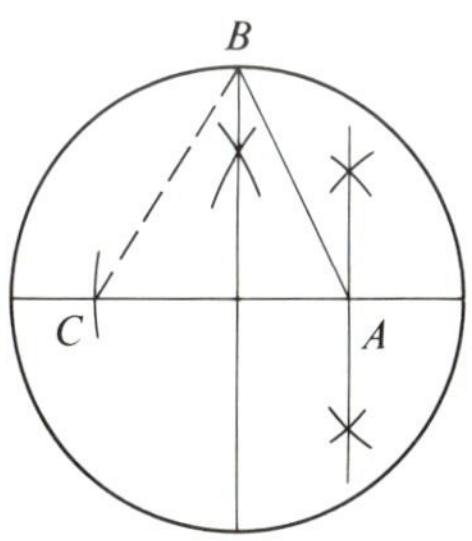

Figure 7.42

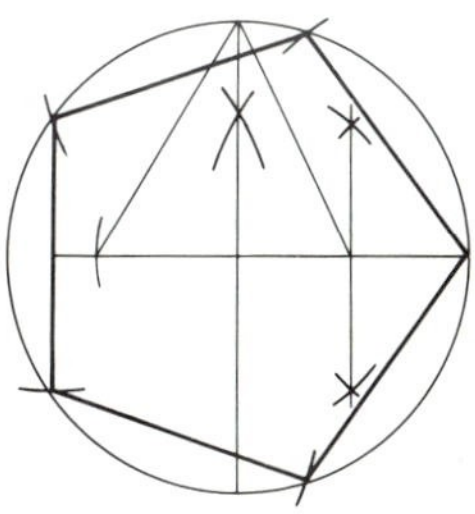

Figure 7.43

Most geometry textbooks give other constructions of regular inscribed pentagons, and although some of these constructions are easier to prove geometrically, their construction requires material not discussed in this text.

Exercise 44

Construct the following regular polygons.

1) square
2) pentagon
3) hexagon
4) octagon

5) decagon (10 sides)
6) 20-sided polygon
7) 16-sided polygon
8) 32-sided polygon
9) 12-sided polygon
10) 24-sided polygon
11) Connect every other vertex of a regular pentagon. What figure is formed?
12) If polygons of 4, 6, 8, 12, 16, and 24 sides are constructed in the same circle, what is true of the area of the polygon as the number of sides increases?
13) Using Problem 12, what is the largest value that the area of any inscribed polygon can have?
14) What is the size of each central angle formed in the construction of a regular polygon having 4 sides? 8 sides? 12 sides? 16 sides? n sides?
15) Construct the Maltese cross with the angle shown. (Hint: begin with a regular octagon.)

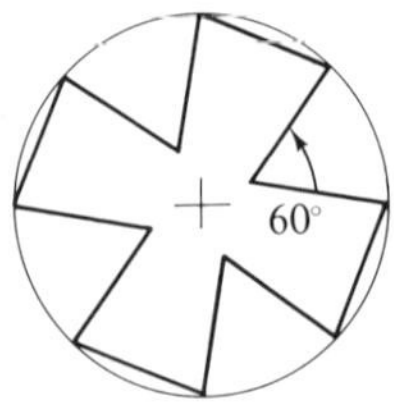

Review Exercise Chapter 7

1) Define the following terms:

polygon	angle
hexagon	ray
pentagon	parallelogram
rhombus	rectangle
quadrilateral	square
area	volume

2) Find the volume of a solid whose bases are trapezoids and whose sides are perpendicular to the bases if the area of the base is 25 square inches and the altitude is 40 inches.
3) A tank truck is in the form of a cylinder with a diameter of 8 feet and a length of 35 feet. How many gallons of liquid will the tank hold?
4) In triangles ABC and DEF, $\angle A = \angle D$, $\angle B = \angle E$, and $\angle C = \angle F$. What can be said about the two triangles?

5) If a canyon is measured as shown below, how far across is it, that is, what is distance x?

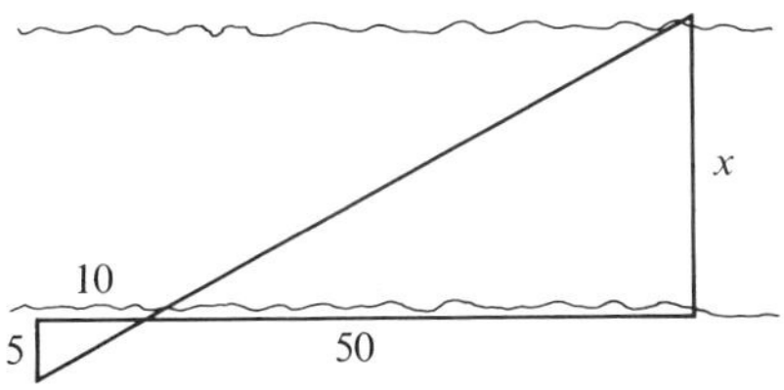

6) Find the hypotenuse of a right triangle with legs of 9 inches and 18 inches.

7) Find the total surface area of a rectangular solid having $l = 10$, $w = 8$, $h = 15$.

8) Find the total area of a pyramid having a square base with sides of 8 inches and an altitude of 12 inches.

9) Concentric circles are circles with the same center. Find the area between two concentric circles having radii of 15 and 10 inches.

10) Find the volume of a right circular cone with an altitude of 15 inches and a base radius of 5 inches.

11) A machine bolt is formed as shown. The top is a hemisphere. Find the volume of the bolt before threading.

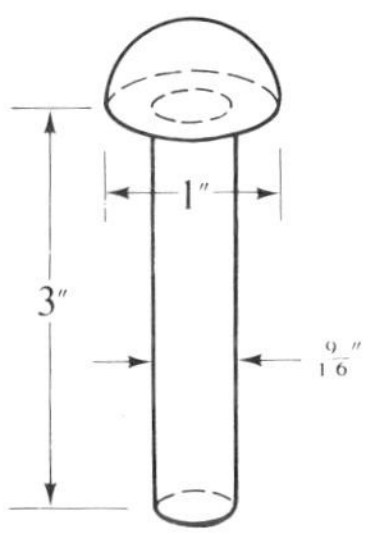

12) Find the length of the belt around the pulleys as shown.

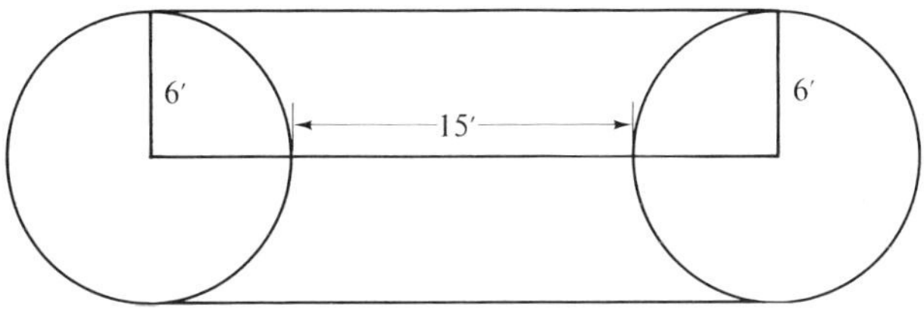

13) Find the length of the diagonal of a square with a 7-inch side.

14) What can be said about the diagonals of a rhombus?

15) How many cubic feet of water would fill a swimming pool with a circular form having a diameter of 36 feet and an average depth of 6 feet?

16) Find the area of the triangle shown below.

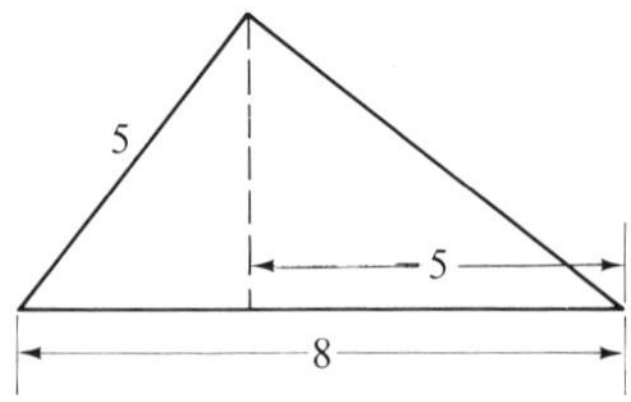

17) If the diameter of a barrel in the shape of a right circular cylinder is twice the height, and if the barrel holds 55 gallons, what is the radius of the base and the height of the barrel?

18) Construct the Star of David (a six-pointed star). Begin with a regular inscribed hexagon.

19) Find the area of a regular hexagon with a 6-inch side.

20) Construct a triangle with base angles of 30° and 60°. How large is the third angle? Measure the short leg and the longest side of the triangle. What observations can you make about their respective lengths?

8

Systems of Equations and Algebraic Operations

8.1 Systems of Linear Equations

A common problem encountered by the practicing technician is one in which he knows two different relationships between two varying quantities, and needs a solution set that will satisfy both relationships. At this point we examine a few of the methods used to solve such problems.

In Section 5.11 we learned that the graph of a linear equation is a straight line and that any point which lies on this line corresponds to a solution of the linear equation. In sketching the graph we consider a single relationship between the two different quantities (variables). Suppose we now add a second relationship between the same two variables and examine the possibilities that such a relationship presents. If we have two linear equations in the same two variables and if we consider them from a graphical standpoint, only three conditions can possibly exist: a) the lines are parallel (they never intersect); or b) the lines coincide (the equations are represented by the same line); or c) the lines intersect in a single point. (See Figure 8.1.)

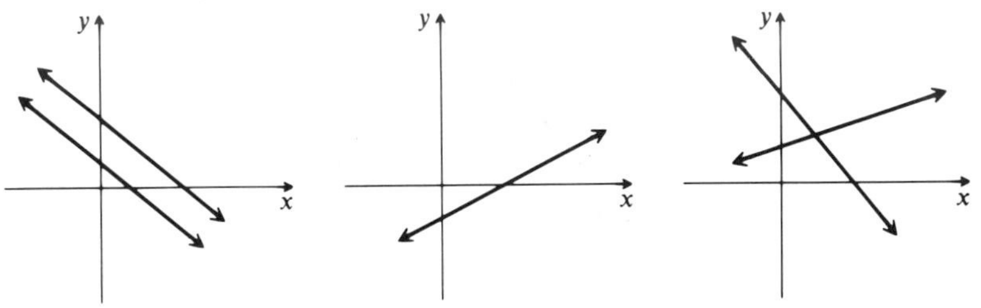

Figure 8.1

The first of these possibilities indicates that no common solution exists, since the two lines do not have a common point. Such a system is said to be *inconsistent.*

The second possibility is that any point on one line is also a point on the other line. This system is said to be *dependent.*

The third possibility is the one that will be of greatest interest to us, since the point of intersection represents a common solution to the system of equations. In this case the system is said to be *consistent* and *independent.*

All such pairs of equations are called *simultaneous equations.*

Example 1. The graph of the system of equations $3x + y = 3$ and $3x + y = 9$ is shown below. This system is inconsistent, since no common solution exists.

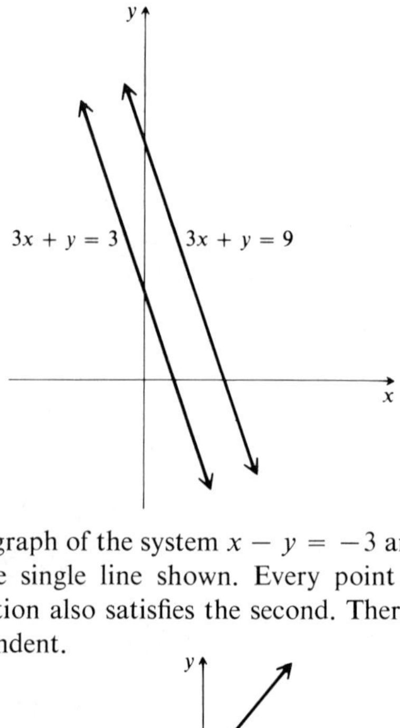

Example 2. The graph of the system $x - y = -3$ and $4x - 4y = -12$ is the single line shown. Every point satisfying the first equation also satisfies the second. Therefore, the system is dependent.

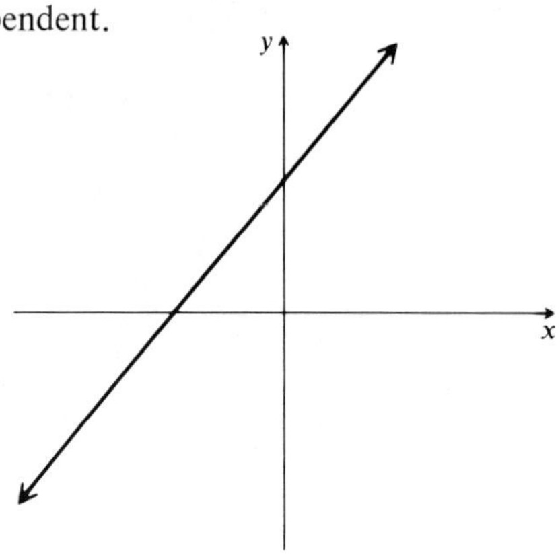

Example 3. The system of equations $3x - y = 1$ and $x + y = 3$ have a common point (1,2) as shown. This system is consistent and independent.

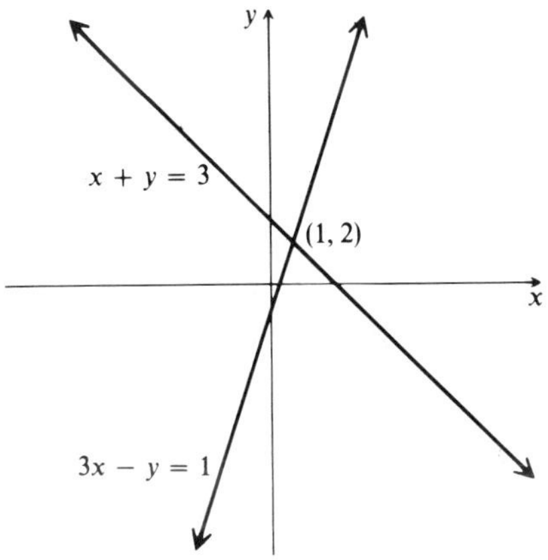

8.2 Solution of Systems of Linear Equations by Graphing

Suppose that the design of a balanced rocker on an oil well requires two braces with a combined weight of two hundred pounds and with a difference in weight of 40 pounds. If we let x be the weight of the heavier brace and y the weight of the lighter brace, then we know two things about x and y. First, the sum of their weights must be 200 pounds, which can be represented as $x + y = 200$; second, the difference of their weights must be 40 pounds, which can be represented as $x - y = 40$. We now have the system of equations

$$\begin{aligned} x + y &= 200, \\ x - y &= 40. \end{aligned}$$

From previous discussions, you probably have concluded that one method of solving this system would be to sketch the graphs of the two equations and read their common solution from their point of intersection on the graph. If we sketch both of these lines on the same graph (Figure 8.2), it can be seen that the one point that lies on both lines is the one where $x = 120$ and $y = 80$. This point (120,80) is the common solution to the system since these coordinates satisfy both equations.

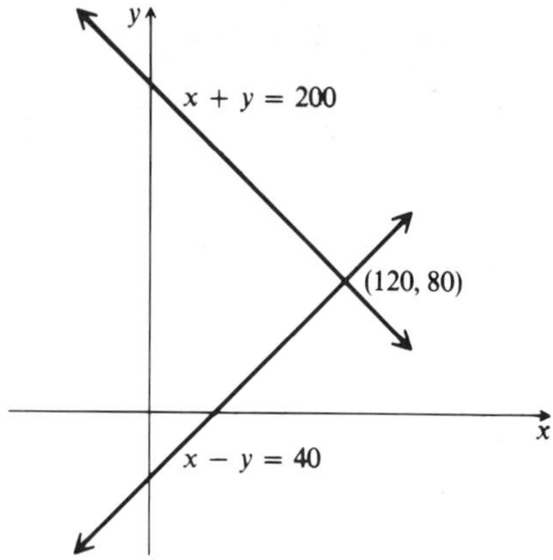

Figure 8.2

Solving systems of equations by graphing has one major disadvantage. If the lines do not intersect in a point that is easily read from the graph, the graphical solution can only be estimated. We now consider two more efficient methods of solving simultaneous linear equations.

8.3 Solution of Systems of Linear Equations by Addition or Subtraction

Let us return to the system of equations discussed in the previous section.

$$x + y = 200, \quad (1)$$
$$x - y = 40. \quad (2)$$

Since we are working with two equations we can employ an extension of Axiom 4. This extension states that equal quantities added to or subtracted from equal quantities produce equal quantities. Thus, when the two equations of the above system are added, we obtain the new equation

$$2x = 240$$

which has the solution

$$x = 120.$$

We now substitute this value for x into either equation (1) or (2) and solve for y. Selecting equation (1) we have

$$120 + y = 200,$$

which has the solution

$$y = 80.$$

With this as background, we now consider several examples of solutions to simultaneous linear equations and also introduce a means of checking the correctness of these solutions.

Example 1. Solve the following system of equations.

$$x + y = 6, \quad (1)$$
$$x - y = 4. \quad (2)$$

Subtracting equation (2) from equation (1) we obtain

$$2y = 2$$

which has the solution

$$y = 1.$$

We can now substitute this value into equation (1), obtaining

$$x + 1 = 6 \quad \text{or} \quad x = 5.$$

If we had substituted $y = 1$ into equation (2), we would have obtained

$$x - 1 = 4,$$

which has the same solution, $x = 5$. The solution to this system of equations is the ordered pair (5,1). To check this result, we substitute 5 for x and 1 for y in both equations (1) and (2). Substituting in equation (1) yields

$$5 + 1 = 6 \quad \text{or} \quad 6 = 6$$

which checks. Substituting into equation (2) yields

$$5 - 1 = 4 \quad \text{or} \quad 4 = 4$$

which also checks. The fact that both equations are satisfied verifies that (5,1) is the solution of the system of equations.

Each of the two previous problems was constructed in such a way that one of the two variables could be eliminated by a direct addition or subtraction of the two equations. The fact that this is not always possible is illustrated by the following example.

Example 2. Solve the following system of equations.

$$2x + y = 4, \quad (1)$$
$$3x - 2y = 5. \quad (2)$$

If we decide to eliminate the terms involving y by addition or subtraction, their numerical coefficients have to be the same. If we multiply both the left-hand and right-hand members of equation (1) by 2, utilizing the axiom for multiplying equals by equals, we obtain the new but equivalent equation

$$2(2x + y) = 2(4)$$

or

$$4x + 2y = 8.$$

Adding this new equation to equation (2), we obtain

$$7x = 13,$$

which has the solution

$$x = \frac{13}{7}.$$

We now substitute this value for x into equation (1) and solve for y.

$$2\left(\frac{13}{7}\right) + y = 4,$$

$$\frac{26}{7} + y = 4,$$

$$y = \frac{2}{7}.$$

Notice that the solution (13/7,2/7) would not be easy to obtain graphically. The check of this problem is left to the reader.

After reading through the examples above, you may wonder what would happen if the addition-subtraction method were applied to an inconsistent system of equations (parallel lines). The lines represented by the following system are parallel (Figure 8.3).

$$x + y = 6, \qquad (1)$$

$$x + y = 4. \qquad (2)$$

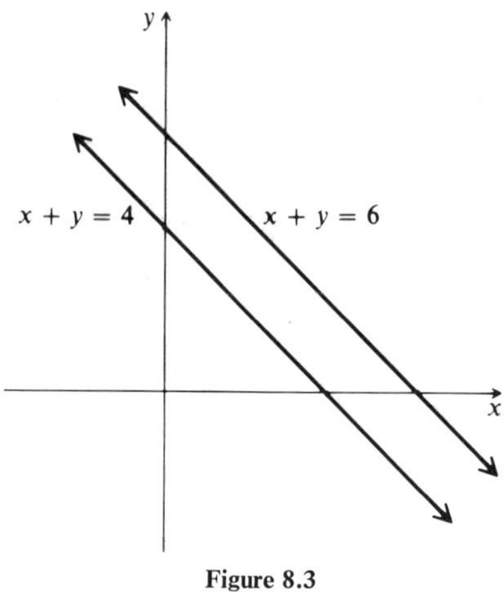

Figure 8.3

If we subtract equation (2) from equation (1) to eliminate x, we find that y is eliminated also and that the result is the contradictory statement

$$0 = 2.$$

A statement such as this always results from attempts to solve inconsistent systems.

Example 3. Solve the following system of equations.

$$3x + 2y = 4, \qquad (1)$$
$$2x + 5y = 10. \qquad (2)$$

Multiplying equation (1) by 2 and equation (2) by 3 we obtain

$$6x + 4y = 8, \qquad (3)$$

and

$$6x + 15y = 30. \qquad (4)$$

Notice that the coefficient of x is the same in both equations, and subtracting equation (4) from equation (3) we obtain

$$-11y = -22$$

which has the solution

$$y = 2.$$

Substituting 2 for y in equation (1) to obtain the solution for x we have

$$3x + 4 = 4,$$
$$3x = 0,$$
$$x = 0.$$

The solution to the system is therefore (0,2). Checking in each equation,

$$3(0) + 2(2) = 4, \qquad (1)$$
$$4 = 4;$$

and

$$2(0) + 5(2) = 10, \qquad (2)$$
$$10 = 10.$$

8.4 Solution of Systems of Linear Equations by Substitution

The addition-subtraction method is not the only technique for solving linear equations. One of the many other methods is the method of substitution. It is this method that we will treat briefly in this section.

If we return for a moment to the graph of the system of equations shown in Figure 8.2, we see that the point of intersection of the two lines (the common solution to the two equations) is the only point on the graph where the value of y in each equation is represented by the same number. This means that we can solve one equation for y and then substitute this value into the other equation. The following illustrates this technique.

Example 1. Solve the following system of equations.

$$x + y = 200. \tag{1}$$
$$x - y = 40. \tag{2}$$

First we solve equation (1) for y to obtain $y = 200 - x$ We now substitute this quantity for y into equation (2), which yields

$$\begin{aligned} x - (200 - x) &= 40 \\ x - 200 + x &= 40 \\ 2x &= 240 \\ x &= 120. \end{aligned}$$

We obtain the value for y in the same manner as we did when we solved this system by the addition-subtraction method in the previous section. Putting this value for x into equation (2) we obtain

$$\begin{aligned} 120 - y &= 40 \\ -y &= -80 \\ y &= 80. \end{aligned}$$

Therefore, the solution is (120,80).

The process of solving simultaneous linear equations by the substitution method may be briefly summarized as follows.

1. Solve one of the equations for one variable in terms of the other.
2. Substitute the expression obtained in this manner into the other equation.
3. Solve the resulting equation for the remaining variable.
4. Substitute the result of Step 3 into the equation of Step 1 and solve for the remaining variable.
5. Check the results on both original equations.

Some equations are particularly well adapted to solution by substitution. The following example illustrates such a system.

Example 2. Solve the following system of equations.

$$2x - y = 4, \tag{1}$$
$$3y = -2x \tag{2}$$

Solving equation (2) for y we have

$$y = \frac{-2}{3}x.$$

Replacing y in equation (1) by this quantity yields

$$\begin{aligned} 2x + \frac{2}{3}x &= 4, \\ 6x + 2x &= 12, \\ 8x &= 12, \\ x &= \frac{3}{2}. \end{aligned}$$

Substituting this quantity into equation (2) we find that

$$3y = -3 \quad \text{or} \quad y = -1.$$

Therefore, the solution of the system is $\left(\frac{3}{2}, -1\right)$.

Exercise 45

Solve each of the following systems of linear equations by the addition-subtraction method and check.

1) $3x + y = 20,$ $2x - y = 10.$

2) $x + 4y = 18,$ $3x - 4y = 22.$

3) $x + 4y = 9,$ $x + 3y = 7.$

4) $3x + y = 12,$ $8x - y = 10.$

5) $3x + y = 25,$ $3x - y = 10.$

6) $x + y = 6,$ $y - x = 1.$

7)–12) Solve Problems 1) – 6) by the method of graphing.

Solve the following systems of linear equations by the method of substitution.

13) $5y = 30x,$ $x + y = 7.$

14) $-2x + y = 7,$ $x + y = 10.$

15) $3x + 2y = 16,$ $x = -\frac{1}{2}y + 2.$

16) $x - 3y = 5,$ $2x + y = 17.$

17) $3x - y = 25,$ $x - 3y = 3.$

18) $x = 4y + 2,$ $y = x - 5.$

Solve the following systems of linear equations by a method of your choosing.

19) $3x + 4y = 16,$ $2x - 2y = 5.$

20) $12x - 3y = 2,$ $6x - y = 5.$

21) $4.5x - 1.5y = 3.5,$ $1.5x - .5y = 2.5.$

22) $\frac{1}{2}x - \frac{1}{3}y = 1,$ $\frac{2}{3}x - \frac{1}{2}y = \frac{2}{3}.$

23) $\frac{1}{3}x + y = 0,$ $x + y = 4.$

24) $7x + 5y = 1,$ $3x - y = 2.$

Solve the following problems using two unknowns.

25) The sum of two resistors in a series is 320 ohms. If the smaller resistor has a rating that is 3/5 the rating of the larger resistor, what must be their respective ratings?

26) In a certain triangle two of the sides are equal. The perimeter is 80 inches and each of the equal sides is 4 inches longer than the third side. Find the lengths of the sides of the triangle.

27) The sum of two angles is 90°. If one angle is 34° more than the other, what is the size of the two angles?

28) Upon examining two credit card statements from his service station, a man noticed that he had been charged \$5.34 for 10 gallons of gas and 3 quarts of oil on one occasion, and \$6.64 for 10 gallons of gas and 5 quarts of oil on another. Assuming that the price of gas and oil remained constant between the two statement dates, what was the cost of one gallon of gas and one quart of oil?

8.5 Stated Problems

Many of the problems to be solved by the technician are encountered in stated form. It is then up to the technician to interpret the statements and find an equation that describes the conditions set forth. Some stated problems have been interspersed through the exercise sets in previous chapters. We will now discuss a more formal method of solution.

The greatest difficulty encountered in this type of problem is usually a result of failure to organize one's work logically or failure to devise a general plan of attack. Both of these problems can be overcome if the following outline is used.

1. Read the entire problem through carefully.
2. Read the problem a second time, noting what is given and what is to be found. Note each given quantity on your paper and let some variable, such as x, represent the unknown quantity.
3. Reread the problem again, expressing all unknown quantities in terms of the variable.
4. From the problem, decide what quantities are equal and formulate an equation from this decision.
5. Solve the equation.
6. Check the solution with the statement of the problem.

Step 6 is particularly important, since a correct solution to the wrong equation would satisfy the equation but not the problem.

Example 1. The sum of three consecutive natural numbers is 33. Find the numbers.

$$\begin{aligned} \text{Let the first number} &= n. \\ \text{Let the second number} &= n + 1. \\ \text{Let the third number} &= n + 2. \\ \text{Sum} &= 33. \end{aligned}$$

The equation is

$$\begin{aligned} n + (n + 1) + (n + 2) &= 33. \\ n + n + 1 + n + 2 &= 33. \\ 3n + 3 &= 33. \\ 3n &= 30. \\ n &= 10. \\ n + 1 &= 11. \\ n + 2 &= 12. \end{aligned}$$

Thus the three numbers are 10, 11, and 12. Since $10 + 11 + 12 = 33$ and since 10, 11, and 12 are consecutive natural numbers, the solution checks.

Example 2. What are the dimensions of a rectangular piece of sheet metal if its perimeter is 480 inches and its length is three times its width?
Let width $= x$. Let length $= 3x$. The formula for the perimeter of a rectangle is $p = 2(L + W)$.
The equation then is

$$\begin{aligned} 2(3x + x) &= 480. \\ 2(4x) &= 480. \\ 8x &= 480. \\ x &= 60 \quad \text{and} \quad 3x = 180. \end{aligned}$$

Thus the dimensions of the sheet metal are 60 x 180 inches. To check, the perimeter is

$$\begin{aligned} 2 \cdot 60 + 2 \cdot 180 &= 120 + 360 \\ &= 480. \end{aligned}$$

The length is 180, which is three times the width, 60.

Example 3. A collection of nickels, dimes, and quarters has a value of \$5.25. How many nickels are there if there are twice as many dimes as nickels and six more quarters than dimes? Let the number of nickels $= x$. Let the number of dimes $= 2x$. Let the number of quarters $= 2x + 6$. The total worth is \$5.25. The equation is

$$5x + 10(2x) + 25(2x + 6) = 525.$$

Notice that the values of the different coins are equated with their total value.

$$\begin{aligned} 5x + 20x + 50x + 150 &= 525. \\ 75x + 150 &= 525. \\ 75x &= 375. \\ x &= 5. \end{aligned}$$

Thus there are 5 nickels in the collection of coins. The check is left to the reader.

A common type of stated problem encountered, particularly in mechanics, is the work problem dealing with the *lever*, a simple machine. The pivot point of a lever or the point at which the lever is balanced is called the *fulcrum*. The distance along the lever from the fulcrum to the point at which the force is applied is called the *lever arm*. (See Figure 8.4.) In general, lever problems can be solved by remembering that in order to balance (to be in equilibrium) the length of one lever arm times the force applied on that side of the fulcrum must equal the length of the other lever arm times the force applied on the other side of the fulcrum, or

$$F_1 \times L_1 = F_2 \times L_2.$$

In this case, as in the case of many stated problems, a sketch of the given situation can be very helpful.*

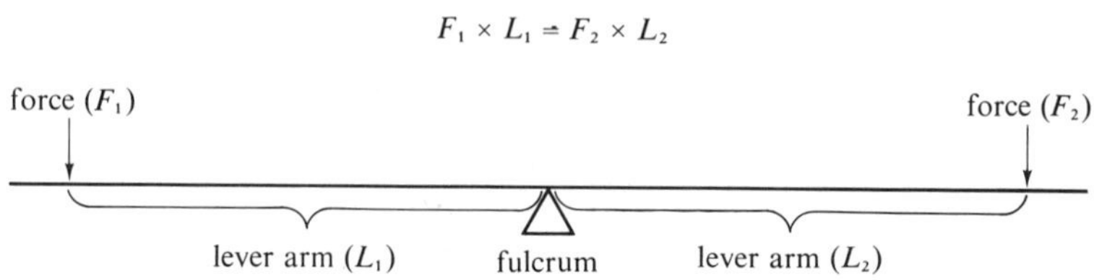

Figure 8.4

Example 4. A weight of 500 pounds is placed 20 feet from the fulcrum of a lever. How far from the fulcrum must a 300-pound weight be placed for the lever to balance?

Let the length of the unknown lever arm equal x. The equation for the problem is then

$$\begin{aligned} 300x &= 500 \times 20 \\ \frac{300x}{300} &= \frac{500 \times 20}{300} \\ x &= 33\tfrac{1}{3}. \end{aligned}$$

So the desired distance is $33\frac{1}{3}$ feet.

* For our purposes in this text we will neglect the weight of the lever itself in dealing with this type of problem.

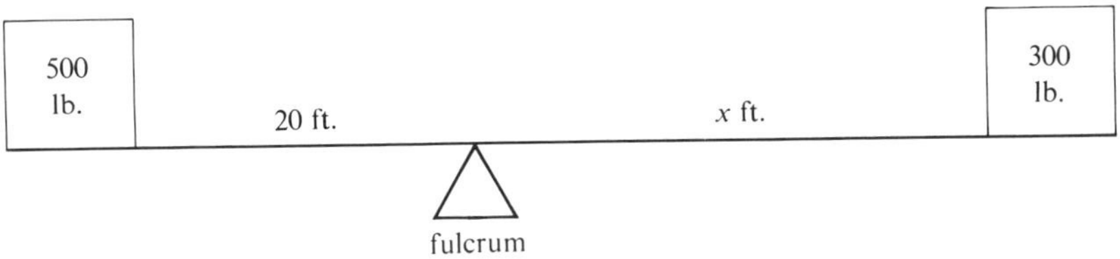

Example 5. How many pounds of force must be exerted at one end of a lever 12 feet long to lift a 450-pound rock if the fulcrum is 3 feet from the rock?

Let x be the unknown force. Then the equation becomes

$$3 \times 450 = (12 - 3)x$$
$$1350 = 9x.$$
$$150 = x.$$

Therefore, 150 pounds of force must be exerted to lift the rock. Since $3 \times 450 = 9 \times 150$, the solution checks.

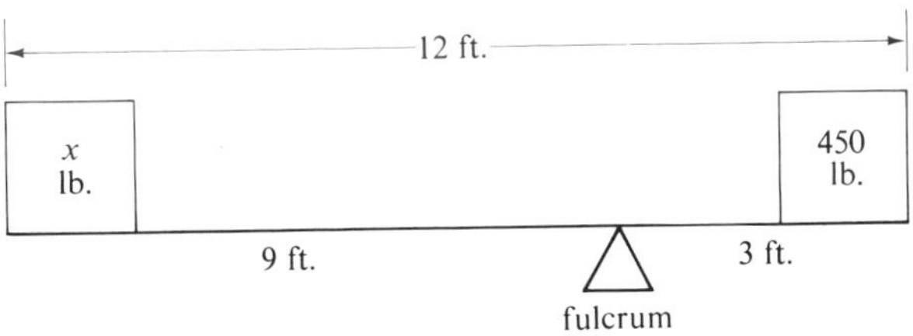

Exercise 46

Solve the following.

1) Divide 83 into two parts such that one part is three greater than three times the other.

2) Find three consecutive even integers whose sum is 126.

3) The numerator of a fraction is 4 less than the denominator. The fraction when reduced is 5/6. Find the original fraction.

4) How many gallons of water should be added to 5 gallons of a 25% alcohol solution in order to get a 10% solution?

5) A collection of coins contains 3 dimes and twice as many nickels as quarters. How many of each kind of coin is in the collection if its value is $3.80?

6) Three-fourths of a number is 5/6 more than 1/2 of the number. Find the number.

7) Mortar is a mixture of cement and sand. How many pounds of cement worth 24 cents a pound and how many pounds of sand worth 2 cents a pound are mixed to obtain 120 pounds of mortar worth 12 cents a pound?

8) The sum of a number and its square is 30. Find the number.

9) A man is paid \$25 for each day that he works and forfeits \$5 for each day he is idle. If his paycheck is \$275 at the end of 17 days, how many days did he work?

10) Two motorists start toward each other at the same time from towns 279 miles apart. If their average speeds are 42 and 51 mph, how much time elapsed before they met? (Hint: distance = rate × time.)

11) The cost of parts for a tune-up is one-half as much as labor. Find the cost of parts and labor if the tune-up costs \$15.00.

12) An auto salvage yard received \$864.00 from the sale of 900 junk tires and wheels. If the tires sold for \$.70 each and the wheels for \$2.50 each, how many of each did the salvage yard sell?

13) A metal pipe 10 feet long is to be divided into two pieces. Find the length of each piece if one is two-thirds as long as the other.

14) If a sheet metal box is made in the shape of a cube having a side area of 25 square inches, how much metal is used for the entire box if it has no lid?

15) The sum of two consecutive integers is 15. Find the numbers.

16) A rectangular field is 35 feet longer than it is wide. The length of the fence around the field is 380 feet. Find the dimensions of the field.

17) How many cubic centimeters of water must be added to 1000 cubic centimeters of a 75% acid solution to make a 60% solution?

18) Two planes start from the same airport and fly in opposite directions at the rates of 350 mph and 500 mph. When will the distance between the two planes be 2075 miles?

19) When a certain casting is machined the weight is decreased by 10%. Find the finished weight of a 9-pound casting.

20) Six less than one-half of a number equals four. Find the number.

21) The sum of the perimeters of two squares is 80 inches. If the side of one square is 4 inches more than the other, find the length of a side of each.

22) If a weight of 325 pounds is 15 feet from the fulcrum of a lever, how far from the fulcrum must a weight of 475 pounds be placed to balance the lever?

23) A man purchased 200 acres of land for \$150,000. If part of the land cost \$500 per acre and part of it cost twice as much, how many acres did he purchase at each price?

24) A 75-pound weight is on a lever 8 feet on one side of the fulcrum. Where should a weight of 60 pounds be placed to give equilibrium?

25) Two automobiles start from the same point, one going north and the other going south. One travels 2/3 as fast as the other. In 8 hours they are 600 miles apart. Find the speed of each.

26) The length of a rectangle is twice its width. If the perimeter is 96 inches, find the area of the rectangle.

27) A student has quiz grades of 75 and 85. What grade must he make on a third quiz to have an average of 80?

28) Admission to a football game was \$1.25 for adults and \$.50 for children. How many adults and children were in the audience of 806 if the box office receipts were \$781.00?

29) One machine can stamp 1800 mounting brackets in 1 hour and another machine can stamp 1500 brackets in 1 hour. If 20,000 brackets must be stamped and the slower machine is started 2 hours after the fast machine, how long will it take to finish the job?

30) A radiator which has a 24-quart capacity is filled with a 20% antifreeze solution. If it requires 45% antifreeze to give adequate protection, how much pure antifreeze must be added to obtain the desired mixture?

8.6 Equations in More Than Two Unknowns

Many applied problems require finding solutions to equations in three or more variables. We shall consider only those equations that are of the form

$$Ax + By + Cz = D,$$

where A, B, C and D are arbitrary constants. Equations of this type are frequently called linear equations even though their graph is a plane surface as opposed to a straight line.

Consider for a moment the equation

$$3x + 2y + 4z = 11.$$

If $x = 1$, $y = -2$ and $z = 3$, the equation is satisfied. We say that the ordered triple $(1, -2, 3)$ satisfies, or is a solution to, the equation. Since the above equation represents a plane, there are infinitely many points which lie on it and therefore infinitely many ordered triples that satisfy the equation.

8.7 Solutions of Systems of Equations in More Than Two Unknowns

Recall from earlier sections of this chapter that two equations in two unknowns could be solved by three different methods provided that a solution existed. These methods were addition-subtraction, graphing, and substitution. In this section we shall pursue the solution to a system of three equations in three unknowns by the addition-subtraction or by the substitution method. Sketching the graphs of equations in three unknowns requires techniques that are beyond the scope of this text, and the solutions obtained by graphing have the same drawbacks that were noted earlier.

Example 1. Solve the following system of equations.

$$2x + y - z = -1, \tag{1}$$

$$x + y + z = 2, \tag{2}$$

$$3x - 2y + z = 7. \tag{3}$$

If we add equation (1) to equation (2) we obtain

$$3x + 2y = 1. \tag{4}$$

We can obtain another equation in the same two unknowns by subtracting equation (2) from equation (3) to obtain

$$2x - 3y = 5. \tag{5}$$

We have now reduced the problem of solving a system in three unknowns to that of solving one in two unknowns.

$$3x + 2y = 1, \tag{4}$$

$$2x - 3y = 5. \tag{5}$$

Multiplying (4) by 3 and (5) by 2 yields

$$\begin{aligned} 9x + 6y &= 3, \\ 4x - 6y &= 10. \end{aligned}$$

From this we find, by adding the two equations, that

$$\begin{aligned} 13x &= 13, \\ x &= 1. \end{aligned}$$

Substituting $x = 1$ in (4) we find

$$\begin{aligned} 3 + 2y &= 1 \\ 2y &= -2 \\ y &= -1. \end{aligned}$$

Substituting $x = 1$, $y = -1$ in (2) yields

$$\begin{aligned} 1 - 1 + z &= 2 \\ z &= 2. \end{aligned}$$

The solution is $(1,-1,2)$.
The substitutions that were made to find y and z could have been made in any of the appropriate equations. Equation (2) was chosen since it had the simplest coefficients.

Example 2. Solve the system by substitution.

$$x + 2y - z = 4 \qquad (1)$$

$$2x + y - z = 7 \qquad (2)$$

$$x - y + z = 5 \qquad (3)$$

Solving equation (3) for x yields

$$x = 5 + y - z.$$

We now substitute this quantity into equations (1) and (2) for x and simplify

$$\begin{aligned} 5 + y - z + 2y - z &= 4 \\ 3y - 2z &= -1 \qquad (4) \\ 2(5 + y - z) + y - z &= 7 \\ 3y - 3z &= -3 \qquad (5) \end{aligned}$$

Solving (4) and (5) simultaneously we find $y = 1$ and $z = 2$ which yields $x = 4$ when substituted into equation (1). Thus the solution to the system is the ordered triple $(4,1,2)$.
To check this solution we must verify that it satisfies each of the three original equations.

$$\begin{aligned} 4 + 2 \cdot 1 - 2 &= 4; \\ 4 &= 4. \\ 2 \cdot 4 + 1 - 2 &= 7; \\ 7 &= 7. \\ 4 - 1 + 2 &= 5 \\ 5 &= 5. \end{aligned}$$

The techniques used to solve three equations in three variables can be extended to any number of equations in any number of variables.

Exercise 47

Solve each of the following systems. Write the solutions as ordered triples.

1) $\begin{aligned} x + y + z &= 6 \\ 2x + 3y + z &= 9 \\ 2x - y + z &= 3 \end{aligned}$

2) $\begin{aligned} x + y &= 3 \\ x + y - z &= -4 \\ y + z &= 6 \end{aligned}$

3) $\begin{aligned} 3M + N + R &= 7 \\ 2N - R &= 5 \\ M + 3N + 2R &= 11 \end{aligned}$

4) $\begin{aligned} R_1 + R_2 + R_3 &= 11 \\ R_1 - R_2 + R_3 &= 3 \\ R_1 + 2R_2 + R_3 &= 15 \end{aligned}$

5) $\begin{aligned} .1M + .2N + .3P &= 1.3 \\ .6M + .3N + .2P &= 2.7 \\ .5M - .2N + .7P &= 2.9 \end{aligned}$

6) $\begin{aligned} a + b &= 1 \\ a + c &= 4 \\ b + c &= 2 \end{aligned}$

7) $\begin{aligned} L + R &= 6 \\ L - S &= 3 \\ R + S &= 5 \end{aligned}$

8) $\begin{aligned} x + 4y - 3z &= 0 \\ x - y - z &= 7 \\ 2x + 3y &= 5 - z \end{aligned}$

9) $\begin{aligned} x + y + z &= 3 \\ x + y - z &= 1 \\ x - y + z &= 1 \end{aligned}$

10) $\begin{aligned} x + y + z &= 1 \\ 4x + 2y + z &= 3 \\ 100x + 10y + z &= 4 \end{aligned}$

8.8 Matrices and Determinants

A *matrix* is a rectangular array of symbols enclosed by parentheses or brackets and usually denoted by capital letters. A matrix may appear in many forms such as the *row matrix*, $A = (a,b,c,d)$, or a *column matrix*

$B = \begin{pmatrix} a \\ b \\ c \\ d \end{pmatrix}$, or a matrix having many rows and columns

of which $C = \begin{pmatrix} a & e & i \\ b & f & j \\ c & g & k \\ d & h & l \end{pmatrix}$ is an example.

The entries in a matrix are called the *elements* of the matrix and the size or *dimensions* of a matrix is determined by the number of rows (m) and columns (n) that the matrix possesses. Thus the matrices above have dimensions 1 x 4, 4 x 1 and 4 x 3 respectively. Thus the dimensions of a general matrix having m rows and n columns are denoted as $m \times n$. If the numbers of rows and columns is the same the matrix is said to be *square*.

Two matrices are said to be equal if they have the same dimensions and if their corresponding elements are the same. Thus

$$\begin{pmatrix} 1 & 3 \\ 2 & 4 \end{pmatrix} = \begin{pmatrix} \frac{2}{2} & \frac{6}{2} \\ \sqrt{4} & \sqrt[3]{64} \end{pmatrix}$$

but

$$\begin{pmatrix} 1 & 3 \\ 2 & 4 \end{pmatrix} \neq \begin{pmatrix} 1 & 3 \\ -2 & 4 \end{pmatrix}$$

or

$$\begin{pmatrix} 1 & 2 \\ 3 & 4 \\ 5 & 6 \end{pmatrix} \neq \begin{pmatrix} 1 & 2 & 3 \\ 4 & 5 & 6 \end{pmatrix}.$$

A *zero matrix* is any matrix in which all entries are zero.

8.9 Operations on Matrices

If two matrices have the same dimensions, their sum can be found. This sum is defined to be the matrix which is obtained by adding the corresponding elements of the two matrices. For example the sum of two matrices A and B denoted $A + B$ is shown below.

$$\begin{pmatrix} 1 & 2 \\ 3 & 4 \\ 5 & 6 \end{pmatrix} + \begin{pmatrix} -1 & 0 \\ 2 & 3 \\ 1 & 2 \end{pmatrix} = \begin{pmatrix} 1 + (-1) & 2 + 0 \\ 3 + 2 & 4 + 3 \\ 5 + 1 & 6 + 2 \end{pmatrix} = \begin{pmatrix} 0 & 2 \\ 5 & 7 \\ 6 & 8 \end{pmatrix}.$$

Similarly two matrices of the same dimensions can be subtracted by subtracting their corresponding entries. Using the matrices of the preceding example, $A - B$ can be determined as

$$\begin{pmatrix} 1 & 2 \\ 3 & 4 \\ 5 & 6 \end{pmatrix} - \begin{pmatrix} -1 & 0 \\ 2 & 3 \\ 1 & 2 \end{pmatrix} = \begin{pmatrix} 1 - (-1) & 2 - 0 \\ 3 - 2 & 4 - 3 \\ 5 - 1 & 6 - 2 \end{pmatrix} = \begin{pmatrix} 2 & 2 \\ 1 & 1 \\ 4 & 4 \end{pmatrix}.$$

If a matrix is multiplied by a constant, k, each entry in the matrix is multiplied by the constant. That is,

$$\text{if } A = \begin{pmatrix} 5 & 3 \\ 2 & 4 \end{pmatrix}, \text{ then } kA = \begin{pmatrix} 5k & 3k \\ 2k & 4k \end{pmatrix}.$$

Two additional matrices worth noting are the *diagonal matrix* which is a square matrix having nonzero elements only along the principal diagonal (the diagonal from upper left to lower right). The second is the *unit* or *identity matrix* which is a square matrix having entries of one along the principal diagonal and zeros elsewhere.

8.10 Multiplication of Matrices

Two matrices may be multiplied if they are *conformable*. This means that to find the product of matrix A times matrix B, the number of columns in matrix A must equal the number of rows in matrix B. Thus a 3×2 matrix is conformable to any $2 \times n$ matrix. The dimensions of the product matrix will be $3 \times n$.

An element in the product matrix, say the element in the 1st row, 2nd column, can be found by taking the sum of all the products of each element in row 1 of the first matrix and column 2 of the second matrix. This multiplication is more easily shown by an example such as the following.

$$A = \begin{pmatrix} 1 & 2 & 1 \\ 2 & 1 & 3 \end{pmatrix} \qquad B = \begin{pmatrix} 1 & 1 \\ 2 & 0 \\ 3 & 1 \end{pmatrix}$$

$$A \cdot B = \begin{pmatrix} 1 \cdot 1 + 2 \cdot 2 + 1 \cdot 3 & 1 \cdot 1 + 2 \cdot 0 + 1 \cdot 1 \\ 2 \cdot 1 + 1 \cdot 2 + 3 \cdot 3 & 2 \cdot 1 + 1 \cdot 0 + 3 \cdot 1 \end{pmatrix} = \begin{pmatrix} 8 & 2 \\ 13 & 5 \end{pmatrix}$$

Notice that matrix addition is commutative and associative and that matrix multiplication is associative but not commutative. $B \times A$ cannot be found, since B and A are not conformable.

Matrices are frequently used as a means of representing a series of linear equations. For example, the system

$$\begin{aligned} x + y + z &= 3 \\ 2x - y + z &= 7 \\ x + y + 3z &= 9 \end{aligned}$$

can be represented as the product of the matrix made up of the coefficients of the variables and the column matrix consisting of the variables set equal to the column matrix made up of the constants.

$$\begin{pmatrix} 1 & 1 & 1 \\ 2 & -1 & 1 \\ 1 & 1 & 3 \end{pmatrix} \cdot \begin{pmatrix} x \\ y \\ z \end{pmatrix} = \begin{pmatrix} 3 \\ 7 \\ 9 \end{pmatrix}$$

Exercise 48

Carry out each of the indicated operations. If the operation is not possible, state the reasons why.

1) $\begin{pmatrix} 1 & 3 \\ 2 & 4 \end{pmatrix} + \begin{pmatrix} 2 & 3 \\ 1 & 5 \end{pmatrix}$

2) $\begin{pmatrix} 2 & 1 & 3 \\ 0 & 2 & 1 \\ 0 & 1 & 2 \end{pmatrix} + \begin{pmatrix} 6 & -1 & 0 \\ 1 & 3 & 2 \\ 2 & 0 & 0 \end{pmatrix}$

3) $3\begin{pmatrix} 1 & 3 \\ 2 & 7 \end{pmatrix}$

4) $2\begin{pmatrix} 1 & 5 \\ 2 & 3 \end{pmatrix} + \begin{pmatrix} 6 & 1 \\ 1 & 2 \end{pmatrix}$

5) $2\begin{pmatrix}1\\5\\6\end{pmatrix} - 3\begin{pmatrix}2\\1\\3\end{pmatrix}$

6) $\begin{pmatrix}3 & 2\\1 & 4\end{pmatrix} - \begin{pmatrix}2 & 3\\1 & 4\end{pmatrix}$

7) $(1\quad 2\quad 3)\begin{pmatrix}3\\2\\1\end{pmatrix}$

8) $\begin{pmatrix}1 & 3\\2 & 4\end{pmatrix} \cdot \begin{pmatrix}1 & 2 & 1\\2 & 1 & 1\\1 & 3 & 1\end{pmatrix}$

9) $\begin{pmatrix}1 & 1\\2 & 1\end{pmatrix} \cdot \begin{pmatrix}1 & 3 & 4\\2 & 2 & 2\end{pmatrix}$

10) $\begin{pmatrix}1 & 0\\0 & 1\end{pmatrix} \cdot \begin{pmatrix}1 & 3 & 4\\2 & 2 & 2\end{pmatrix}$

11) $\begin{pmatrix}1 & 0 & 0\\0 & 1 & 0\\0 & 0 & 1\end{pmatrix} \cdot \begin{pmatrix}2 & 0 & 7\\1 & 3 & 9\\2 & 4 & 7\end{pmatrix}$

12) What name is assigned to the first matrix in both problems 10) and 11)?

Find the values of x and y that will make the following true.

13) $\begin{pmatrix}x & 1\\3 & y\end{pmatrix} = \begin{pmatrix}2 & 7\\4 & 9\end{pmatrix}$

14) $\begin{pmatrix}x+1\\y-1\end{pmatrix}\begin{pmatrix}2\\3\end{pmatrix} = \begin{pmatrix}2 & 7\\1 & 4\end{pmatrix}$

15) $\begin{pmatrix}x+1 & -y\\y & x\end{pmatrix} + \begin{pmatrix}1 & 1\\2 & 3\end{pmatrix} = \begin{pmatrix}2 & 4\\-1 & 3\end{pmatrix}$

16) $\begin{pmatrix}y & 2x-y\\2 & 3\end{pmatrix} + \begin{pmatrix}1 & x\\2 & 5\end{pmatrix} = \begin{pmatrix}1 & 1\\4 & 8\end{pmatrix}$

17) $(1\quad 2\quad 3)\begin{pmatrix}x & y & z\\2 & 2 & 1\\3 & 1 & 0\end{pmatrix} = (1\quad 8\quad 12)$

18) If $A = \begin{pmatrix}1 & 3\\2 & 4\end{pmatrix}$ and $B = \begin{pmatrix}2 & 0\\1 & 3\end{pmatrix}$ show that $A \cdot B \neq B \cdot A$.

8.11 Determinants

Associated with each square matrix there is a real number called the *determinant* of the matrix. For the matrix $A = \begin{pmatrix}a & c\\b & d\end{pmatrix}$ we define its determinant to be

$$D = \begin{vmatrix}a & c\\b & d\end{vmatrix} = ad - bc.$$

The *order* of the determinant is the number of entries in any row or column. This determinant is of order two. The determinant of a 2 x 2 matrix is obtained by taking the product of the entries on the principal diagonal and subtracting the product of the entries on the other diagonal. Notice that determinants are enclosed by vertical bars as opposed to parentheses.

Example 1. $\begin{vmatrix} 1 & 3 \\ 2 & 4 \end{vmatrix} = 1 \cdot 4 - 3 \cdot 2 = 4 - 6 = -2.$

Example 2. $\begin{vmatrix} 2 & -1 \\ 3 & 7 \end{vmatrix} = 2 \cdot 7 - (-1) \cdot 3 = 14 + 3 = 17.$

If the matrix under consideration has dimensions 3 x 3, then its determinant is of order three and is defined to have the value

$$\begin{vmatrix} a_1 & b_1 & c_1 \\ a_2 & b_2 & c_2 \\ a_3 & b_3 & c_3 \end{vmatrix} = a_1b_2c_3 + a_2b_3c_1 + a_3b_1c_2 - a_1b_3c_2 - a_2b_1c_3 - a_3b_2c_1.$$

A convenient scheme for obtaining this product is to copy the first two columns of the determinant to the right of the last column. Then to evaluate, we compute the sum of the products of each diagonal, downward to the right, containing three elements as we move from left to right. Then we subtract the sum of the products of each diagonal containing three elements as we move downward from right to left. These products are indicated below.

$$D = \begin{vmatrix} a_1 & b_1 & c_1 \\ a_2 & b_2 & c_2 \\ a_3 & b_3 & c_3 \end{vmatrix} = \left|\begin{array}{ccc|cc} a_1 & b_1 & c_1 & a_1 & b_1 \\ a_2 & b_2 & c_2 & a_2 & b_2 \\ a_3 & b_3 & c_3 & a_3 & b_3 \end{array}\right.$$

When a determinant is evaluated, we say that it has been *expanded*. For example,

$$\text{evaluate} \begin{vmatrix} 1 & 0 & 2 \\ 2 & 1 & 1 \\ 1 & 3 & 4 \end{vmatrix}$$

We have

$$\left|\begin{array}{ccc|cc} 1 & 0 & 2 & 1 & 0 \\ 2 & 1 & 1 & 2 & 1 \\ 1 & 3 & 4 & 1 & 3 \end{array}\right. = 4 + 0 + 12 - (0 + 3 + 2) = 11.$$

The above process is useful only for 3rd order determinants. Other methods will be developed for determinants of a higher order.

8.12 Evaluation of Determinants by Minors and Cofactors

The *minor* of an element of a determinant is defined to be the determinant that results when the row and column containing that element is deleted. For example, consider the minors of each element in row 1 of

$$A = \begin{vmatrix} 1 & 4 & 7 \\ 2 & 5 & 8 \\ 3 & 6 & 9 \end{vmatrix}.$$

The minor of 1 is $\begin{vmatrix} 5 & 8 \\ 6 & 9 \end{vmatrix}$. The minor of 4 is $\begin{vmatrix} 2 & 8 \\ 3 & 9 \end{vmatrix}$. The minor of 7 is $\begin{vmatrix} 2 & 5 \\ 3 & 6 \end{vmatrix}$. If we multiply each minor by $(-1)^{i+j}$ where i is the row number and j the column number of the generating element, the result is called the *cofactor of the element*. In the above example the cofactor of 1 is

$$(-1)^{1+1} \cdot \begin{vmatrix} 5 & 8 \\ 6 & 9 \end{vmatrix} = \begin{vmatrix} 5 & 8 \\ 6 & 9 \end{vmatrix}.$$

The cofactor of 4 is

$$(-1)^{1+2} \cdot \begin{vmatrix} 2 & 8 \\ 3 & 9 \end{vmatrix} = -\begin{vmatrix} 2 & 8 \\ 3 & 9 \end{vmatrix}.$$

The cofactor of 7 is

$$(-1)^{1+3} \cdot \begin{vmatrix} 2 & 5 \\ 3 & 6 \end{vmatrix} = \begin{vmatrix} 2 & 5 \\ 3 & 6 \end{vmatrix}.$$

If we multiply each cofactor by its generating element and form the sum of all such quantities generated from any one row or column, the result is the value of the determinant.

Example 1. Evaluate the following determinant by expanding it about row 1.

$$A = \begin{vmatrix} 2 & 3 & 2 \\ 1 & -2 & 1 \\ 4 & -1 & 2 \end{vmatrix}$$

$$= 2\begin{vmatrix} -2 & 1 \\ -1 & 2 \end{vmatrix} - 3\begin{vmatrix} 1 & 1 \\ 4 & 2 \end{vmatrix} + 2\begin{vmatrix} 1 & -2 \\ 4 & -1 \end{vmatrix}$$

$$= 2(-4+1) - 3(2-4) + 2(-1+8)$$

$$= -6 + 6 + 14 = 14.$$

Example 2. Evaluate the following determinants by expanding about column 2.

$$A = \begin{vmatrix} 2 & 3 & 2 \\ 1 & -2 & 1 \\ 4 & -1 & 2 \end{vmatrix}$$

$$= -3 \begin{vmatrix} 1 & 1 \\ 4 & 2 \end{vmatrix} + (-2) \begin{vmatrix} 2 & 2 \\ 4 & 2 \end{vmatrix} - (-1) \begin{vmatrix} 2 & 2 \\ 1 & 1 \end{vmatrix}$$

$$= -3(2 - 4) - 2(4 - 8) + (2 - 2)$$

$$= 6 + 8 = 14.$$

Notice that Example 2 and Example 1 are the same, and that the value of the determinant is independent of the selection of the row or column used for expansion.

The method of evaluating determinants by minors and cofactors applies to determinants of any order. For example, a 4 x 4 determinant can be expanded about any row or column.

Example 3. Find the value of the determinant.
We choose to expand about column 4 since it contains two zero elements.

$$\begin{vmatrix} 1 & 0 & 7 & 0 \\ 2 & 1 & 2 & 1 \\ 1 & 2 & 0 & 2 \\ 1 & 1 & 3 & 0 \end{vmatrix} = 1 \begin{vmatrix} 1 & 0 & 7 \\ 1 & 2 & 0 \\ 1 & 1 & 3 \end{vmatrix} - 2 \begin{vmatrix} 1 & 0 & 7 \\ 2 & 1 & 2 \\ 1 & 1 & 3 \end{vmatrix} = -17.$$

If we had expanded about a row or column with four nonzero elements, we would have had four 3×3 determinants to expand.

From Example 3, we see that a 5th order determinant is expanded as five 4th order determinants which in turn are each expanded as four 3rd order determinants. Thus, to expand by minors, it is quite useful to choose a row or column containing as many zeros as possible. The next section lists some properties of determinants that can be used to make the expansion easier.

8.13 Properties of Determinants

The following is a partial listing of the properties of determinants. Each property is illustrated, as an example, for a second order determinant even though each applies to a determinant of any order.

I) *Any row or column may be added to or subtracted from any other row or column without changing the value of the determinant.*

Example:

$$\begin{vmatrix} 1 & 2 \\ 3 & 4 \end{vmatrix} = \begin{vmatrix} 4 & 6 \\ 3 & 4 \end{vmatrix} = -2.$$

Here we added row 2 to row 1.

II) *If any row or column is multiplied by a nonzero constant, the value of the determinant is multiplied by the same constant.*

Example:

$$\begin{vmatrix} 1 & 3 \\ 2 & 4 \end{vmatrix} = 4 - 6 = -2.$$

If we multiply column 2 by k and expand we have

$$\begin{vmatrix} 1 & 3k \\ 2 & 4k \end{vmatrix} = 4k - 6k = k(4 - 6) = -2 \cdot k.$$

III) *If any two rows or columns are interchanged, the resulting determinant is the negative of the original determinant.*

Example:

$$\begin{vmatrix} 1 & 3 \\ 2 & 4 \end{vmatrix} = 4 - 6 = -2.$$

If we interchange row 1 and row 2 and expand, we have

$$\begin{vmatrix} 2 & 4 \\ 1 & 3 \end{vmatrix} = 6 - 4 = 2.$$

IV) *If any row or column is zero, the value of the determinant is zero.*

Example:

$$\begin{vmatrix} 1 & 2 \\ 0 & 0 \end{vmatrix} = 0 - 0 = 0.$$

VI) *If any two rows or columns of a determinant are equal, the value of the determinant is zero.*

Example:

$$\begin{vmatrix} 2 & 2 \\ 2 & 2 \end{vmatrix} = 4 - 4 = 0.$$

VII) *If the product of a constant and any row or column is added to any other row or column, the value of the determinant remains unchanged.*

Example:

$$\begin{vmatrix} 1 & 3 \\ 2 & 4 \end{vmatrix} = 4 - 6 = -2.$$

If we add the product of 3 and row 2 to row 1, we have

$$\begin{vmatrix} 7 & 15 \\ 2 & 4 \end{vmatrix} = 28 - 30 = -2.$$

The reader is encouraged to verify each of the properties for a determinant of order three.

In Example 3 of the previous section we expanded the 4th order determinant about the 4th column, since it contained two zeros. We can and should use the properties of determinants as an aid in evaluation of determinants.

Example 1. Expand $\begin{vmatrix} 1 & 1 & 3 & 1 \\ 2 & 1 & 2 & 2 \\ 1 & 2 & 3 & 4 \\ 3 & 2 & -1 & 1 \end{vmatrix}$

If we subtract column 1 from columns 2 and 4 and three times column 1 from column 3, we have

$$\begin{vmatrix} 1 & 0 & 0 & 0 \\ 2 & -1 & -4 & 0 \\ 1 & 1 & 0 & 3 \\ 3 & -1 & -10 & -2 \end{vmatrix} = 1 \begin{vmatrix} -1 & -4 & 0 \\ 1 & 0 & 3 \\ -1 & -10 & -2 \end{vmatrix}$$

$$= 1[12 - 8 - 30] = 34.$$

In this example we expanded about row one since it contained three zeros.

Exercise 49

Expand each of the following determinants.

1) $\begin{vmatrix} 1 & 1 \\ 3 & 6 \end{vmatrix}$

2) $\begin{vmatrix} a & c \\ 3 & 2 \end{vmatrix}$

3) $\begin{vmatrix} 2 & -7 \\ 1 & 4 \end{vmatrix}$

4) $\begin{vmatrix} -1 & 0 \\ -3 & 6 \end{vmatrix}$

5) $\begin{vmatrix} a & c \\ b & d \end{vmatrix}$

6) $\begin{vmatrix} 2 & 7 \\ 1 & 3 \end{vmatrix}$

Expand each of the following without using minors.

7) $\begin{vmatrix} 1 & 4 & 1 \\ 2 & 1 & 2 \\ 3 & 2 & 2 \end{vmatrix}$

8) $\begin{vmatrix} 0 & 2 & 0 \\ 1 & 1 & 0 \\ 0 & 3 & 2 \end{vmatrix}$

9) $\begin{vmatrix} 5 & 3 & 2 \\ 1 & 0 & -3 \\ 4 & 2 & 3 \end{vmatrix}$

10) $\begin{vmatrix} 2 & 6 & 1 \\ 1 & 2 & 0 \\ 3 & 4 & 0 \end{vmatrix}$

11)–14) Expand each of the problems 7)–10) by use of minors and cofactors. Do not use the properties of determinants before expanding.

Use the properties of determinants to introduce zeros into each of the following. Then expand.

15) $\begin{vmatrix} 1 & 1 & 2 \\ 2 & 1 & 1 \\ 3 & 4 & 3 \end{vmatrix}$

16) $\begin{vmatrix} 1 & 3 & 2 \\ 1 & 4 & 1 \\ 5 & -1 & 6 \end{vmatrix}$

17) $\begin{vmatrix} 1 & 2 & 1 & 3 \\ 2 & 1 & 4 & 1 \\ 1 & 3 & 2 & 1 \\ 3 & 2 & 1 & 2 \end{vmatrix}$

18) $\begin{vmatrix} 1 & 2 & -1 & 1 \\ 3 & 1 & 2 & 4 \\ 1 & 3 & 3 & 2 \\ 2 & -4 & 4 & 1 \end{vmatrix}$

19) $\begin{vmatrix} 1 & 1 & 0 & 2 & 1 \\ 2 & 0 & 1 & 0 & 2 \\ 1 & 2 & 2 & 3 & 4 \\ 0 & 1 & 1 & 4 & 1 \\ 1 & 0 & 0 & 2 & 3 \end{vmatrix}$

20) $\begin{vmatrix} 5 & 2 & 5 & 3 & 3 \\ 1 & -1 & 1 & -6 & 2 \\ 4 & 3 & 3 & 2 & 1 \\ 2 & 2 & 2 & 1 & -4 \\ 6 & 4 & -1 & 5 & 2 \end{vmatrix}$

8.14 Systems of Equations and Cramer's* Rule

Determinants can be used to solve a system of equations in which there are n equations in n variables. To see how they are applied we first consider the solution to two general equations in two variables.

$$\begin{aligned} a_1x + b_1y &= c_1 \\ a_2x + b_2y &= c_2 \end{aligned}$$

To solve for y by the addition-subtraction method we multiply the first equation by a_2 and the second equation by a_1, and then subtract.

$$\begin{aligned} a_1a_2x + a_2b_1y &= a_2c_1 \\ a_1a_2x + a_1b_2y &= a_1c_2 \\ a_2b_1y - a_1b_2y &= a_2c_1 - a_1c_2 \end{aligned}$$

Factoring y from the left-hand member we have

$$y(a_2b_1 - a_1b_2) = a_2c_1 - a_1c_2.$$

Dividing by $a_2b_1 - a_1b_2$ yields

$$y = \frac{a_2c_1 - a_1c_2}{a_2b_1 - a_1b_2}$$

or

$$y = \frac{a_1c_2 - a_2c_1}{a_1b_2 - a_2b_1}.$$

* Eighteenth century mathematician credited with reviving interest in determinants.

If we now multiply the first equation by b_1 and the second equation by b_2 and then follow the pattern outlined for y we find

$$x = \frac{b_2c_1 - b_1c_2}{a_1b_2 - a_2b_1}.$$

These results can be expressed by determinants as

$$x = \frac{\begin{vmatrix} c_1 & b_1 \\ c_2 & b_2 \end{vmatrix}}{\begin{vmatrix} a_1 & b_1 \\ a_2 & b_2 \end{vmatrix}} \qquad y = \frac{\begin{vmatrix} a_1 & c_1 \\ a_2 & c_2 \end{vmatrix}}{\begin{vmatrix} a_1 & b_1 \\ a_2 & b_2 \end{vmatrix}}$$

Notice that the determinant in the denominator in each case is formed from the coefficients of the variables taken in order. The determinant in the numerator is obtained by replacing the coefficients of the variable to be evaluated by the constants and then forming the determinant of the coefficients. This process is known as *Cramer's Rule*, and it applies to determinants of any order.

Example 1. Solve the system

$$2x + 3y = 6$$
$$x + 3y = 4$$

$$x = \frac{\begin{vmatrix} 6 & 3 \\ 4 & 3 \end{vmatrix}}{\begin{vmatrix} 2 & 3 \\ 1 & 3 \end{vmatrix}} = \frac{6}{3} = 2, \quad y = \frac{\begin{vmatrix} 2 & 6 \\ 1 & 4 \end{vmatrix}}{\begin{vmatrix} 2 & 3 \\ 1 & 3 \end{vmatrix}} = \frac{2}{3}.$$

Example 2. Solve the system

$$x + y + 2z = 4$$
$$2x - y + 3z = 9$$
$$x + 4y = 0$$

$$x = \frac{\begin{vmatrix} 4 & 1 & 2 \\ 9 & -1 & 3 \\ 0 & 4 & 0 \end{vmatrix}}{\begin{vmatrix} 1 & 1 & 2 \\ 2 & -1 & 3 \\ 1 & 4 & 0 \end{vmatrix}} \quad y = \frac{\begin{vmatrix} 1 & 4 & 2 \\ 2 & 9 & 3 \\ 1 & 0 & 0 \end{vmatrix}}{D} \quad z = \frac{\begin{vmatrix} 1 & 1 & 4 \\ 2 & -1 & 9 \\ 1 & 4 & 0 \end{vmatrix}}{D}$$

Here D represents the determinant of the coefficients of the variables which is the same as found for x. Expanding we find

$$x = \frac{24}{9} = \frac{8}{3}, \; y = \frac{-6}{9} = \frac{-2}{3}, \; z = \frac{9}{9} = 1.$$

Notice that when a variable is missing from an equation, as z from the third equation, its coefficient is assumed to be zero.

Exercise 50

Solve each of the following systems by use of Cramer's Rule.

1) $\begin{aligned} 3x + y &= 20 \\ 2x - y &= 10 \end{aligned}$

2) $\begin{aligned} x + y &= 6 \\ 4x - 2y &= 3 \end{aligned}$

3) $\begin{aligned} 7x - 11y &= 0 \\ 19x - 3y &= 4 \end{aligned}$

4) $\begin{aligned} x + 4y &= 18 \\ 3x - 4y &= 22 \end{aligned}$

5) $\begin{aligned} 12x - 3y &= 2 \\ 6x - y &= 5 \end{aligned}$

6) $\begin{aligned} x &= 4y + 2 \\ y &= x - 5 \end{aligned}$

7) $\begin{aligned} x + y + z &= 6 \\ 2x + 3y + z &= 9 \\ 2x - y + z &= 3 \end{aligned}$

8) $\begin{aligned} x + y \phantom{{}-z} &= 3 \\ x + y - z &= -4 \\ y + z &= -6 \end{aligned}$

9) $\begin{aligned} a + b &= 1 \\ a + c &= 4 \\ b + c &= 2 \end{aligned}$

10) $\begin{aligned} 3M + N + R &= 7 \\ 2N - R &= 5 \\ M + 3N + 2R &= 11 \end{aligned}$

11) $\begin{aligned} x + y + z &= 3 \\ x + y + z + w &= 5 \\ 2x - y + 3w &= 4 \\ 3x - 2y + 3z - w &= 0 \end{aligned}$

12) $\begin{aligned} s + t + u - v &= -1 \\ u - v &= -1 \\ 2s - 2t - 2v + u &= 3 \\ 4s - t - 3v + u &= 2 \end{aligned}$

13) $\begin{aligned} 2x + 2y - 2z - 2w &= 0 \\ x + y - 2z + 2w &= 4 \\ x + y &= 4 \\ y - z + w &= 2 \end{aligned}$

14) $\begin{aligned} 2x - y + z + w &= 8 \\ x - y + z + w &= 3 \\ y - 5z + w &= 8 \\ x + z - w &= 0 \end{aligned}$

8.15 The Inverse of a Matrix

Recall from an earlier section that if two matrices A and B are conformable then $A \cdot B$ can be found. If A is the identity matrix, then $A \cdot B = B$.

Example 1. $\begin{pmatrix} 1 & 0 \\ 0 & 1 \end{pmatrix}\begin{pmatrix} x & z \\ y & w \end{pmatrix} = \begin{pmatrix} x + 0 & z + 0 \\ 0 + y & 0 + w \end{pmatrix} = \begin{pmatrix} x & z \\ y & w \end{pmatrix}.$

Example 2. $\begin{pmatrix} 1 & 0 & 0 \\ 0 & 1 & 0 \\ 0 & 0 & 1 \end{pmatrix}\begin{pmatrix} x \\ y \\ z \end{pmatrix} = \begin{pmatrix} x + 0 + 0 \\ 0 + y + 0 \\ 0 + 0 + z \end{pmatrix} = \begin{pmatrix} x \\ y \\ z \end{pmatrix}.$

This fact provides us with an interesting and useful method of solving simultaneous linear equations. However, we first need to develop the concept of the inverse of a matrix.

The inverse of a matrix A is matrix A^{-1}, if it exists, such that $A \cdot A^{-1} = I$, where I represents the identity matrix.

Example 3. Let $A = \begin{pmatrix} 1 & 3 \\ 2 & 4 \end{pmatrix}$.

$$A^{-1} = \frac{1}{-2}\begin{pmatrix} 4 & -3 \\ -2 & 1 \end{pmatrix}.$$

Now

$$A^{-1} \cdot A = \frac{1}{-2}\begin{pmatrix} 4 & -3 \\ -2 & 1 \end{pmatrix}\begin{pmatrix} 1 & 3 \\ 2 & 4 \end{pmatrix}$$

$$= \frac{1}{-2}\begin{pmatrix} -2 & 0 \\ 0 & -2 \end{pmatrix} = \begin{pmatrix} 1 & 0 \\ 0 & 1 \end{pmatrix}.$$

We now examine the process by which the inverse of a 2×2 matrix can be found as well as the condition that is necessary for its existence.

Let
$$A = \begin{pmatrix} a_1 & a_2 \\ b_1 & b_2 \end{pmatrix} \quad \text{and} \quad A^{-1} = \begin{pmatrix} x & y \\ z & w \end{pmatrix}.$$

Then $A \cdot A^{-1} = I$, or

$$\begin{pmatrix} a_1 & a_2 \\ b_1 & b_2 \end{pmatrix}\begin{pmatrix} x & y \\ z & w \end{pmatrix} = \begin{pmatrix} 1 & 0 \\ 0 & 1 \end{pmatrix}.$$

This yields

$$\begin{pmatrix} a_1x + a_2z & a_1y + a_2w \\ b_1x + b_2z & b_1y + b_2z \end{pmatrix} = \begin{pmatrix} 1 & 0 \\ 0 & 1 \end{pmatrix},$$

from which we see

$$\begin{aligned} a_1x + a_2z &= 1 \quad & a_1y + a_2w &= 0 \\ b_1x + b_2z &= 0 \quad & b_1y + b_2w &= 1 \end{aligned}$$

Solving these pairs of equations for x, y, z and w by Cramer's rule, we find

$$x = \frac{b_2}{a_1b_2 - a_2b_1} \qquad y = \frac{-a_2}{a_1b_2 - a_2b_1}$$

$$z = \frac{-b_1}{a_1b_2 - a_2b_1} \qquad w = \frac{a_1}{a_1b_2 - a_2b_1}$$

The denominator of each of these fractions is the determinant of A. Thus we see that

$$A^{-1} = \frac{1}{D}\begin{pmatrix} b_2 & -a_2 \\ -b_1 & a_1 \end{pmatrix}.$$

Thus the inverse of a 2×2 matrix is obtained by:

1) interchanging the elements on the principal diagonal;
2) multiplying the other two elements by -1;
3) multiplying this matrix by the reciprocal of the determinant of A.

From the last equation we see that the inverse of a matrix exists, provided that $D \neq 0$. A matrix whose determinant is zero is said to be *singular*. Thus, only the inverse of a *nonsingular* matrix can be found.

Example 4. Find the inverse of

$$A = \begin{pmatrix} 1 & -5 \\ 3 & 7 \end{pmatrix}.$$

$$D = 7 + 15 = 22.$$

Interchanging the elements on the principal diagonal and negating the other elements we have

$$\begin{pmatrix} 7 & 5 \\ -3 & 1 \end{pmatrix}.$$

Thus
$$A^{-1} = \frac{1}{22}\begin{pmatrix} 7 & 5 \\ -3 & 1 \end{pmatrix}.$$

Example 5. Find the inverse of

$$A = \begin{pmatrix} 5 & 10 \\ 2 & 4 \end{pmatrix}.$$

Since $D = 0$ the matrix is singular and no inverse exists.

The inverse of a $n \times n$ matrix is somewhat more difficult to find than that of a 2×2 matrix. For this reason, we state, without proof, one method that can be used to find it.

$$\text{If } A = \begin{pmatrix} a_{11} & a_{12} & a_{13} & \cdots & a_{1n} \\ a_{21} & a_{22} & a_{23} & \cdots & a_{2n} \\ \cdot & & & & \\ \cdot & & & & \\ \cdot & & & & \\ a_{n1} & a_{n2} & a_{n3} & \cdots & a_{nn} \end{pmatrix},$$

then
$$A^{-1} = \frac{1}{D}\begin{pmatrix} A_{11} & A_{21} & A_{31} & \cdots & A_{n1} \\ A_{12} & A_{22} & A_{32} & \cdots & A_{n2} \\ \cdot & & & & \\ \cdot & & & & \\ \cdot & & & & \\ A_{1n} & A_{2n} & A_{3n} & \cdots & A_{nn} \end{pmatrix},$$

where A_{11} is the cofactor of a_{11}, A_{12} the cofactor of a_{12}, and so on.

Notice that the rows and columns have been interchanged. The matrix created in this fashion is called the *transpose* of the original matrix.

Example 6. Find the inverse of

$$A = \begin{pmatrix} 1 & 0 & 1 \\ 2 & 2 & 2 \\ 3 & 1 & -1 \end{pmatrix}.$$

We first find $D = -8 \neq 0$. Now we obtain the matrix

$$\begin{pmatrix} -4 & 8 & -4 \\ 1 & -4 & -1 \\ -2 & 0 & 2 \end{pmatrix}$$

whose transpose is

$$\begin{pmatrix} -4 & 1 & -2 \\ 8 & -4 & 0 \\ -4 & -1 & 2 \end{pmatrix}.$$

Thus

$$A^{-1} = -\frac{1}{8}\begin{pmatrix} -4 & 1 & -2 \\ 8 & -4 & 0 \\ -4 & -1 & 2 \end{pmatrix}.$$

Notice that we have found inverses only for square matrices.

8.16 Solving Equations by Use of Matrices

The product of a matrix and an inverse provides us with a method of solving simultaneous equations, which is computer-oriented. Computer-oriented means that the drudgery of the arithmetic calculations can be programmed for computer execution. Other methods than the one which follows are frequently used to solve simultaneous equations by use of matrices with computers, but they are beyond the scope of this book.

Example 1. Solve the system of equations using matrices.

$$3x + 6y = 9,$$
$$5x - 2y = 4.$$

These equations can be written as

$$\begin{pmatrix} 3 & 6 \\ 5 & -2 \end{pmatrix}\begin{pmatrix} x \\ y \end{pmatrix} = \begin{pmatrix} 9 \\ 4 \end{pmatrix}.$$

Multiplying both sides of the equation by the inverse of $\begin{pmatrix} 3 & 6 \\ 5 & -2 \end{pmatrix}$ yields

$$-\frac{1}{36}\begin{pmatrix} -2 & -6 \\ -5 & 3 \end{pmatrix}\begin{pmatrix} 3 & 6 \\ 5 & -2 \end{pmatrix}\begin{pmatrix} x \\ y \end{pmatrix} = \frac{-1}{36}\begin{pmatrix} -2 & -6 \\ -5 & 3 \end{pmatrix}\begin{pmatrix} 9 \\ 4 \end{pmatrix}$$

$$\begin{pmatrix} 1 & 0 \\ 0 & 1 \end{pmatrix}\begin{pmatrix} x \\ y \end{pmatrix} = \begin{pmatrix} \frac{7}{6} \\ 33/36 \end{pmatrix}$$

The latter product yields $\begin{pmatrix} x \\ y \end{pmatrix} = \begin{pmatrix} \frac{7}{6} \\ \frac{33}{36} \end{pmatrix}$.

Thus we see $x = \frac{7}{6},\ y = \frac{33}{36} = \frac{11}{12}.$

Example 2. Solve the system of equations using matrices.

$$\begin{aligned} x + z &= 4, \\ 2x + 2y + 2z &= 9, \\ 3x + y - z &= 1. \end{aligned}$$

These equations can be written as

$$\begin{pmatrix} 1 & 0 & 1 \\ 2 & 2 & 2 \\ 3 & 1 & -1 \end{pmatrix}\begin{pmatrix} x \\ y \\ z \end{pmatrix} = \begin{pmatrix} 4 \\ 9 \\ 1 \end{pmatrix}.$$

Multiplying both sides of the equation by the inverse of the 3×3 matrix yields

$$-\frac{1}{8}\begin{pmatrix} -4 & 1 & -2 \\ 8 & -4 & 0 \\ -4 & -1 & 2 \end{pmatrix}\begin{pmatrix} 1 & 0 & 1 \\ 2 & 2 & 2 \\ 3 & 1 & -1 \end{pmatrix}\begin{pmatrix} x \\ y \\ z \end{pmatrix}$$

$$= -\frac{1}{8}\begin{pmatrix} -4 & 1 & -2 \\ 8 & -4 & 0 \\ -4 & -1 & 2 \end{pmatrix}\begin{pmatrix} 4 \\ 9 \\ 1 \end{pmatrix}$$

$$\begin{pmatrix} 1 & 0 & 0 \\ 0 & 1 & 0 \\ 0 & 0 & 1 \end{pmatrix}\begin{pmatrix} x \\ y \\ z \end{pmatrix} = \begin{pmatrix} \frac{9}{8} \\ \frac{1}{2} \\ \frac{23}{8} \end{pmatrix}$$

so that $x = \frac{9}{8},\ y = \frac{1}{2},\ z = \frac{23}{8}.$

Observe that if the determinant formed by the coefficient of the variables is zero, no solution exists.

Exercise 51

Solve each of the following systems by use of matrices. (Caution: Watch order of variables).

1) $3x + y = 20$
$2x - y = 10$

2) $x + 4y = 18$
$3x - 4y = 22$

3) $x + 4y = 9$
$x + 3y = 7$

4) $3x + y = 12$
$8x - y = 10$

5) $x + y = 6$
$y - x = 1$

6) $x + 3y = 11$
$2x - y = 7$

7) $3x + 2y = 16$
$x + \frac{1}{2}y = 2$

8) $5x - 7y = 9$
$2x - 3y = 6$

9) $7x + 5y = 1$
$3x - y = 2$

10) $2x - 3y = -4$
$x - y = 7$

11) $x + y + z = 6$
$2x + 3y + z = 9$
$2x - y + z = 3$

12) $3M + N + R = 7$
$2N - R = 5$
$M + 3N + 2R = 11$

13) $a + b + c = 3$
$a + b - c = 1$
$a - b + c = 1$

14) $L + R = 6$
$L - S = 3$
$R + S = 5$

15) $2x + 2y - 2z - 2w = 0$
$x + y - 2z + 2w = 4$
$x + y = 4$
$y - z + w = 2$

16) $2x - y + z + w = 8$
$x - y + z + w = 3$
$y - 5z + w = 8$
$x + z - w = 0$

Review Exercise Chapter 8

Solve the following systems of equations by graphing.

1) $3x - y = 6$
$x + y = 2$

2) $2x - 3y = 5$
$4x = 10 + 6y$

Solve the following systems of equations by addition or subtraction.

3) $2x - 3y = 6$
$x - y = 4$

4) $\frac{1}{3}x + \frac{1}{4}y = 1$
$x - y = 2$

5) $5x - y = -3$
$x + 3y = 4$

Solve the following systems of equations by substitution.

6) $y = 5x$
$x - y = 3$

7) $3y = 6x + 2$
$2x - y = -1$

8) $4x - 3y = 16$
$x = y + 2$

Solve the following systems of equations by a method of your choosing.

9) $x = 3y + 5$
$y = 2x/3$

10) $6x = y + 7$
$y = -1$

11) $x + y = 1$
$2x - z = 0$
$y + 2z = -2$

12) $x + 4y - z = 3$
$-2x - 8y + 2z = -6$
$3x + 12y - 3z = 9$

Evaluate the following determinants.

13) $\begin{vmatrix} 1 & 3 & 5 \\ 2 & 0 & -3 \\ 1 & -1 & -2 \end{vmatrix}$

14) $\begin{vmatrix} -3 & 0 & 0 \\ -1 & 1 & 3 \\ 2 & 2 & 3 \end{vmatrix}$

Solve the following systems of equations by Cramer's rule.

15) $3x - 4y = 6$
$x - 2y = -1$

16) $2x - 3y + 2z = -10$
$5x + 2y + 3z = -2$
$4x + 6y + 5z = 2$

Multiply the following matrices.

17) $\begin{pmatrix} 3 & 2 \\ 1 & 6 \end{pmatrix} \cdot \begin{pmatrix} 2 \\ 1 \end{pmatrix}$

18) $\begin{pmatrix} -3 & -1 & 0 \end{pmatrix} \begin{pmatrix} -3 \\ -1 \\ 0 \end{pmatrix}$

Find the inverse of each of the following matrices.

19) $\begin{pmatrix} -1 & 2 \\ 1 & -2 \end{pmatrix}$

20) $\begin{pmatrix} 0 & -1 & 3 \\ -2 & 1 & 1 \\ 0 & 2 & 1 \end{pmatrix}$

9 Higher Order Equations

9.1 Second Degree Equation

In the earlier chapters of this book you learned how to solve first-degree equations of various types. Numerous applications for the skills acquired are found in the form of word problems or in the form of applications to formulas from industry. However, many formula and word problems result in equations that are of a degree greater than one. Problems involving area, volume, chemical analysis, and acceleration are but a few of these. In the following sections we will study equations that are of degree two. These equations are called *quadratic equations*.

9.2 Standard Form of a Quadratic Equation

The standard form of a quadratic equation is

$$ax^2 + bx + c = 0,$$

where a, b, c are any real numbers with the restriction that $a \neq 0$. If $a = 0$, the equation reduces to $bx + c = 0$ which is not of degree two and therefore is not quadratic.

As you recall, first-degree equations were solved by applying the axioms of equality which are stated in Section 5.2. These axioms are equally applicable to quadratic equations, and they will be used to transform the equations into equivalent forms where desirable. For example, we may want to transform $3x^2 = 6x + 9$ into an equivalent equation that has zero as the right-hand member. To do this, we subtract $6x + 9$ from each side of the equation.

$$3x^2 = 6x + 9$$
$$3x^2 - 6x - 9 = 6x + 9 - 6x - 9$$
$$3x^2 - 6x - 9 = 0$$

9.3 Solutions to Quadratic Equations

Several methods for solving quadratic equations exist, and each one has its distinct advantages and disadvantages. Some of the more common ones are factoring, extraction of roots, and completion of the square. Each one of these will be considered separately even though more than one technique may be used on any one problem.

9.4 Solution by Factoring

Solutions of quadratic equations by factoring is based on the rather simple observation that if the product of two or more factors is zero, then at least one of the factors must have been zero. To state this symbolically, we have

if $A \cdot B = 0$, then
$A = 0$ or $B = 0$ or both $A = 0$ and
$B = 0$.

Applying this observation to equation solving is known as *solution by factoring.*

Example 1. Solve $x^2 + 3x + 2 = 0$.
Factoring, we have $(x + 2)(x + 1) = 0$. Now either $x + 2 = 0$ or $x + 1 = 0$. Therefore, $x = -2$ or $x = -1$. We check these solutions by substituting these results back into the original equation to see if it is satisfied by them.

Check:
$$(-2)^2 + 3(-2) + 2 = 0$$
$$4 - 6 + 2 = 0$$
$$0 = 0$$

Check:
$$(-1)^2 + 3(-1) + 2 = 0$$
$$1 - 3 + 2 = 0$$
$$0 = 0$$

Example 2. Solve $x^2 - 7x - 8 = 0$.
Factoring, $(x - 8)(x + 1) = 0$. Now either $x - 8 = 0$ or $x + 1 = 0$. Therefore, $x = 8$ or $x = -1$.

Check:

$$(8)^2 - 7(8) - 8 = 0$$
$$64 - 56 - 8 = 0$$
$$0 = 0$$

Check:

$$(-1)^2 - 7(-1) - 8 = 0$$
$$1 + 7 - 8 = 0$$
$$0 = 0$$

Example 3. Solve $x^2 + 2x = 0$.
Factoring,

$$x(x + 2) = 0.$$

Now

$$x = 0 \quad \text{or} \quad x + 2 = 0.$$
$$x = 0 \quad \text{or} \quad x = -2.$$

The temptation here is to divide through by x losing the solution $x = 0$.

Example 4. Solve $x^2 + 5x = 4$.
There is always a temptation to handle this equation by factoring the left-hand member as $x(x + 5) = 4$ and then set $x = 4$ or $x + 5 = 4$. Substituting these results into $x(x + 5) = 4$, we have $4 \cdot 4 = 4$. This obviously is not correct. Instead, the equation must be written as

$$x^2 + 5x - 4 = 0$$

which does not factor. Hence, the solution to this equation cannot be found by this method.

Example 5. Solve $3x^2 - 4x = 4$.
First we write the equation in standard form as

$$3x^2 - 4x - 4 = 0.$$

Factoring,

$$(3x + 2)(x - 2) = 0.$$

Now either $3x + 2 = 0$ or $x - 2 = 0$, $3x = -2$ or $x = 2$,

which yields $\quad x = -\dfrac{2}{3} \quad$ or $\quad x = 2.$

Check:

$$3\left(-\frac{2}{3}\right)^2 - 4\left(-\frac{2}{3}\right) - 4 = 0$$
$$3\left(\frac{4}{9}\right) + \frac{8}{3} - \frac{12}{3} = 0$$
$$\frac{4}{3} + \frac{8}{3} - \frac{12}{3} = 0$$
$$0 = 0$$

Check:

$$3(2)^2 - 4(2) - 4 = 0$$
$$3(4) - 8 - 4 = 0$$
$$12 - 8 - 4 = 0$$
$$0 = 0$$

Exercise 52

Solve each of the following quadratic equations by the method of factoring. Check as directed by the instructor.

1) $x^2 - 3x - 4 = 0$

2) $x^2 + 3x + 2 = 0$

3) $x^2 + 5x - 14 = 0$

4) $x^2 + 11x + 10 = 0$

5) $x^2 + x - 12 = 0$

6) $x^2 - (-2x) - 15 = 0$

7) $x^2 - 9 = 0$

8) $x^2 + 1x - 42 = 0$

9) $x^2 - 5x = 0$

10) $3r^2 - 21r = 0$

11) $n^2 + 4n + 4 = 0$

12) $n^2 = 6n$

13) $v^2 + 2v - 63 = 0$

14) $v^2 - v - 12 = 0$

15) $x^2 + 12x + 35 = 0$

16) $x^2 - 10x = -16$

17) $x^2 + 3x - 88 = 0$

18) $x^2 = -3x + 108$

19) $x^2 - 7x - 98 = 0$

20) $2x - 1 = x^2$

21) $3x^2 - 13x - 30 = 0$

22) $6x^2 + 25x - 9 = 0$

23) $6x^2 - 17x - 14 = 0$

24) $6x^2 - 7x - 90 = 0$

25) The distance an object falls under the influence of gravity in a vacuum is given by $s = 16t^2$, where s is in feet and t is in seconds. How long would it take the object to fall from a height of 256 ft.?

26) The length of a rectangle is 4 inches more than the width. If the area is 96 square inches, what are the dimensions?

27) Three times the square of a number decreased by twice the number is 40. What is the number?

28) A sheet metal worker has need to form an open gutter by bending up the sides of a sheet of tin 30 feet long and 12 inches wide. If the cross-sectional area must be 18 square inches, how much is bent up on each side?

9.5 Extraction of Roots

A quadratic equation of the form $x^2 = c$ can be solved by a method which is known as *extraction of roots*. This method employs the concept that if two quantities are equal, then their square roots would be equal. To apply this idea we extract the square root of each side of the equation. For example, consider the equation

$$x^2 = 4.$$

If we extract the square root of each side, we find the solutions to be

$$x = 2 \quad \text{or} \quad x = -2,$$

which is frequently written as

$$x = \pm 2.$$

To verify that both 2 and -2 are solutions we check the results by substitution.

Check:	Check:
$(-2)^2 = 4$	$(2)^2 = 4$
$4 = 4$	$4 = 4$

Example 1. Solve $y^2 = 9$ by extraction of roots.
Extracting the square root of each side we have

$$y = \pm 3.$$

This equation could also have been solved by factoring as follows.

$$y^2 = 9$$
$$y^2 - 9 = 0$$
$$(y - 3)(y + 3) = 0$$

$$y - 3 = 0 \qquad y + 3 = 0$$
$$y = 3 \qquad y = -3$$

Other forms of equations exist that can be solved by extraction of roots. To illustrate what is meant by this statement consider the equation

$$(y + 2)^2 = 4.$$

If we let $y + 2 = x$ and then substitute x for $y + 2$ in the original equation, we have

$$x^2 = 4,$$

which has the solution of

$$x = \pm 2.$$

If we now replace x by $y + 2$, we have

$$y + 2 = \pm 2,$$

which we consider as

$$y + 2 = 2 \quad \text{or} \quad y + 2 = -2.$$

Solving these, we find that

$$y = 0 \quad \text{or} \quad y = -4.$$

The reader should observe at this point that there is no real need for the substitution of x for $y + 2$. If we extract the square root of each side directly, we have

$$y + 2 = \pm 2,$$

which is precisely the same result we achieved by substitution.

Example 2. Solve $(y + 5)^2 = 16$.
Extracting roots, $y + 5 = \pm 4$

$$y + 5 = 4 \quad \text{or} \quad y + 5 = -4$$
$$y = -1 \qquad\qquad y = -9$$

Example 3. Solve $(n - 2)^2 = 7$.
Extracting roots, $n - 2 = \pm\sqrt{7}$.
This yields $n - 2 = +\sqrt{7}$ or $n - 2 = -\sqrt{7}$. We now have $n = 2 + \sqrt{7}$ or $n = 2 - \sqrt{7}$. Since $\sqrt{7}$ cannot be removed exactly from the radical, we can only indicate the sum or difference.

Decimal approximations can be found in the table on the end papers, should it be desirable to have an answer correct to the nearest tenth, hundredth or thousandth.

Example 4. Solve $2(x + 6)^2 = 72$.
Dividing each side by 2 we have

$$(x + 6)^2 = 36.$$

Extracting roots

$$x + 6 = \pm 6$$
$$x = -6 + 6 \quad \text{or} \quad x = -6 - 6$$
$$x = 0 \quad \text{or} \quad x = -12.$$

Exercise 53

Solve each of the following equations by extraction of roots.

1) $x^2 = 9$	2) $x^2 = 4$
3) $y^2 = 16$	4) $y^2 = 25$
5) $x^2 = 11$	6) $n^2 = 36$
7) $r^2 = 19$	8) $s^2 = 45$
9) $(x + 1)^2 = 16$	10) $(x - 3)^2 = 25$
11) $(x + 4)^2 = 36$	12) $(x - 11)^2 = 12$
13) $(x + 13)^2 = 18$	14) $(x + 2)^2 = 21$
15) $(y - 6)^2 = 25$	16) $(s + 2)^2 = 225$

17) $2x^2 = 18$

18) $(3y^2)/2 = 24$

19) $5(x + 3)^2 = 125$

20) $(3x + 4)^2 = 25$

21) $(3y - 2)^2 = 25$

22) $2(x + 7)^2 = 50$

23) $6(y + 4)^2 = 24$

24) $2(3y + 7)^2 = 72$

25) $7(x + 2)^2 = 77$

9.6 Solutions by Completing the Square

In the previous section you saw that an equation of the form $(x + a)^2 = c$ could be solved by extracting the square root of each side of the equation. In fact, any quadratic equation can be solved by extraction of roots, if we apply the additional operation of completing the square. Let us illustrate this by considering

$$x^2 - 6x + 8 = 0$$

which can be written

$$x^2 - 6x = -8.$$

If we add the square of one half the coefficient of the linear term (first degree term) to each side, we obtain

$$x^2 - 6x + 9 = -8 + 9.$$

In this equation the left-hand side is the square of $(x - 3)$. This means the equation can be written as

$$(x - 3)^2 = 1,$$

and the solution set is obtained by extraction of roots.

Example 1. Solve $x^2 + 4x + 1 = 0$ by completion of the square. We first write the equation in the form

$$x^2 + 4x = -1.$$

Adding the square of one half the coefficient of the linear term to each member, we have

$$x^2 + 4x + 4 = 4 - 1.$$

The left-hand member is equivalent to $(x + 2)^2$. This yields $(x + 2)^2 = 3$, from which we obtain solutions $x = -2 + \sqrt{3}$ and $x = -2 - \sqrt{3}$.

Example 2. Solve $2x^2 + 5x + 1 = 0$ by completion of the square. We first write the equation in the form

$$2x^2 + 5x^2 = -1.$$

Since it is easier to complete the square when the coefficient of the 2nd degree term is one, we divide both members by 2. This yields

$$x^2 + \frac{5}{2}x = -\frac{1}{2}.$$

Adding the square of one-half the coefficient of the linear term to each member, we have

$$x^2 + \frac{5}{2}x + \frac{25}{16} = -\frac{1}{2} + \frac{25}{16}.$$

This can be written as

$$\left(x + \frac{5}{4}\right)^2 = \frac{17}{16},$$

from which we obtain the solutions

$$x = \frac{-5}{4} \pm \frac{\sqrt{17}}{4} \text{ or } \frac{-5 + \sqrt{17}}{4}, \frac{-5 - \sqrt{17}}{4}.$$

9.7 Solutions by Formula

In each of the preceding examples we solved a quadratic equation by completing the square. This method of solving a quadratic equation is easily applied to the general quadratic equation. Such an application results in a formula that can be used for solving any quadratic equation. This formula is known as the *quadratic formula*. Consider the general quadratic equation

$$ax^2 + bx + c = 0.$$

If we follow the procedure for completing the square, we first transform the equation to the form

$$ax^2 + bx = -c.$$

Dividing both members by a we have

$$x^2 + \frac{b}{a}x = -\frac{c}{a}.$$

Adding the square of one-half of the coefficient of the linear term to both members yields

$$x^2 + \frac{b}{a}x + \frac{b^2}{4a^2} = \frac{b^2}{4a^2} - \frac{c}{a}$$

or

$$\left(x + \frac{b}{2a}\right)^2 = \frac{b^2 - 4ac}{4a^2},$$

from which we obtain

$$x + \frac{b}{2a} = \pm\frac{\sqrt{b^2 - 4ac}}{2a}.$$

Finally, solving for x yields

$$x = \frac{-b \pm \sqrt{b^2 - 4ac}}{2a}.$$

This quadratic formula is of such importance that it should be committed to memory along with the general quadratic equation to which it is applied.

Example 1. Solve $x^2 + 3x + 2 = 0$ by use of the quadratic formula. We first identify the a, b and c values to be $a = 1$, $b = 3$ and $c = 2$. Substituting these directly into the formula we have

$$x = \frac{-3 \pm \sqrt{(3)^2 - 4(1)(2)}}{2(1)}.$$

Simplifying,

$$x = \frac{-3 \pm \sqrt{9 - 8}}{2} \text{ or } \frac{-3 \pm 1}{2}.$$

Considering each combination yields

$$x = \frac{-3 + 1}{2}, \frac{-3 - 1}{2}$$

$$x = -1, -2.$$

Example 2. Solve $3x^2 + 6x - 10 = 0$ by use of the quadratic formula. Here $a = 3$, $b = 6$ and $c = -10$. Substituting into the formula we have

$$x = \frac{-6 \pm \sqrt{(6)^2 - 4(3)(-10)}}{6}$$

$$x = \frac{-6 \pm \sqrt{36 + 120}}{6}$$

$$x = \frac{-6 \pm 2\sqrt{39}}{6}$$

$$x = \frac{-3 + \sqrt{39}}{3}, \frac{-3 - \sqrt{39}}{3}.$$

To check these solutions as usual we substitute each of them into the original equation and verify that the equation is satisfied. Checking only the first of the two solutions, we have

$$3\left(\frac{-3+\sqrt{39}}{3}\right)^2 + 6\left(\frac{-3+\sqrt{39}}{3}\right) - 10 = 0$$

$$3\left(\frac{9-6\sqrt{39}+39}{9}\right) + \frac{-18+6\sqrt{39}}{3} - 10 = 0.$$

Simplifying and collecting over the lowest common denominator yields

$$\frac{48-6\sqrt{39}-18+6\sqrt{39}-30}{3} = 0$$

$$0 = 0.$$

Example 3. Solve for n by use of the quadratic formula if

$$n^2 + 2n + 4 = 0.$$

Here $a = 1$, $b = 2$, $c = 4$. Substituting we have

$$\begin{aligned} n &= \frac{-2 \pm \sqrt{(2)^2 - 4(1)(4)}}{2} \\ &= \frac{-2 \pm \sqrt{4-16}}{2} \\ &= \frac{-2 \pm \sqrt{-12}}{2} \\ &= \frac{-2+2i\sqrt{3}}{2}, \frac{-2-2i\sqrt{3}}{2}. \end{aligned}$$

Notice that the roots here are complex conjugates. This will always be the case when solutions to quadratic equations involve imaginary numbers.

Example 4. Solve for y by use of the quadratic formula if $2y^2 - 2y + 3 = 0$, where $a = 2$, $b = -2$, $c = 3$.

$$\begin{aligned} y &= \frac{-(-2) \pm \sqrt{(-2)^2 - 4(2)(3)}}{2 \cdot 2} \\ &= \frac{2 \pm \sqrt{-20}}{4} \\ &= \frac{2 \pm 2i\sqrt{5}}{4} \\ &= \frac{2(1 \pm i\sqrt{5})}{4} \\ &= \frac{1+i\sqrt{5}}{2}, \frac{1-i\sqrt{5}}{2}. \end{aligned}$$

Exercise 54

Solve each of the following by completing the square. Check your solutions.

1) $x^2 + 4x = -2$
2) $n^2 + 6n = 7$
3) $x^2 + 8x - 9 = 0$
4) $2x^2 + 4x - 3 = 0$

Solve each of the following by use of the quadratic formula.

5) $x^2 - 5x + 4 = 0$
6) $x^2 + 4x - 5 = 0$
7) $2x^2 - 7x = 3$
8) $x^2 + x - 6 = 0$
9) $s^2 - 4s + 5 = 0$
10) $6n^2 = 5n + 6$
11) $s^2 + s = 1$
12) $2n^2 - n + 1 = 0$
13) $2x^2 - 7x + 6 = 0$
14) $x^2 + 13x + 12 = 0$
15) $x^2 + 2x + 5 = 0$
16) $3x^2 + 12x + 9 = 0$
17) $6x = x^2 - 2$
18) $8x^2 = 1 + x$
19) $x^2 - 3 = 4x^2 + 2x$
20) $6x^2 + 4x = 0$
21) $x^2 + 12x + 9 = 0$
22) $3x^2 = 7x - 2$
23) $\frac{x^2}{4} + 2x - 2 = 0$
24) $(x^2 - 3x)/2 = (x^2 + 1)/3$
25) $2x(x + 4) = 7$
26) $(5x + 1)(x + 2) = 4$
27) Two positive numbers differ by 7 and their product is 70. Find the numbers.
28) If each side of a square piece of sheet metal is trimmed by 3 inches the area is decreased by 63 square inches. Find the original dimensions.
29) A formula from electrical circuitry relating the current i (amps), the voltage E (volts) and resistance R (ohms) states for a specified circuit that $i^2R + iE = 4000$. If $R = 20$ ohms and $E = 50$ volts, what current flows in the circuit?

9.8 The Discriminant of the Quadratic Formula

In Example 4 of the previous section we found the roots of the equation $2y^2 - 2y + 3 = 0$ to be complex conjugates. Actually this discovery could have been made without actually solving the equation by using a segment of the quadratic formula which is known as the *discriminant*. This segment is that portion of

$$x = \frac{-b \pm \sqrt{b^2 - 4ac}}{2a}$$

which appears under the radical, namely $b^2 - 4ac$.

In fact, the discriminant tells us far more about the nature of the roots of a quadratic equation than whether or not the roots are complex conjugates. This information is illustrated by the following analysis.

I)	$b^2 - 4ac = 0$	roots are real and equal.
II)	$b^2 - 4ac > 0$	roots are real and unequal.
III)	$b^2 - 4ac < 0$	roots are complex conjugates.

We can carry the analysis to even greater extent by considering whether the discriminant is a perfect square. If it is, and if it is also greater than zero, we can also say that the roots will be rational. If it is not a perfect square, the roots will be irrational.

Example 1. Without solving the equation, determine the nature of the roots.

$$3x^2 + 6x + 5 = 0.$$
$$b^2 - 4ac = 6^2 - 4 \cdot 3 \cdot 5 = 36 - 60 = -24 < 0.$$

Therefore, the roots are complex conjugates.

Example 2. Find the discriminant and determine the nature of the roots of the equation $4x^2 - 2x - 1 = 0$.

$$b^2 - 4ac = (-2)^2 - 4 \cdot 4(-1) = 4 + 16 = 20 > 0.$$

Therefore, the roots are real and unequal. Further, since 20 is not a perfect square, the roots are irrational.

Example 3. Use the discriminant to determine the nature of the roots of the equation $4x^2 + 12x + 9 = 0$.

$$b^2 - 4ac = 144 - 144 = 0.$$

Thus the roots are real and equal.

On occasion it is desirable to find a value or range of values for either a, b, or c in the quadratic equation

$$ax^2 + bx + c = 0$$

so that the roots will be of a particular nature. This can be accomplished in certain cases by use of the discriminant. The following examples illustrate the procedure.

Example 4. Determine the value of b so the roots of $3x^2 + bx + 1$ are real and unequal.

If the roots are to be real and unequal, then

$$b^2 - 4ac > 0.$$

Thus

$$b^2 - 4 \cdot 3(1) > 0$$
$$b^2 - 12 > 0.$$

From this we see that either $b > \sqrt{12}$ or $b < -\sqrt{12}$ may satisfy the conditions.

Example 5. Determine the value of c so that the roots of $2x^2 - 6x + c = 0$ are complex conjugates.

If the roots are complex conjugates, $b^2 - 4ac < 0$. Thus we have

$$36 - 8c < 0,$$

or

$$c > \frac{36}{8}$$
$$c > \frac{9}{2}.$$

9.9 Roots and Coefficients: Their Relationship

You have already seen how the discriminant can be used to determine the nature of the roots of a quadratic equation and how it is directly related to the coefficients of the general quadratic equation. In this section we will discover a further relationship between the coefficients of the general quadratic.

Consider the roots of the general quadratic equation

$$x_1 = \frac{-b + \sqrt{b^2 - 4ac}}{2a} \quad \text{and} \quad x_2 = \frac{-b - \sqrt{b^2 - 4ac}}{2a}.$$

If we form the sum of these two solutions $x_1 + x_2$, we find that the sum of the roots of a quadratic equation is $-b/a$. If we multiply x_1 by x_2, we find the product of the roots is c/a. These observations lead us to the following theorem.

Theorem: *If* r_1 *and* r_2 *are the roots of the general quadratic equation, then*

$$r_1 + r_2 = \frac{-b}{a}$$
$$r_1 \cdot r_2 = \frac{c}{a}.$$

Example 1. Find the sum and product of the roots of $x^2 + x - 6 = 0$ without solving, where $a = 1$, $b = 1$, $c = -6$.

$$r_1 + r_2 = \frac{-1}{1} \qquad r_1 \cdot r_2 = \frac{-6}{1}.$$

Thus the sum is -1, and the product is -6.

Example 2. Verify that the roots of $2x^2 - x + 4 = 0$ are $\frac{1 \pm i\sqrt{31}}{4}$, where $a = 2$, $b = -1$, $c = 4$.

The sum is $1/2$. The product is 2. These results are obtained when

$$\frac{1 + i\sqrt{31}}{4} \quad \text{and} \quad \frac{1 - i\sqrt{31}}{4}$$

are added and multiplied.

Exercise 55

Without solving, determine the nature of the roots of each of the following equations. Classify as a) real and equal, b) real and unequal, or c) complex conjugates.

1) $x^2 - 3x + 2 = 0$
2) $2x^2 + 12x + 9 = 0$
3) $2x = x^2 + 2$
4) $4x^2 - 3x = 0$
5) $x^2 + x - 5 = 0$
6) $y^2 - 4y + 5 = 0$
7) $2x^2 + 8x - 3 = 0$
8) $6x^2 = 5x + 6$
9) $3x(x - 2) = 7$
10) $(x + 1)3x = 7(x - 1)$
11) $\frac{x^2 - 3x}{2} = \frac{x^2 + 1}{3}$
12) $(4x + 2)(x - 3) = -2$
13) $3x^2 = 7x^2 - 2x$

Find the values of a, b or c, so that the indicated conditions are met.

14) $ax^2 + 4x - 2 = 0$ Roots are complex conjugates.
15) $9x^2 + 24x + c + 3 = 0$ Roots are real and equal.
16) $x^2 + bx + 5$ Roots are real and unequal.
17) Find k if the roots of $kx^2 + 3x + 2k = 0$ are to be real and unequal.

Find the sum and product of the roots of each of the following equations.

18) $2x^2 + 3x - 4 = 0$
19) $4x^2 - x + 6 = 0$
20) $3x^2 = 2x - 9$

9.10 The Graphs of Quadratic Equations

Just as with the linear equation in two variables, the quadratic equation in two variables possesses a solution set whose members are ordered pairs. To see this is the case, we consider the quadratic equation

$$y = x^2 - 5.$$

In order to find ordered pairs satisfying the equation we arbitrarily assign values to x and compute the related values for y. For instance, if we let x assume the value 2 in the equation above we have

$$\begin{aligned} y &= (2)^2 - 5 \\ y &= 4 - 5 \\ y &= -1 \end{aligned}$$

so that $(2,-1)$ is a solution. We could also let x assume the value -4. We then have

$$\begin{aligned} y &= (-4)^2 - 5 \\ y &= 16 - 5 \\ y &= 11 \end{aligned}$$

so that $(-4,11)$ is a solution. In a similar manner we find that $(1,-4)$, $(-1,-4)$, $(-2,-1)$, $(-3,4)$, $(3,4)$, $(0,-5)$ are also solutions to the equation. In fact, the solution set contains an infinite number of ordered pairs, each obtained by assigning an arbitrary real number value to x and computing y. If we now plot the corresponding points on the rectangular coordinate system, we have the graph shown in Figure 9.1.

It is easy to see that these points do not lie on a straight line, but the question is, do they form any meaningful pattern. If we join these points with a smooth curve, moving from left to right a definite pattern seems to be present. The graph is shown in Figure 9.2.

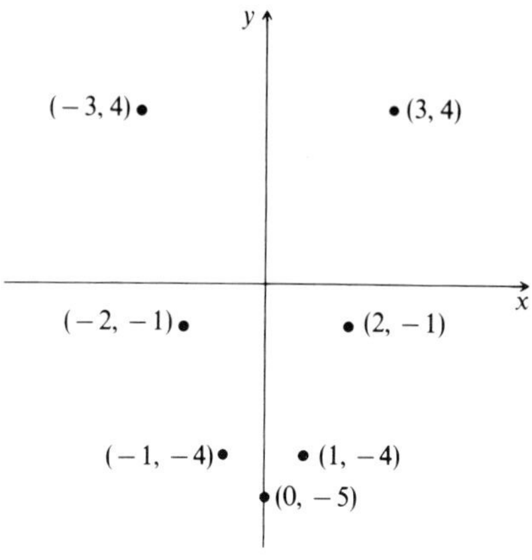

Figure 9.1

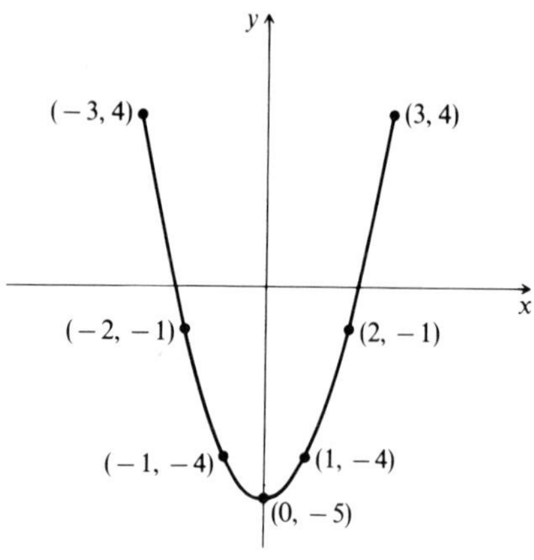

Figure 9.2

To check the reasonableness of this interpretation, we will plot one additional point between each of those shown on Figure 9.2. To do this we will let x assume the values $-5/2$, $-3/2$, $-1/2$, $1/2$, $3/2$, $5/2$ obtaining the ordered pairs

$$\left(-\frac{5}{2},\frac{5}{4}\right), \left(-\frac{3}{2},-\frac{11}{4}\right), \left(-\frac{1}{2},-\frac{19}{4}\right), \left(\frac{1}{2},-\frac{19}{4}\right), \left(\frac{3}{2},-\frac{11}{4}\right), \left(\frac{5}{2},\frac{5}{4}\right).$$

The result of plotting the additional points is shown in Figure 9.3.

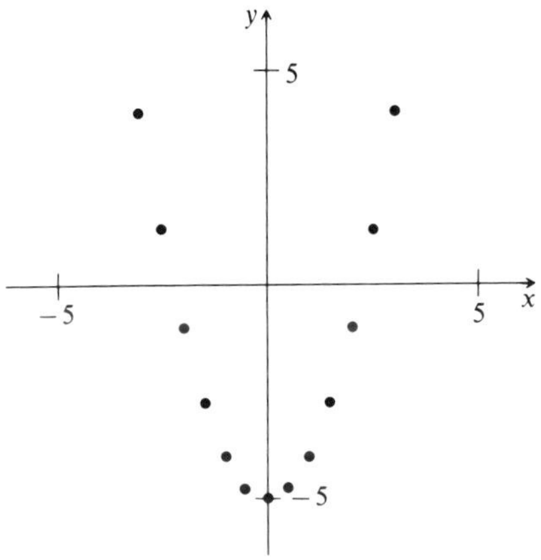

Figure 9.3

The location of these new points supports the contention that the points all lie on a smooth curve obtained by joining the points from left to right. Curves of this general type are called *parabolas*, and such curves have seemingly endless applications in industry. Radar antennas are parabolic; reflectors on most flashlights and automobile headlights are parabolic; and many of the largest telescopes employ parabolic mirrors, to name just a few. Additional treatment of this extremely important curve will be carried out in the following discussion of conic sections.

9.11 The Conic Sections

Conic Sections is a name that is given generally to classes of quadratic equations whose graphs are similar to the surfaces obtained when a right circular cone is cut by a plane. The most common of these sections are known as the *circle*, the *ellipse*, the *parabola*, and the *hyperbola*.

To define each of the conics in terms of a cone and a generating plane we first need to consider some terminology associated with the cone. The terms *base*, *axis of symmetry*, *nappe*, and *lateral element* are defined by illustration in Figure 9.4. The portion of the cone above T is called the *upper nappe* while that portion below is the *lower nappe*. The *axis of symmetry* or *axis of rotation* is the straight line ATB, where A and B are the centers of the circular bases of each nappe.

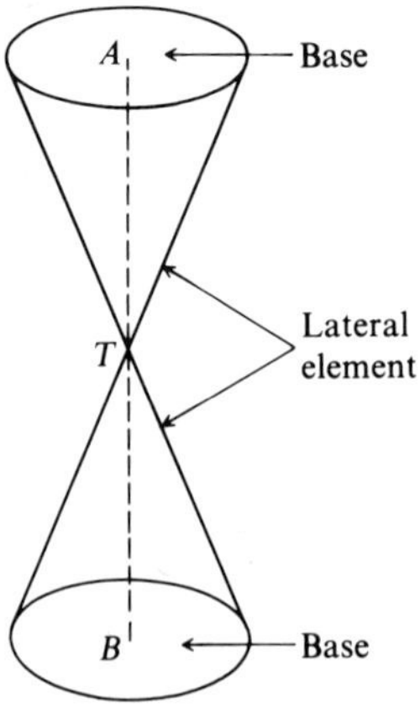

Figure 9.4

In the following sections we will examine the equations of each of the conic sections. No formal derivations of the equations will be shown.

9.12 The Circle

A *circle* is that section of the cone that is obtained when a plane is passed through the cone parallel to the base. (Figure 9.5.) The circle may also be defined as *the location of all points in a plane of equal distance from a fixed point.*

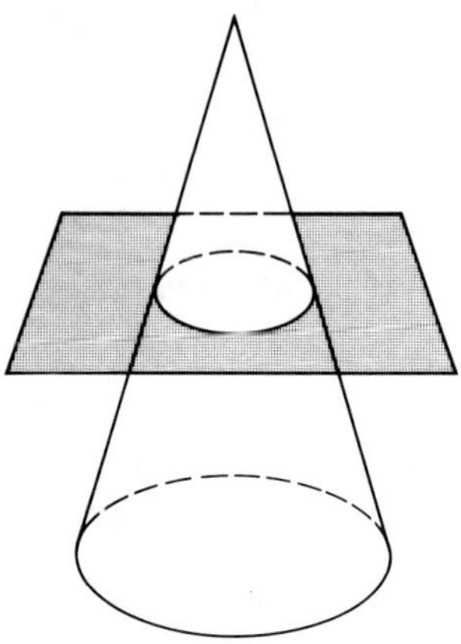

Figure 9.5

The equation of a circle with its center at the origin is given by

$$x^2 + y^2 = r^2, \tag{1}$$

where r is the radius of the circle.

If the center of the circle is located at a point (h,k) other than the origin, its equation is given by

$$(x - h)^2 + (y - k)^2 = r^2. \tag{2}$$

If $(h,k) = (0,0)$, Equation (2) reduces to Equation (1). Thus Equation (1) is a special case of Equation (2).

Example 1. Find the center and radius of the circle

$$x^2 + y^2 = 9.$$

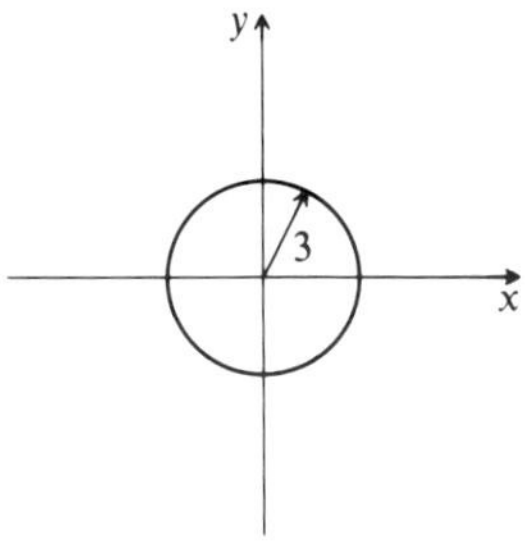

Sketch its graph.

The center is at the origin, and the radius is 3 units, as shown in the graph.

Example 2. Find the center and radius of the circle

$$(x - 1)^2 + (y - 2)^2 = 25.$$

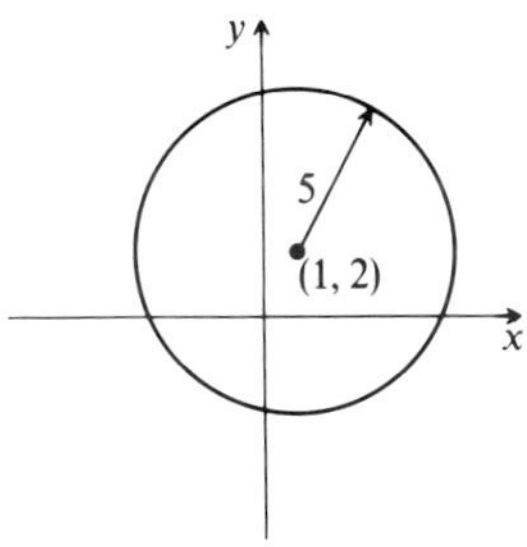

Sketch its graph.

Since $(x - h) = (x - 1)$ and $(y - k) = (y - 2)$, $h = 1$ and $k = 2$. The center is at the point $(1,2)$. The radius is 5.

Example 3. Find the center and radius of the circle

$$(x + 2)^2 + (y - 1)^2 = 4.$$

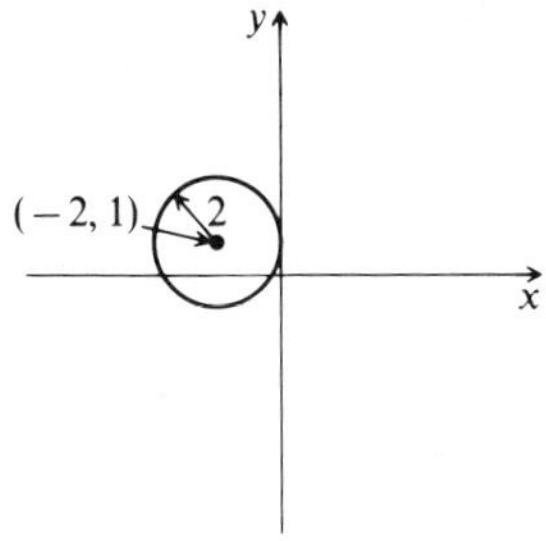

Sketch its graph.
Here $(h,k) = (-2,1)$, $r = 2$. The center is at $(-2,1)$ and the radius is 2.

Example 4. Find the center and radius of the circle

$$x^2 + 4x + y^2 + 6y = 0$$

We first put the equation in the form $(x - h)^2 + (y - k)^2 = r^2$ by completing the square on each of the variables.

$$x^2 + 4x + 4 + y^2 + 6y + 9 = 13$$
$$(x + 2)^2 + (y + 3)^2 = 13$$

Thus we see the center is at $(-2,-3)$ and the radius is $\sqrt{13}$ units.

9.13 The Ellipse

The *ellipse* is the section of a cone obtained when a plane is passed through the cone in such a way so as to be not parallel to the base, the axis of symmetry, or a lateral element. See Figure 9.6.
An ellipse is defined as the location of all points in a plane the sum of whose distances from two fixed points is a constant. (Figure 9.7.)

The equation of an ellipse with its center at the origin is either

$$\frac{x^2}{a^2} + \frac{y^2}{b^2} = 1 \tag{3}$$

or

$$\frac{y^2}{a^2} + \frac{x^2}{b^2} = 1. \tag{4}$$

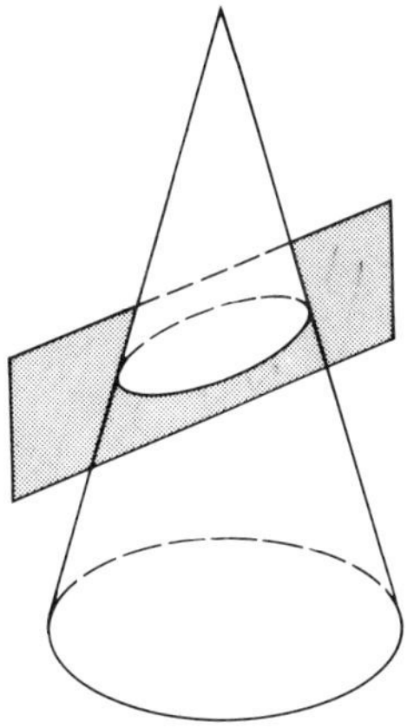

Figure 9.6

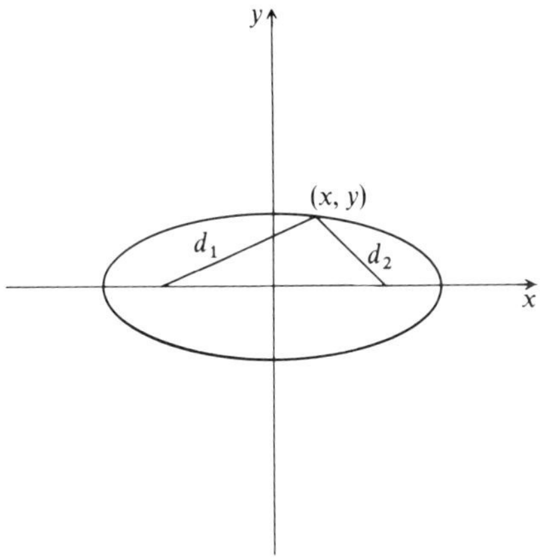

Figure 9.7

In each of these equations, the longer axis or *major axis* is $2a$ units long while the shorter axis or *minor axis* is $2b$ units long. The major axis is along the x axis if a is associated with the x term as in Equation (3), or it is along the y axis if a is associated with the y term as it is in Equation (4). The graphs of these two equations are illustrated in Figure 9.8.

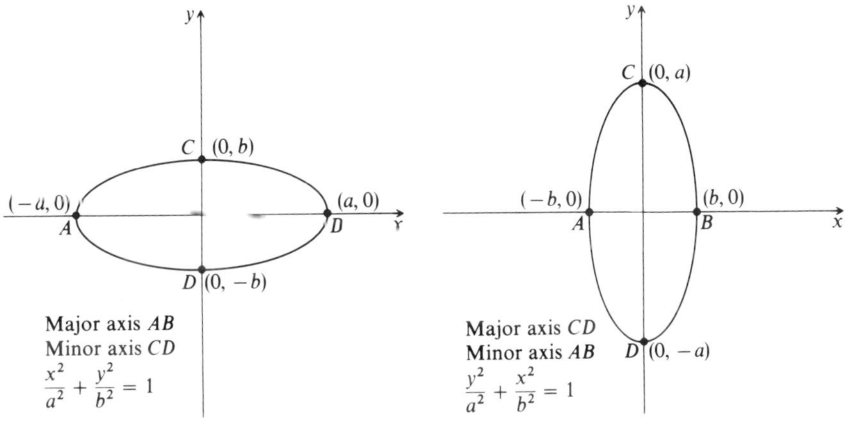

Figure 9.8

If the center of an ellipse is moved to the point (h,k) from the origin, then its equation is either

$$\frac{(x-h)^2}{a^2} + \frac{(y-k)^2}{b^2} = 1 \tag{5}$$

or

$$\frac{(y-k)^2}{a^2} + \frac{(x-h)^2}{b^2} = 1. \tag{6}$$

The major and minor axis are still $2a$ and $2b$ units long respectively, and they lie on lines through (h,k) parallel to the x and y axis. The graphs of Equations (5) and (6) are shown in Figure 9.9.

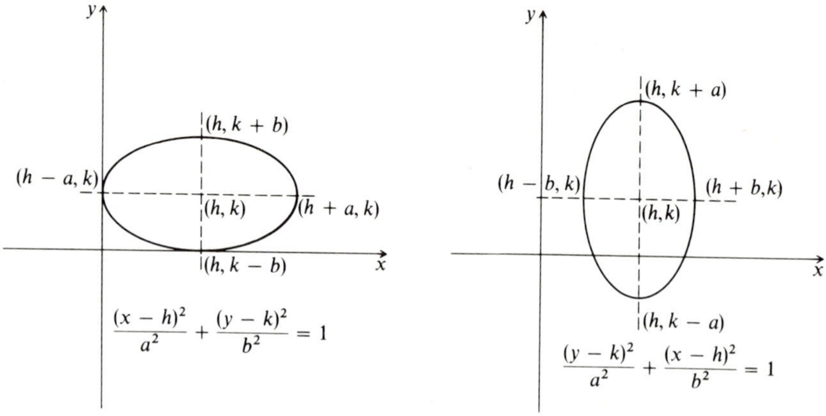

Figure 9.9

If $(h,k) = (0,0)$ then Equations (3) and (4) reduce to the form of Equations (1) and (2). Thus (1) and (2) are special cases of (3) and (4).

Example 1. Find the equation of an x ellipse with its center at the origin, major axis 10 units, minor axis 8 units. Sketch.
a is one-half the major axis and b one-half the minor axis. Thus $a = 5$, and $b = 4$. Since we are seeking an x ellipse, the form is

$$\frac{x^2}{a^2} + \frac{y^2}{b^2} = 1.$$

Substituting $a = 5$, $b = 4$, we have

$$\frac{x^2}{25} + \frac{y^2}{16} = 1.$$

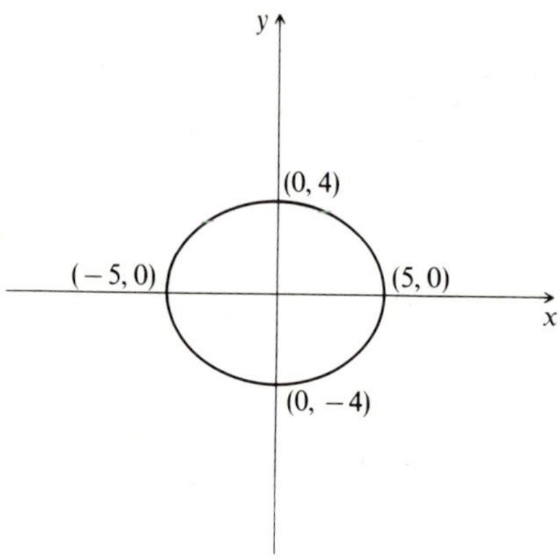

Example 2. Find the equation of the y ellipse with center at $(3, -4)$, $a = 8$, $b = 4$.

Here $h = 3$, $k = -4$. Substituting into the basic form,

$$\frac{(y-k)^2}{a^2} + \frac{(x-h)^2}{b^2} = 1.$$

We obtain

$$\frac{(y+4)^2}{64} + \frac{(x-3)^2}{16} = 1.$$

The graph is shown below.

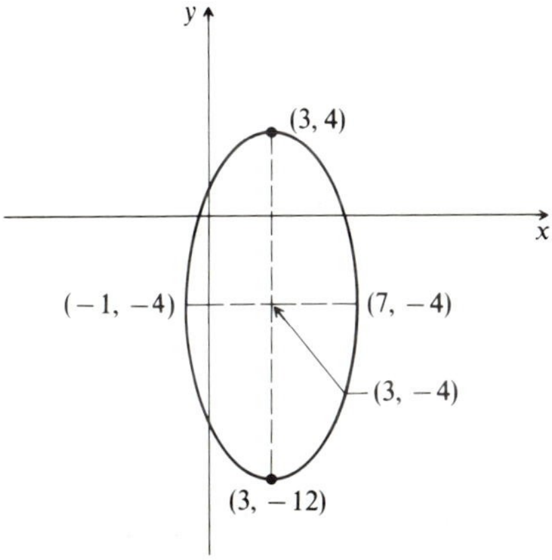

If we compare the equation of an ellipse to that of the circle under the conditions that $a = b$, we can see that a circle is a special case of an ellipse. In this case the major and minor axis have the same length.

Exercise 56

Identify each of the following equations. If it is a circle find its center and radius. If it is an ellipse, find its center and the coordinates of the end-points of the major and minor axis. Sketch.

1) $x^2 + y^2 = 9$
2) $x^2 + y^2 = 16$
3) $(x-2)^2 + (y-4)^2 = 16$
4) $\frac{x^2}{25} + \frac{y^2}{9} = 0$
5) $(x+2)^2 + (y-4)^2 = 9$
6) $(x-7)^2 + (y-2)^2 = 4$
7) $\frac{(x-2)^2}{25} + \frac{(y-3)^2}{16} = 1$
8) $\frac{(x+2)^2}{9} + \frac{(y+3)^2}{25} = 1$
9) $x^2 + 6x + y^2 - 4y - 3 = 0$
10) $x^2 + 2x + y^2 = 3$

11) $\frac{(x-4)^2}{4} + \frac{y^2}{1} = 1$

12) $(x-2)^2 + y^2 = 9$

13) $\frac{x^2}{\frac{1}{9}} + \frac{y^2}{\frac{1}{25}} = 1$

14) $\left(x - \frac{1}{2}\right)^2 + \left(y - \frac{1}{3}\right)^2 = \frac{1}{4}$

15) $(y-2)^2 = (x-3)^2 + 9$

16) $\frac{(x-7)^2}{3} = 3 - \frac{(y-2)^2}{12}$

9.14 The Hyperbola

The *hyperbola* is the conic section obtained when a plane is passed through the cone parallel to the axis of symmetry of the cone. Figure 9.10.

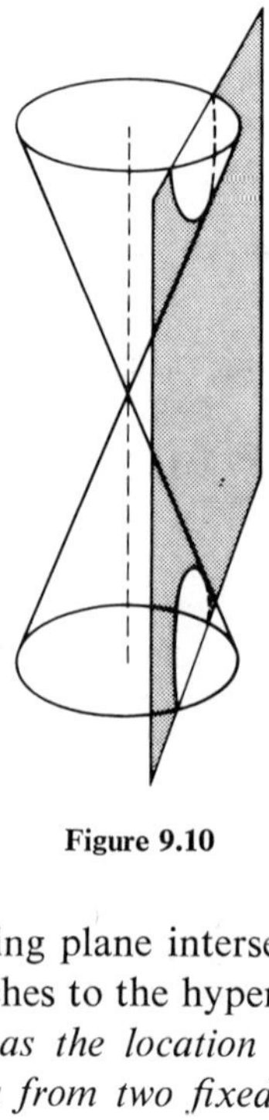

Figure 9.10

Notice that the generating plane intersects both nappes of the cone and that there are two branches to the hyperbola.

A hyperbola is defined as the location of all points in a plane, the difference of whose distances from two fixed points is a constant (Figure 9.11).

The equation of a hyperbola with its center at the origin is either

$$\frac{x^2}{a^2} - \frac{y^2}{b^2} = 1 \tag{7}$$

$$\frac{y^2}{a^2} - \frac{x^2}{b^2} = 1. \tag{8}$$

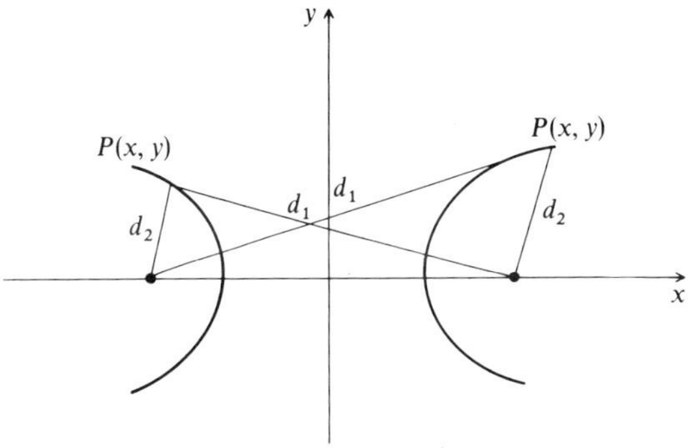

For any point P on the hyperbola, $d_1 - d_2 = k$.

Figure 9.11

In the equations the key to determining the type of hyperbola, x or y is the association of the negative term. Equation (7) is that of an x hyperbola since the x term is positive; while Equation (8) is that of a y hyperbola since the y term is positive. The graphs of these two equations are given in Figure 9.12.

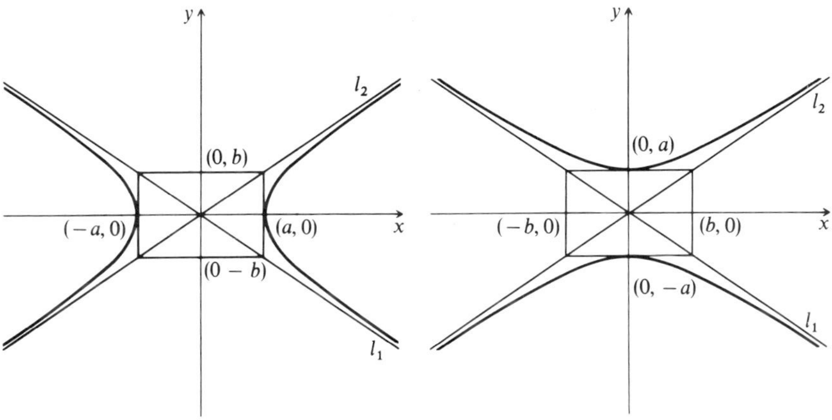

Figure 9.12

Notice that the center of the hyperbola does not lie interior to either of the branches. The lines l_1 and l_2 are called *asymptotes* of the hyperbola since the graph will approach this line as the absolute value of x grows large without bound. The asymptotes are usually sketched by constructing a rectangle whose center is at the intersection of the asymptotes and whose sides are $2a$ and $2b$ units respectively. The asymptotes are the diagonals of this rectangle.

Example 1. Sketch the hyperbola whose equation is

$$\frac{x^2}{9} - \frac{y^2}{16} = 1.$$

Here $a = 3$, $b = 4$. The x term is positive so the equation represents an x hyperbola.

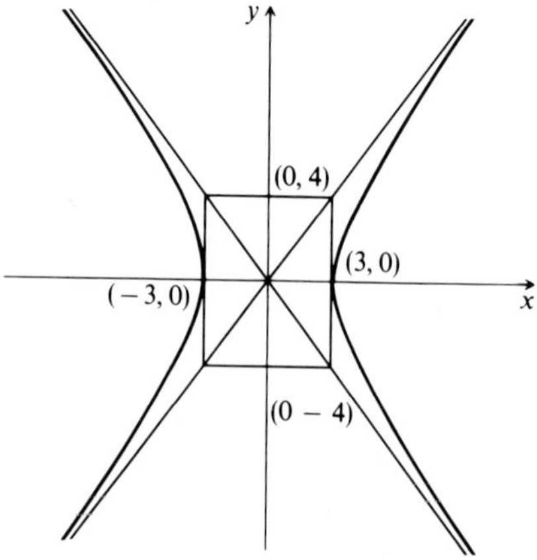

The distance $2a$ is said to be the length of the *transverse axis* and the distance $2b$ the length of the *conjugate axis*. In Example 1 the length of the transverse axis is 6 units; while the length of the conjugate axis is 8.

Example 2. Sketch the graph of the hyperbola whose equation is

$$\frac{y^2}{16} - \frac{x^2}{9} = 1.$$

Here $a = 4$ and $b = 3$. The term involving y is positive so the equation represents a y hyperbola.

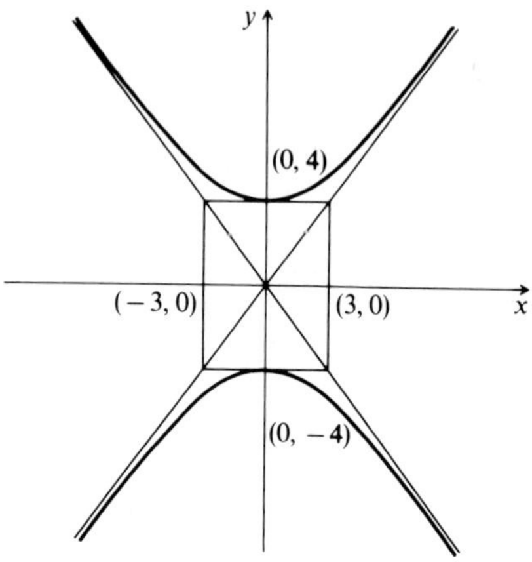

As in the two previous sections we can move the center from the origin to a point (h,k) by replacing x by $(x - h)$ and y by $(y - \text{k})$. Thus the equation of a hyperbola with center at (h,k) is either

$$\frac{(x-h)^2}{a^2} - \frac{(y-k)^2}{b^2} = 1 \tag{9}$$

or

$$\frac{(y-k)^2}{a^2} - \frac{(x-h)^2}{b^2} = 1. \tag{10}$$

Example 3. Sketch the graph of

$$\frac{(y-2)^2}{16} - \frac{(x+1)^2}{9} = 1.$$

$a = 4$, $b = 3$. The center is at $(-1,2)$.

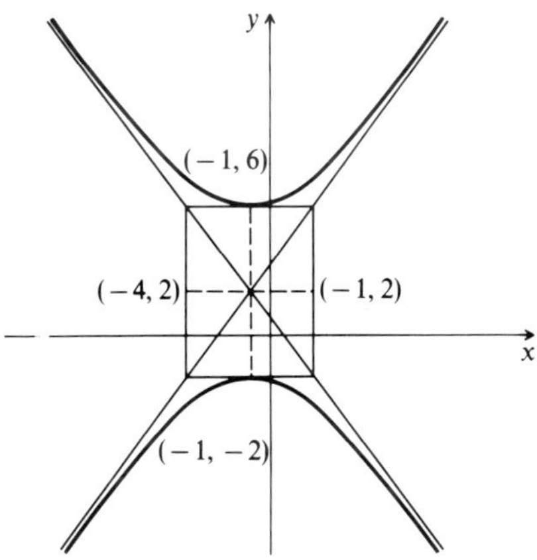

Example 4. Sketch the graph of

$$\frac{(y-1)^2}{10} - \frac{x^2}{4} = -1.$$

We first multiply both sides of the equation by (-1) to put the equation into the form

$$\frac{x^2}{4} - \frac{(y-1)^2}{10} = 1.$$

$a = 2$, $b = \sqrt{10}$. The center is at $(0,1)$.

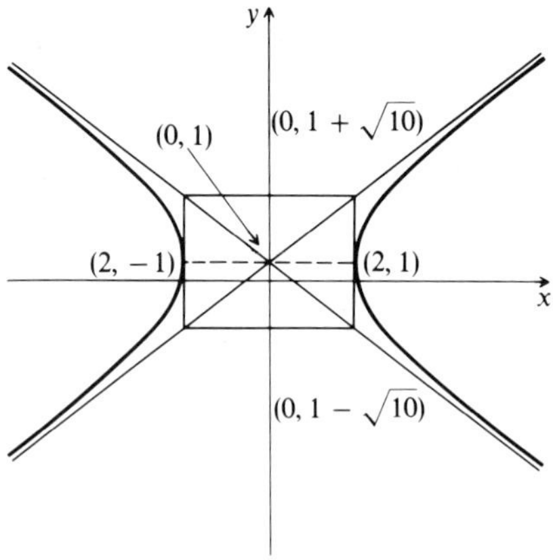

9.15 The Parabola

The *parabola* is that section of a cone obtained when the generating plane passes through the cone parallel to a lateral element. (Figure 9.13.)

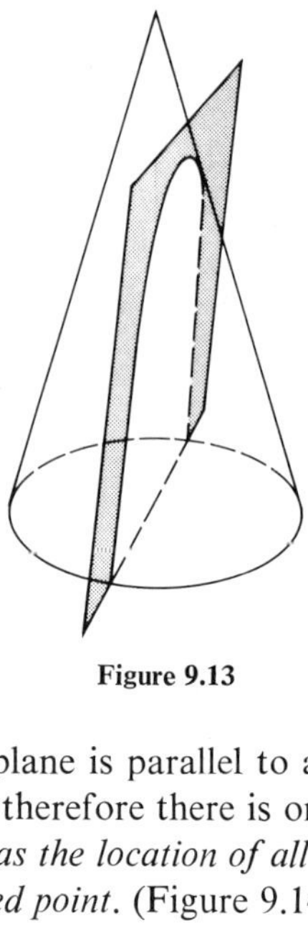

Figure 9.13

Since the generating plane is parallel to a lateral element, it will not intersect both nappes and therefore there is only one branch.

A *parabola is defined as the location of all points in a plane equidistant from a fixed line and a fixed point*. (Figure 9.14.)

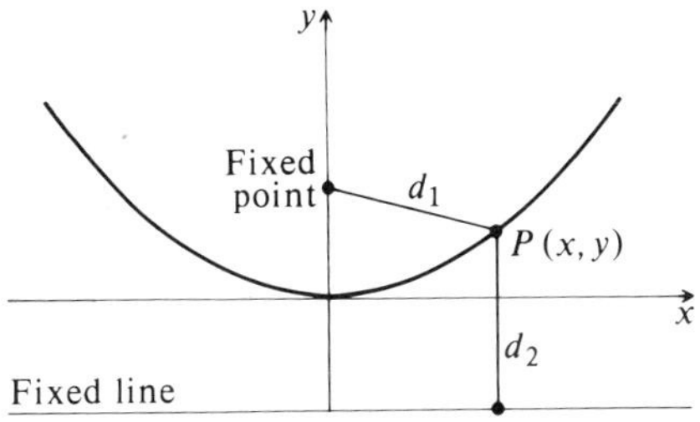

For any point P on the parabola $d_1 = d_2$

Figure 9.14

The general equation of a parabola is

$$y = ax^2 + bx + c, \tag{11}$$

where a, b and c are arbitrary constants, $a \neq 0$. If the turning point or vertex is at the origin, $b = c = 0$. The line drawn through the vertex so that one side of the parabola is an image of the other is known as the *axis of symmetry*. Parabolas open up or down accordingly as a is positive or negative. The location of the vertex depends upon the values of b and c. In fact, it can be shown that a parabola that opens up or down with its vertex at (h,k) has as its equation

$$(y - k) = a(x - h)^2. \tag{12}$$

Example 1. Sketch the parabola $y = 2x^2$.
Here $(h,k) = (0,0)$, and the parabola opens up. The additional points shown were found by letting x take on the values of ± 1, ± 2, ± 3 and computing the corresponding value of y. In this manner we obtain a better idea of the shape of the parabola.

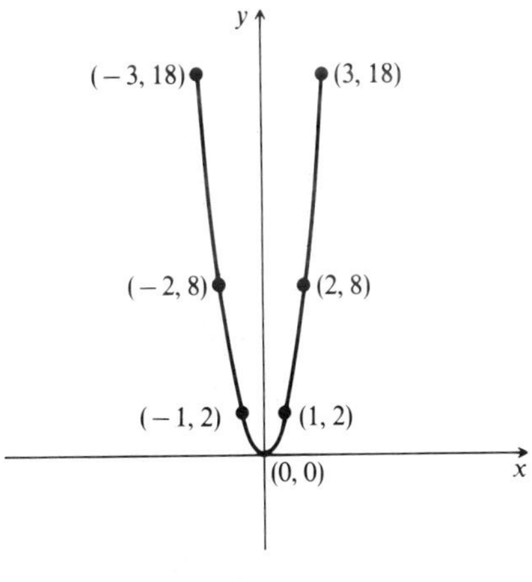

The axis of symmetry is the y axis.

Example 2. Sketch the graph of $y = x^2 + 2x + 1$.
We first rewrite the equation as $y = (x + 1)^2$.
From this form we see that $(h,k) = (-1,0)$. Thus the axis of symmetry is the line $x = -1$. Since $a = 1$, the parabola opens up.

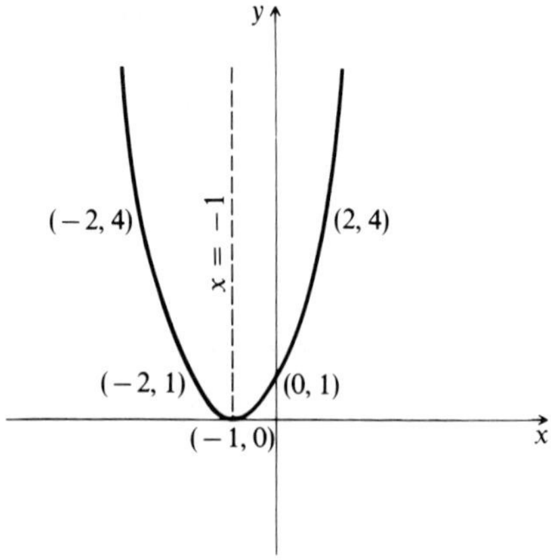

Example 3. Sketch the graph of $y - 2 = -2(x - 3)^2$. Here $(h,k) = (3,2)$ and $a = -2$. The axis of symmetry is the line $x = 3$. The parabola opens down, as indicated by a negative.

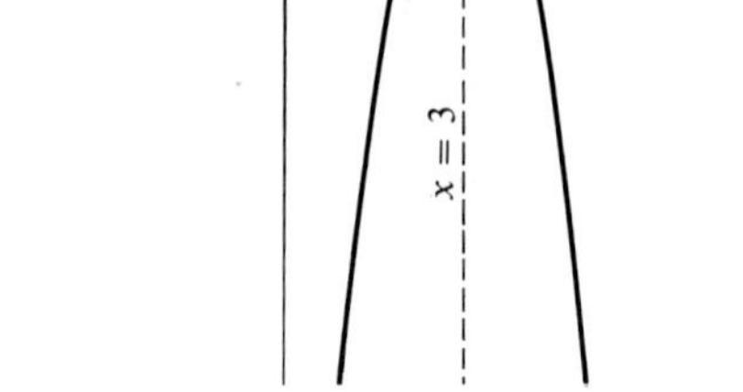

If a parabola opens down, the curve is said to be *concave down*. If it opens up, the curve is *concave up*.

If a parabola does not appear in the form

$$y - k = a(x - h)^2,$$

it can always be put in that form by algebraic methods. However, for the purposes of sketching, point by point plotting usually is sufficient.

Example 4. Sketch the graph of

$$y = \tfrac{1}{2}x^2 - 4x + 8.$$

The ordered pairs (0,8), (2,2), (4,0), (6,2) and (8,8) satisfy the equation. Further we see that the axis of symmetry is the line $x = 4$. The graph is shown below.

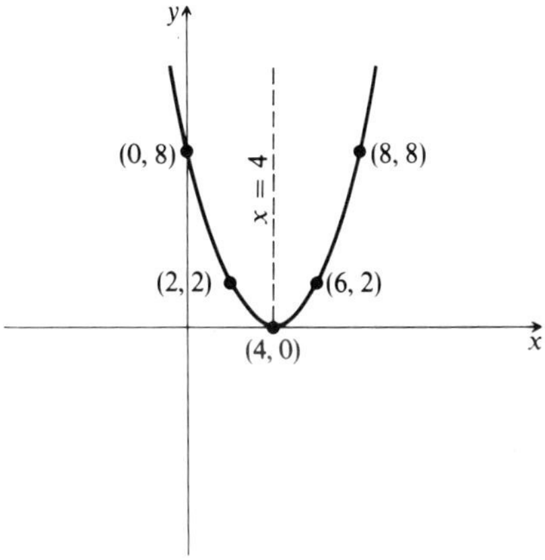

Exercise 57

Sketch the graph of each of the following. In the case of a hyperbola, find its center and identify it as a x or a y hyperbola.

1) $y = x^2$

2) $y = x^2 - 2$

3) $y = x^2 + 3$

4) $y = 2x^2 - 1$

5) $y = -x^2 + 5$

6) $y = x^2 - 2x$

7) $y = x^2 - 3x + 2$

8) $y = 2 - x - x^2$

9) $y = \dfrac{1}{3}x^2 + 9$

10) $y = \dfrac{x^2}{4} - 1$

11) $y = x^2 - 2x + 1$

12) $\dfrac{x^2}{4} - \dfrac{y^2}{9} = 1$

13) $\dfrac{(x-1)^2}{16} - \dfrac{(y+1)^2}{36} = 1$

14) $\dfrac{(y-2)^2}{25} - \dfrac{x^2}{16} = 1$

15) $\dfrac{x^2}{10} - \dfrac{y^2}{15} = 1$

16) $\dfrac{(x-1)^2}{4} - \dfrac{(y-1)^2}{4} = 1$

17) The surface area of a cube is given by $A = 6s^2$. Sketch the graph of the surface area equation for $0 < s < 5$. Use different units of length along each axis.

18) If a circular container is filled with water and rotated, the shape of the surface will be parabolic. The equation of the parabola is

$$y = \left(\frac{\omega^2}{2g}\right)x^2$$

where g represents gravity and ω the angle that a point on the surface passes through in a given period of time. Sketch the shape of the cross section of the surface if $\omega = 15$ and $g = 32$.

19) If an object is thrown vertically upward, the distance that it will rise above the surface of the earth is given by $h = vt - 16t^2$, where v is the velocity at which it is thrown and t is the elapsed time. If v is 120, sketch the graph, letting h be the vertical axis and t the horizontal axis.

20) A man has 100 yards of fencing with which he wants to enclose a maximum rectangular area. Show that the equation describing the area is $A = L\,(50 - L)$, where L is the length, and sketch its graph. Can you guess the diemnsions from this graph? What would be the maximum area?

9.16 Graphs of Other Equations

In the earlier sections of this chapter several important types of equations were graphed. Some were graphed by using standard forms, and others were graphed by assigning a specific value to one variable and then determining the corresponding value of the other variable. Many other types of equations occur in industrial applications of mathematics. We will examine a few in the following section.

The basic procedure for sketching the graphs will be point by point plotting. The following examples illustrate some of the more common types of equations that you may expect to encounter.

Example 1. Sketch the graph of $y = 2^x$.
In this example the variable x appears as an exponent and the equation is called an *exponential equation*. If we let $x = -2$, we have $y = 2^{-2} = \frac{1}{4}$. If we let $x = 0$, then $y = 1$; or if we let $x = 1$, then $y = 2^1 = 2$. We therefore have three ordered pairs that lie on the graph $\left(-2,\frac{1}{4}\right)$, (0,1) and (1,2). Some additional points obtained in a similar manner are

$$\left(-3,\frac{1}{8}\right), \left(-1,\frac{1}{2}\right), (2,4), (3,8).$$

The graph is shown in Figure 9.15.

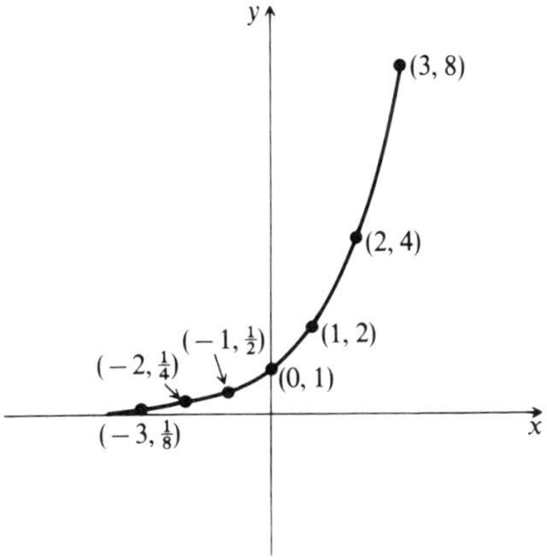

Figure 9.15

Equations of this type play a large role in the theory of logarithms which are found in Chapter 6.

Example 2. Sketch the graph of $y = x^3 + 1$.
An equation of this general type is known as a *cubic equation*. Some ordered pairs satisfying the equation are $(-2,-7)$, $(-1,0)$, $(0,1)$, $(1,2)$,$(2,9)$. The graph is shown in Figure 9.16.

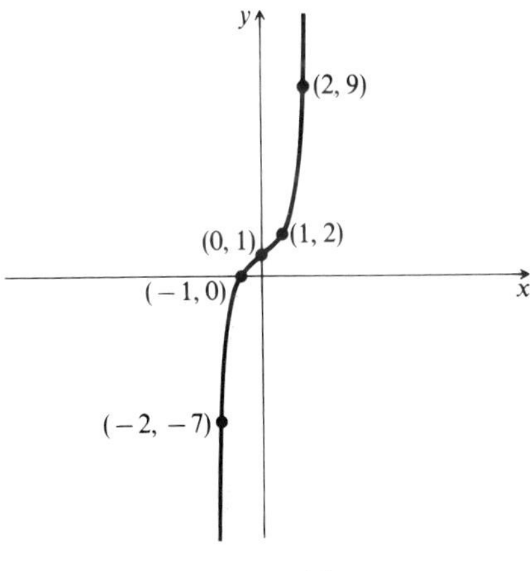

Figure 9.16

Example 3. Sketch the graph of $y = 1/x$.
Notice here that zero cannot be used as a replacement for x since division by zero is not defined. Some ordered pairs satisfying this equation are

$$\left(-3,-\frac{1}{3}\right), \left(-2,-\frac{1}{2}\right), (-1,-1), \left(-\frac{1}{2},-2\right),$$

$$\left(-\frac{1}{3},-3\right), \left(-\frac{1}{4},-4\right); \text{ also } \left(\frac{1}{4},4\right), \left(\frac{1}{2},2\right),$$

$$(1,1), \left(2,\frac{1}{2}\right), \text{ and } \left(3,\frac{1}{3}\right).$$

Since zero is not permissible several fractional values near zero, both positive and negative, were chosen in order to examine the behavior of the graph in that area. The graph is shown in Figure 9.17.

Equations of this type belong to the general class known as *hyperbolas*. Notice that the axes are the asymptotes.

Example 4. Sketch the graph of $y = x^3 - 4x$. Some ordered pairs that satisfy this equation are $(-3,-15)$, $(-2,0)$, $(-1,3)$, $(0,0)$, $(1,-3)$, $(2,0)$, $(3,15)$. The graph is shown in Figure 9.18.

Note that one unit on the x axis is the same as two units on the y axis. This equation, like Example 2 is known as a *cubic equation*.

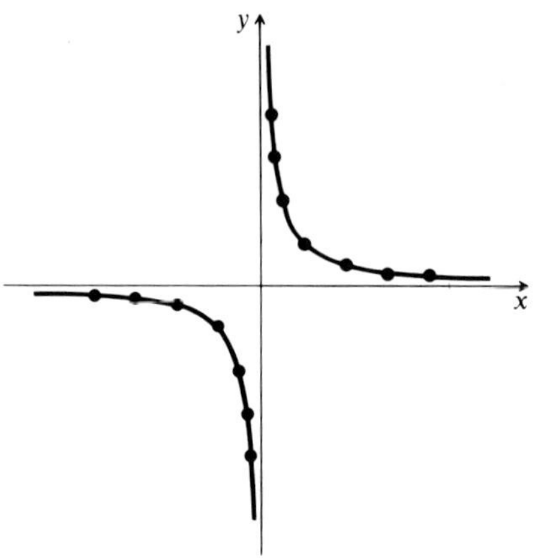

Figure 9.17

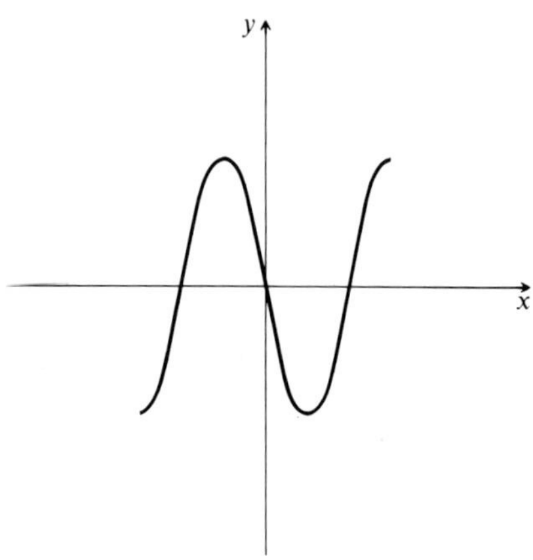

Figure 9.18

Exercise 58

Sketch the graph of each of the following equations.

1) $y = 3^x$

2) $y = x^3 + 2x$

3) $y = x^3$

4) $y = 4^x$

5) $y = 1^x$

6) $xy = 6$

7) $6y = \frac{4}{x}$

8) $x + y^2 = 9x^4$

9) $x^2 + y = 25x^3$

10) $x^3 - y = 6$

11) $x^4 - y = 25$

12) $y = \frac{1}{x + 1}$

13) $y = 3 - \frac{1}{x}$

14) $y = x^3 - x^2$

15) $y = \left(\frac{1}{2}\right)^x$

16) The relationship between resistance R and the diameter of a particular type of copper wire is given by $R = \frac{k}{d^2}$, where d is in mm. If $k = 20$, sketch the graph of R vs. d from $d = 1$ to $d = 4$. Use intervals of .3 mm.

17) At a pressure of 20 pounds the volume of a gas is 40 cubic inches and the formula relating pressure to volume is $V = \frac{800}{P}$. Sketch the graph letting the horizontal axis be the P axis and the vertical axis the V axis. What happens to the volume as the pressure increases?

18) The maximum safe uniformly distributed load L for a horizontal beam is given by the formula $L = \frac{kbd}{l}$, where b is the width, d the depth and l the length. If k remains constant at 600, $b = 2$, and $d = 4$, how does increasing length effect the maximum safe load? Sketch the graph.

Review Exercise Chapter 9

Solve the following quadratic equations by factoring.

1) $x^2 + 3x + 2 = 0$

2) $3x^2 - 5x + 2 = 0$

Solve the following quadratic equations by extraction of roots.

3) $(2x - 3)^2 = 25$

4) $(x + 6)^2 = 3$

Solve the following quadratic equations by completing the square.

5) $x^2 - 6x = 5$

6) $3x^2 + 2x - 2 = 0$

7) $2x^2 = 5x + 6.$

Solve the following quadratic equations by use of the quadratic formula.

8) $m^2 - 5m + 2 = 0$

9) $3x^2 - 2x - 1 = 0$

10) $-x^2 - x + 2 = 0$

Without solving determine a) the nature of the roots, b) the sum of the roots, and c) the product of the roots in each of the following.

11) $3x^2 + \frac{3}{5}x - 9 = 0$

12) $\frac{x}{5} - x^2 = 6$

13) $5x^2 = 3x - 4$

Identify each of the following conics, give its center or vertex and sketch.

14) $\frac{x^2}{4} - \frac{y^2}{9} = 1$

15) $(x - 2)^2 + (y + 6)^2 = 25$

16) $\frac{(x - 3)^2}{25} + \frac{(y + 2)^2}{16} = 1$

17) $(x - 3)^2 = 2(y + 4)$

18) $\frac{(x - 9)^2}{16} - \frac{(y - 2)^2}{9} = 1$

Sketch the following.

19) $y = 5^{2x}$

20) $x^3 + 2x = y$

10 Trigonometric Operations

10.1 Trigonometry

The word *trigonometry* comes from the Greek words *trigonon* meaning triangle and *metria* meaning measurement. The subject of trigonometry is therefore a study of the relationships between the sides of a triangle and the angles related to those sides.

In this chapter we will develop some of the basic relationships that the technical student should be familiar with. As in the previous chapter, much of the development will be given without formal proof. If the student wishes to examine the proofs of the various sections in depth, he can refer to any standard trigonometry text.

10.2 The Trigonometric Ratios

We shall now examine the six ratios that can be formed beween the sides of a right triangle. The student will recall that a right triangle is one that contains a 90° angle. The side opposite the right angle is called the hypotenuse, and the other two sides are called the legs. It is customary to label the side opposite $\angle A$ as leg a, the side opposite $\angle B$ as leg b, and the side opposite $\angle C$ (the right angle) as the hypotenuse c. (See Figure 10.1).

Six ratios can be formed between the sides of the right triangle in Figure 10.1. These are a/c, b/c, a/b, b/a, c/b, and c/a. These ratios are of such great importance that each is given a specific name in relation to certain angles of the triangle. The names and relationships are as follows.

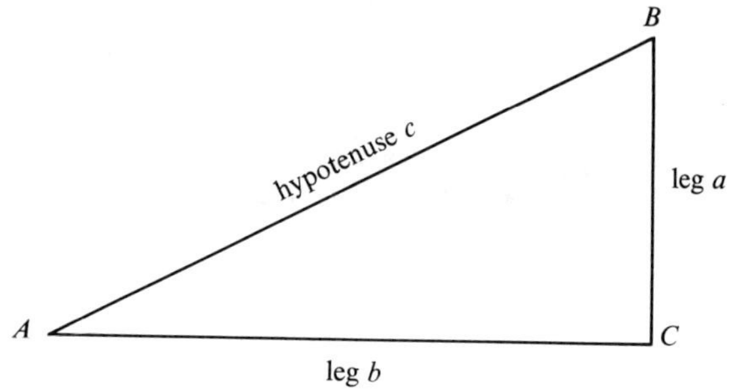

Figure 10.1

$$\sin A \text{ or sine } A = \frac{a}{c} = \frac{\text{side opposite } \angle A}{\text{hypotenuse}},$$
$$\cos A \text{ or cosine } A = \frac{b}{c} = \frac{\text{side adjacent to } \angle A}{\text{hypotenuse}},$$
$$\tan A \text{ or tangent } A = \frac{a}{b} = \frac{\text{side opposite } \angle A}{\text{side adjacent to } \angle A},$$
$$\cot A \text{ or cotangent } A = \frac{b}{a} = \frac{\text{side adjacent to } \angle A}{\text{side opposite } \angle A},$$
$$\sec A \text{ or secant } A = \frac{c}{b} = \frac{\text{hypotenuse}}{\text{side adjacent to } \angle A},$$
$$\csc A \text{ or cosecant } A = \frac{c}{a} = \frac{\text{hypotenuse}}{\text{side opposite } \angle A}.$$

At this point the student should notice that some of the functions are reciprocals of others as follows.

$$\sin A = \frac{1}{\csc A} \quad \text{and} \quad \csc A = \frac{1}{\sin A},$$
$$\cos A = \frac{1}{\sec A} \quad \text{and} \quad \sec A = \frac{1}{\cos A},$$
$$\tan A = \frac{1}{\cot A} \quad \text{and} \quad \cot A = \frac{1}{\tan A}.$$

The student's attention is also called to the fact that the six ratios apply to B as well as A. The trigonometric functions in terms of B are

$$\sin B = \frac{b}{c} \qquad \cot B = \frac{a}{b}$$
$$\cos B = \frac{a}{c} \qquad \sec B = \frac{c}{a}$$
$$\tan B = \frac{b}{a} \qquad \csc B = \frac{c}{b}$$

10.3 Uniqueness of the Ratio

In Chapter 7 we saw that if two triangles are similar their corresponding sides are proportional. This fact allows us to observe the uniqueness of the trigonometric ratios for any angle.

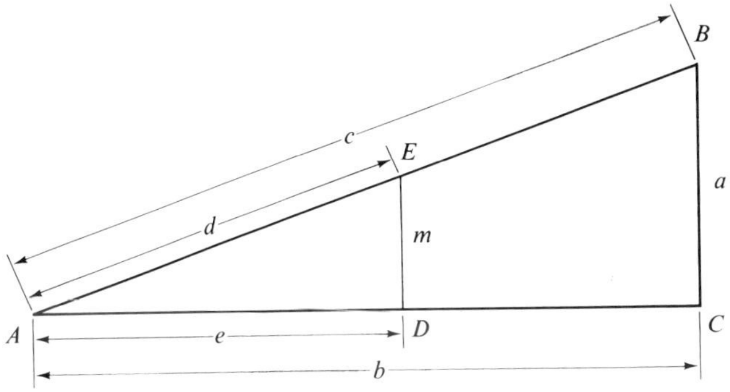

Figure 10.2

In Figure 10.2 any point E on side AB is selected and a perpendicular is constructed to side AC. The foot of this perpendicular is labeled D. Since triangles ABC and AED contain three equal angles, their corresponding sides are proportional. We therefore observe

$$\frac{m}{e} = \frac{a}{b},$$
$$\frac{m}{d} = \frac{a}{c},$$
$$\frac{e}{d} = \frac{b}{c},$$

and so on. This leads us to the conclusion that

$$\tan A = \frac{a}{b} = \frac{m}{e},$$
$$\sin A = \frac{a}{c} = \frac{m}{d},$$
$$\cos A = \frac{b}{c} = \frac{e}{d},$$

and therefore that the length of the sides of two similar triangles does not affect the value of the trigonometric functions of their angles. That is, for any given angle the trigonometric functions of that angle are unique.

In the previous two sections we have introduced the six trigonometric ratios and found them to be unique. In order for one to become familiar enough with these ratios, he needs to use them in enough simple situations that their definitions become part of his vocabulary. The following examples serve to tie these ideas together.

Example 1. Use Triangle I and find the six trigonometric functions of angle A.

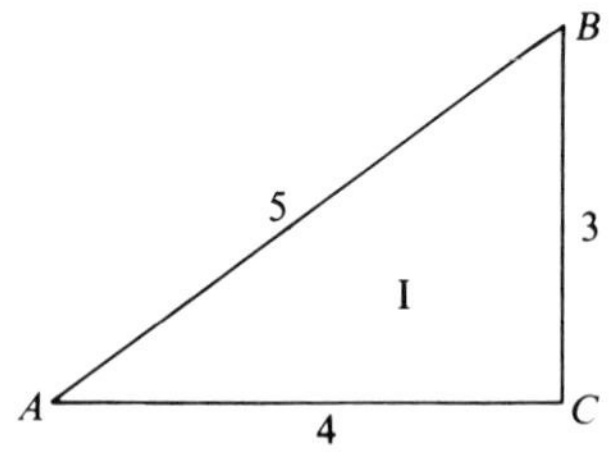

$$\sin A = \frac{\text{opp}}{\text{hyp}} = \frac{3}{5} \qquad \csc A = \frac{\text{hyp}}{\text{opp}} = \frac{5}{3}$$
$$\cos A = \frac{\text{adj}}{\text{hyp}} = \frac{4}{5} \qquad \sec A = \frac{\text{hyp}}{\text{adj}} = \frac{5}{4}$$
$$\tan A = \frac{\text{opp}}{\text{adj}} = \frac{3}{4} \qquad \cot A = \frac{\text{adj}}{\text{opp}} = \frac{4}{3}$$

Example 2. Use Triangle II to find the six trigonometric functions of angle B.

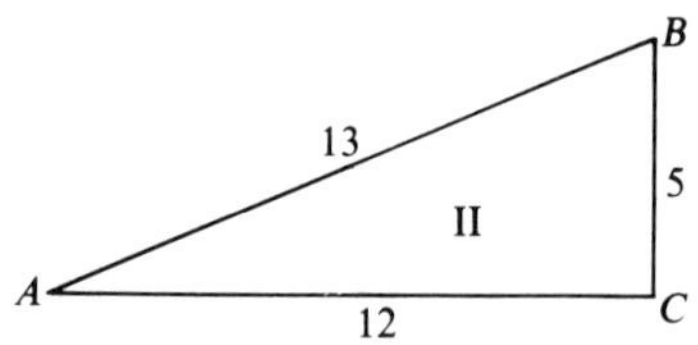

$$\sin B = \frac{\text{opp}}{\text{hyp}} = \frac{12}{13} \qquad \csc B = \frac{\text{hyp}}{\text{opp}} = \frac{13}{12}$$
$$\cos B = \frac{\text{adj}}{\text{hyp}} = \frac{5}{13} \qquad \sec B = \frac{\text{hyp}}{\text{adj}} = \frac{13}{5}$$
$$\tan B = \frac{\text{opp}}{\text{adj}} = \frac{12}{5} \qquad \cot B = \frac{\text{adj}}{\text{opp}} = \frac{5}{12}$$

Example 3. Use the diagram below to find $\cos B$ and $\tan A$.

Since $\cos B = \dfrac{\text{adj}}{\text{hyp}}$ and the hypotenuse is not known, we will first need to determine its length. To do this, we can use the Pythagorean theorem.

$$c^2 = a^2 + b^2$$
$$c^2 = 10^2 + 5^2 = 100 + 25$$
$$c^2 = 125$$
$$c = \sqrt{125} = 11.180.$$

We now have

$$\cos B = \frac{\text{adj}}{\text{hyp}} = \frac{5}{11.18} = .4472,$$
$$\tan A = \frac{\text{opp}}{\text{adj}} = \frac{5}{10} = \frac{1}{2} = .5$$

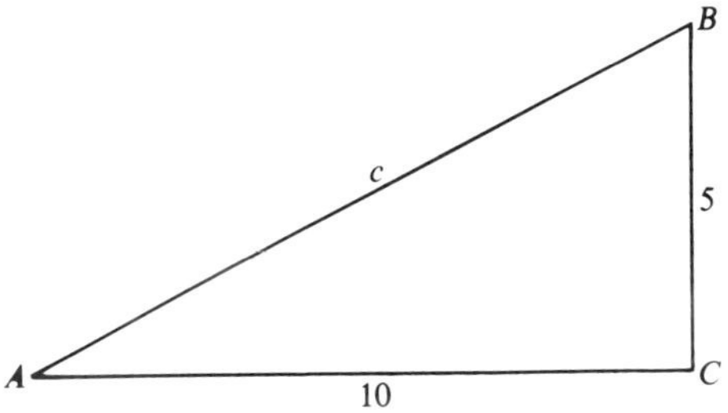

10.4 Standard Reference Triangles

Several angles occur so frequently in applied mathematics that it is worth one's time to study and memorize the relationships between the sides of the right triangles of which they are a part. One such triangle is one in which both acute angles have a measure of 45°. Another important triangle is one in which the acute angles have a measure of 30° and 60° respectively. These two triangles are shown in Figures 10.3 and 10.4.

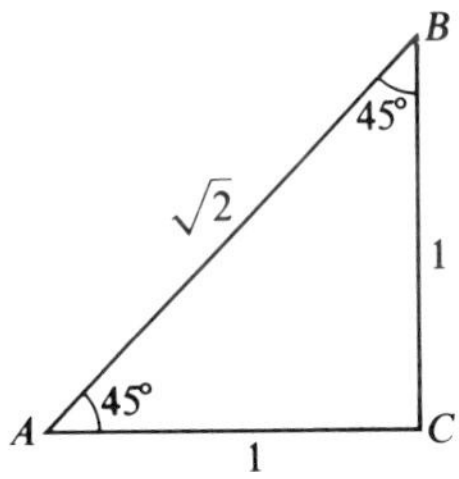

Figure 10.3

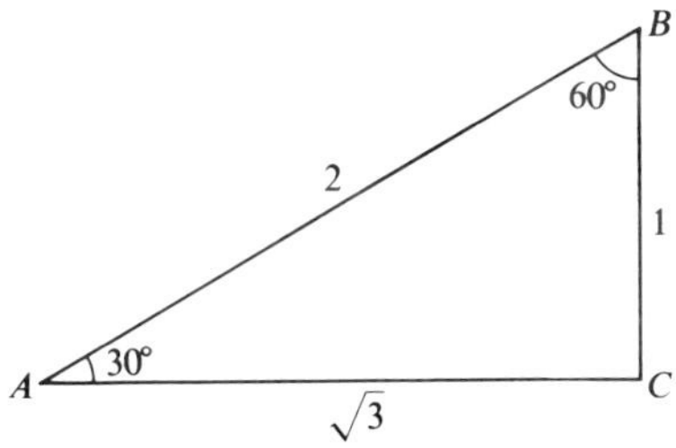

Figure 10.4

The dimensions shown are merely those that the authors feel are most convenient for computation. The 1 on each diagram simply represents one unit and one unit could be 3 inches, 26 cm., 5 yds., or any other unit of measure for that matter. The remaining sides would be proportional to this unit, and the uniqueness of the ratio would be preserved.

Example 4. Find sec 45°.
Referring to Figure 10.3, we have

$$\sec 45^\circ = \frac{\text{hyp}}{\text{adj}} = \frac{\sqrt{2}}{1} = \sqrt{2}.$$

Example 5. Find the six trigonometric functions of 30°.
Referring to Figure 10.4, we have

$$\sin 30^\circ = \frac{1}{2} \qquad \csc 30^\circ = 2$$

$$\cos 30^\circ = \frac{\sqrt{3}}{2} \qquad \sec 30^\circ = \frac{2}{\sqrt{3}} = \frac{2\sqrt{3}}{3}$$

$$\tan 30^\circ = \frac{1}{\sqrt{3}} = \frac{\sqrt{3}}{3} \qquad \cot 30^\circ = \sqrt{3}$$

If a trigonometric ratio for one angle of a right triangle is known, it is possible to find the other five ratios. The following example serves to illustrate the method that would be used.

Example 6. Cos A = .9. Find the other five trigonometric ratios for angle A in the diagram below.
Since $\cos A$ = adj/hyp we have $\cos A = .9 = 9/10$. Using the Pythagorean theorem we find

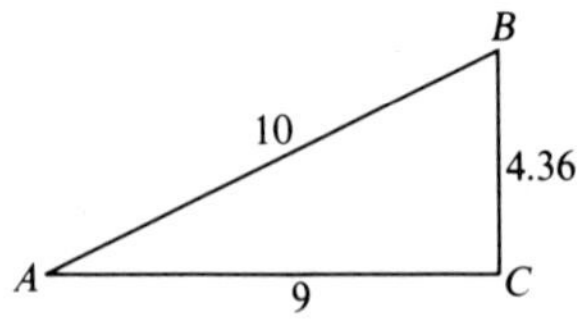

$$\begin{aligned} 10^2 &= 9^2 + a^2 \\ 100 &= 81 + a^2 \\ 19 &= a^2 \\ 4.36 &= a. \end{aligned}$$

Referring to the diagram above, we see that

$$\sin A = \frac{4.36}{10} = \frac{436}{1000} = .436$$

$$\tan A = \frac{4.36}{9} = \frac{436}{900} = .484$$

$$\cot A = \frac{9}{4.36} = \frac{900}{436} = 2.064.$$

$$\sec A = \frac{10}{9} = 1.111$$

$$\csc A = \frac{1000}{436} = 2.294$$

Exercise 59

In each of the following, find the trigonometric ratios from triangle ABC.

1) $\sin A$, $\cos A$
2) $\sec B$, $\tan B$
3) $\sec A$, $\cot B$
4) $\csc A$, $\sec A$
5) $\cot B$, $\csc B$

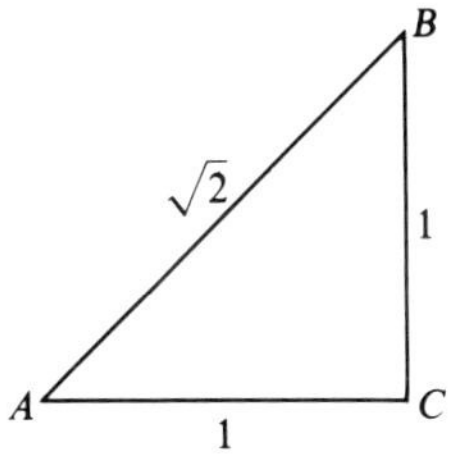

If angle $A = 30°$ and angle $B = 60°$ in a $30° - 60°$ right triangle, find each of the following. Sketch the triangle for reference before you begin.

6) $\cos A$, $\sin A$
7) $\tan A$, $\sec A$
8) $\tan B$, $\cos B$
9) $\sec B$, $\cot A$
10) $\cot B$

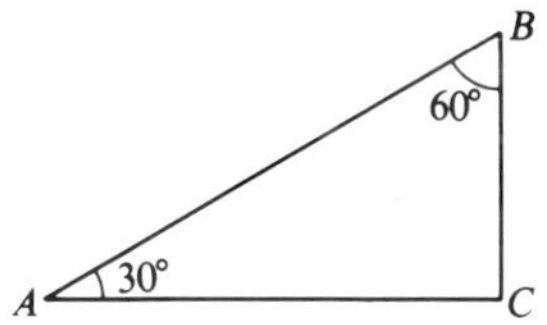

Each of the following problems are based on the diagram for triangle ABC as shown below. Find the indicated trigonometric ratios.

11) $a = 3$, $b = 4$; $\sin A$, $\cos B$
12) $a = 5$, $b = 12$, $\tan A$, $\csc B$
13) $a = 1$, $b = \sqrt{3}$; $\cos A$, $\cot A$

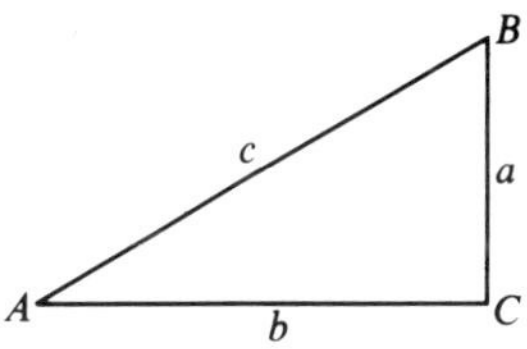

14) $a = 1$, $b = 1$; $\tan A$, $\cot A$
15) $a = 9$, $c = 41$; $\cos A$, $\sin B$
16) $a = 8$, $c = 17$; $\tan A$, $\sin B$
17) $a = 2$, $c = 5$; $\csc A$, $\cot A$
18) $a = 5$, $b = 4$; $\sin B$, $\cos A$
19) $a = 2$, $c = 3$; $\cos B$, $\tan A$
20) If $\cot A = 1$, find $\cos A$.
21) If $\tan B = 3/4$, find $\sin A$.
22) If $\sec A = .3$, find $\cot B$.
23) If $\sin A = 1/2$, find $\cos A$.
24) Without reference to the book, write the definition of each of the trigonometric ratios in terms of opposite, adjacent, and hypotenuse.
25) What relationship exists between $\sin A$ and $\cos B$, $\tan A$ and $\cot B$, $\sec A$ and $\csc B$ for any right triangle?

10.5 Trigonometric Tables and the Values of the Trigonometric Ratios

In section 10.4 the trigonometric ratios for the 30° – 60° triangle as well as the 45° – 45° triangle were computed as based on the standard reference triangle. The correctness of the ratio of the sides for both triangles is easily verified by use of the Pythagorean theorem. Such verification is left to the reader. What is important however, is that we can begin to build our own table of values for each of the trigonometric ratios. Referring to Figures 10.3 and 10.4 (page 279) we have:

	sin	cos	tan	cot	sec	csc
30°	$1/2$	$\sqrt{3}/2$	$\sqrt{3}/3$	$\sqrt{3}$	$2\sqrt{3}/3$	2
45°	$\sqrt{2}/2$	$\sqrt{2}/2$	1	1	$\sqrt{2}$	$\sqrt{2}$
60°	$\sqrt{3}/2$	$1/2$	$\sqrt{3}$	$\sqrt{3}/3$	2	$2\sqrt{3}/3$

If these values are expressed in decimal form and rounded to 4 decimal places we have:

	sin	cos	tan	cot	sec	csc
30°	.5000	.8660	.5774	1.7321	1.1547	2.0000
45°	.7071	.7071	1.0000	1.0000	1.4142	1.4142
60°	.8660	.5000	1.7321	.5774	2.0000	1.1547

In higher mathematics the ratios between the sides of a triangle for any given angle can be computed quite accurately. Approximations of these ratios to four decimal places are found in Table III of the Appendix. The angles from 0° to 45° are found by reading down the left hand column and the angles from 45° to 90° are found by reading up the right hand column.

Note that the column headings are not the same at the top and bottom of the page. The headings at the top are used for angles from 0° to 45°. Those at the bottom are used for angles from 45° to 90°. If we refer to Figure 10.5, we can see the relationship between the angles of a triangle that makes table construction of the above-mentioned type possible.

$$\sin A = \frac{a}{c} \qquad \cos B = \frac{a}{c}$$

$$\tan A = \frac{a}{b} \qquad \cot B = \frac{a}{b}$$

$$\sec A = \frac{c}{b} \qquad \csc B = \frac{c}{b}$$

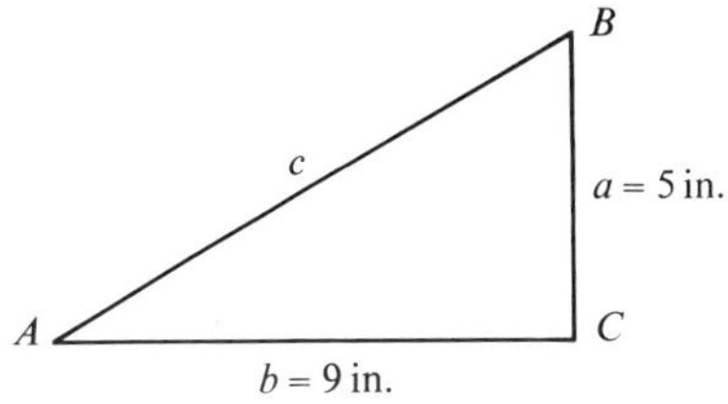

Figure 10.5

We should also notice that $(A + B) = 90°$, and that in general, the trigonometric function of any acute angle is always equal to the cofunction of its complement.

Example 1. To find sin 27° we look under the left hand column for 27° and then trace to the right until we are in the column headed sin. Here we find sin 27° = .4540.

Example 2. To find cot 63° we trace up the right hand column until we find 63°. We then trace to the left until we are in the column headed cot. Here we find cot 63° = 5095. Recall that 63° is between 45° and 90° and that the heading cot would be found at the bottom of the page.

Example 3. Given that tan A = .6009, find angle A. To determine angle A, we search the tables in the column headed tan until we find .6009. We then read to the left the proper angle 31°. Again we must be careful to use the proper angle depending upon whether the heading is at the top or bottom of the page.

10.6 Use of Tables for Solving Triangles

Trigonometric tables can be used to solve for the remaining parts of a right triangle when either one acute angle and a side or two sides are known. They can also be used to find the parts of triangles that do not contain right angles. This use will be developed in a later section.

Example 1. In triangle ABC, $a = 4$ and $b = 6$. Find the angles to the nearest degree.

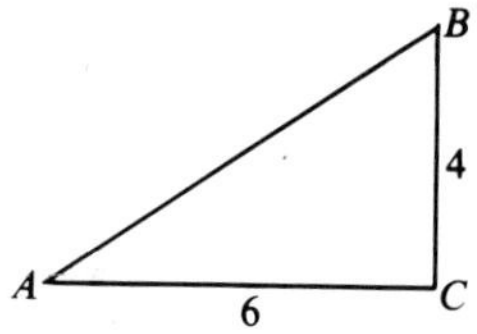

We know the side opposite angle A as well as side adjacent to it. Thus we have

$$\tan A = \frac{4}{6} = .6667.$$

Using the table we find $\tan 33° = .6494$ while $\tan 34° = .6745$. Thus angle $A = 34°$ to the nearest degree. Angle $B = 56°$ since $A + B = 90°$ and $c = 7.2$ by the Pythagorean theorem.

Example 2. In triangle ABC angle $A = 38°$ while side $c = 10$ units. Find side b.

$$\cos A = \frac{\text{adj}}{\text{hyp}} = \frac{b}{c}$$

so

$$\cos 38° = \frac{b}{10}$$

or

$$10 \cos 38° = b.$$

Using the table, we find

$$10(.7880) = b$$
$$7.880 = b.$$

Exercise 60

1) Use Table III to find approximations for the following.

a) sin 12°	b) tan 26°	c) tan 45°
d) cos 45°	e) sin 60°	f) cos 30°
g) sec 19°	h) cot 37°	i) csc 28°
j) sin 90°	k) cos 0°	l) tan 80°

2) Use Table III to find the angle if each of the following functions are known.

a) $\sin x = .3584$ b) $\cos x = .8290$
c) $\tan x = 3.2709$ d) $\cos x = .4384$
e) $\tan x = .9657$ f) $\sin x = .7771$
g) $\cos x = .8910$ h) $\tan x = .1051$

For each of the following problems refer to Triangle ABC; find the indicated angle to the nearest degree or the indicated side to the nearest hundredth. Do not use tables for 30°, 45° or 60° angles.

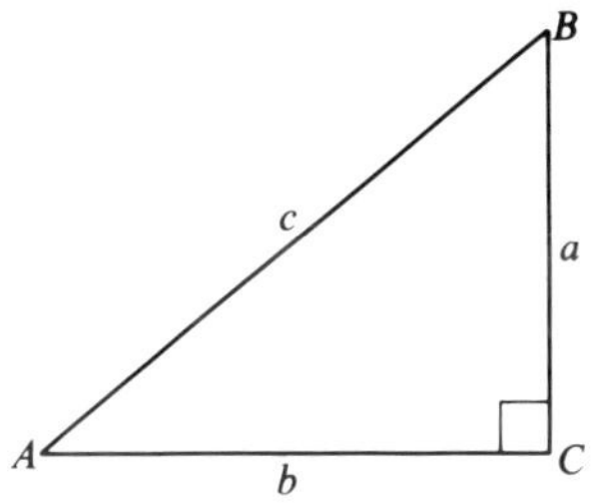

3) $a = 3, b = 5; A =$ ____. 4) $c = 9, b = 2; B =$ ____.
5) $a = 6, c = 12; B =$ ____. 6) $a = 4, b = 8; A =$ ____.
7) $a = 1, c = 2; A =$ ____. 8) $a = 2.12, b = 4.21; B =$ ____.

Assume the angular measure in the following problems to be exact.

9) $A = 60°, b = 2; a =$ ____. 10) $A = 23°, c = 9; a =$ ____.
11) $B = 72°, a = 4.1; b =$ ____. 12) $B = 64°, a = 3.91; a =$ ____.
13) $A = 16°, c = 3.8; b =$ ____. 14) $B = 46°, c = 22; a =$ ____.

10.7 Applications to Right Triangles

Many applied problems from technology and industry relate directly to the right triangle, and their solutions involve using the trigonometric ratios. The following examples show how these triangles are solved.

To solve a right triangle only two parts must be known in addition to the knowledge that one angle is a right angle. One of the known parts must be a side, if all of the remaining parts are to be found.

Example 1. The angle of elevation (the angle to the top) of a power pole from a point 30 feet from the base is observed to be 64°. How high is the pole?
Here we know the angle and the side adjacent to the angle. We want to find the opposite side, and thus use the tangent function.

$$\tan A = \frac{a}{b} = \frac{\text{side opposite } \angle A}{\text{side adjacent to } \angle A}, \quad \tan 64° = \frac{a}{30},$$

$$\begin{aligned} a &= 30 \tan 64° \\ &= 30(2.0503) \\ &= 61.5 \text{ ft.} \end{aligned}$$

Example 2. A stabilizing wire is to be attached to a power pole at a point 25 ft. from the ground. If the anchor is at a point 30 ft. from the base, what will be the angle between the wire and the ground?

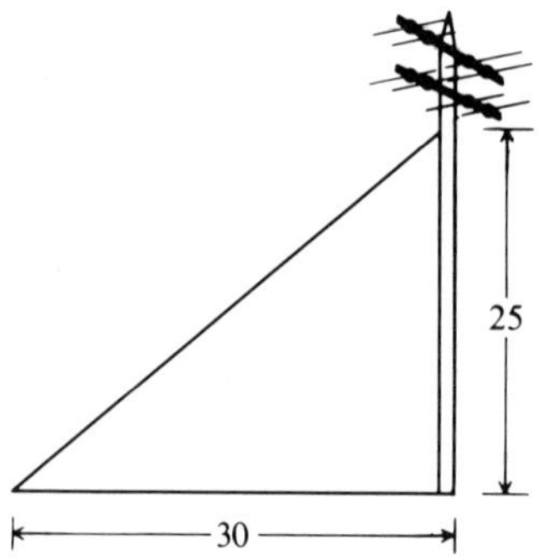

$$\tan A = \frac{\text{opp}}{\text{hyp}} = \frac{25}{30} = .8333.$$

$A = 40°$, to the nearest degree.

Example 3. Given equilateral triangle ABC with altitude BD. If each side is 3 inches, find the length of the altitude.

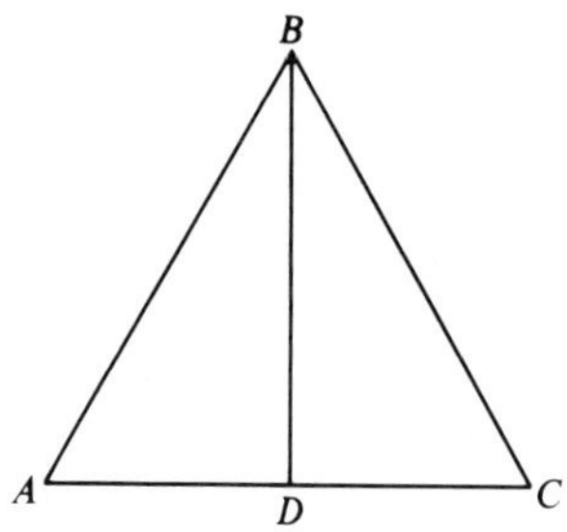

Sin $A = BD/AB$. We know $A = 60^\circ$ since the triangle is equilateral. We also know that $AB = 3$ in. We now have $\sin 60^\circ = BD/3$.

$$\frac{\sqrt{3}}{2} = \frac{BD}{3}; \qquad \frac{3\sqrt{3}}{2} = BD.$$

This may be approximated from tables as

$$\frac{3(1.732)}{2} = 2.598.$$

Exercise 61

1) Find approximations for the side labeled x in each of the following:

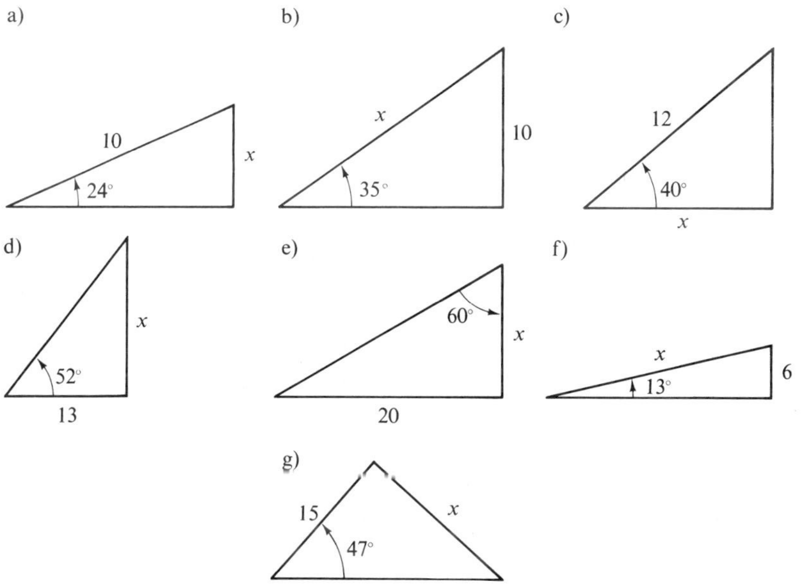

2) A missile was fired at an angle of 28°. How high was it in feet after it had traveled 427 meters? Assume it traveled in a straight line.

3) A 15-foot ladder makes an angle of 18° with a vertical wall. How far is the foot of the ladder from the wall?

4) Two sides of a rectangular piece of sheet metal are 13 inches and 16 inches. What angle does the diagonal make with the sides?

5) A rafter is to span a distance of 17 feet with a rise (vertical distance) of 4 feet. What angle should the carpenter cut the end of the rafter

so it will fit flush against a vertical wall (angle A in the diagram below)?

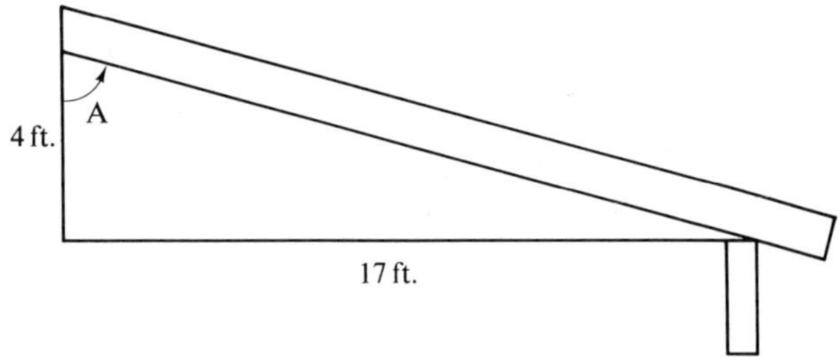

6) Compute the depth d of a screw thread according to the dimensions on the drawing.

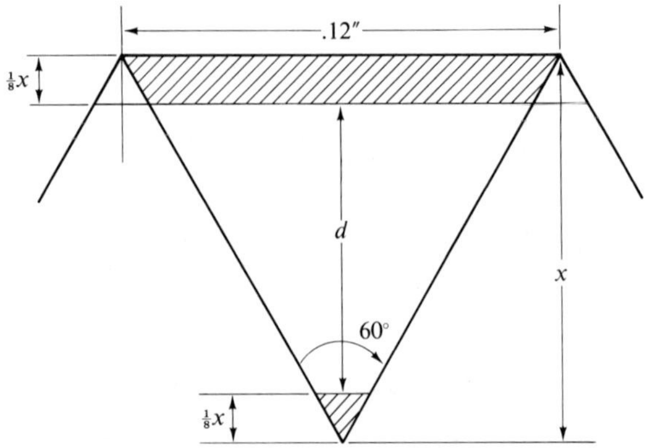

7) A pentagonal rod is to be milled from circular steel stock 2 inches in diameter. If each side of the pentagon is to be the same length, to what depth will the cutting tool need work?

8) Compute the height x of the gauge blocks if $y = 2.951$ in. and the sine bar length $S = 10$ in. Angle M is 12°.

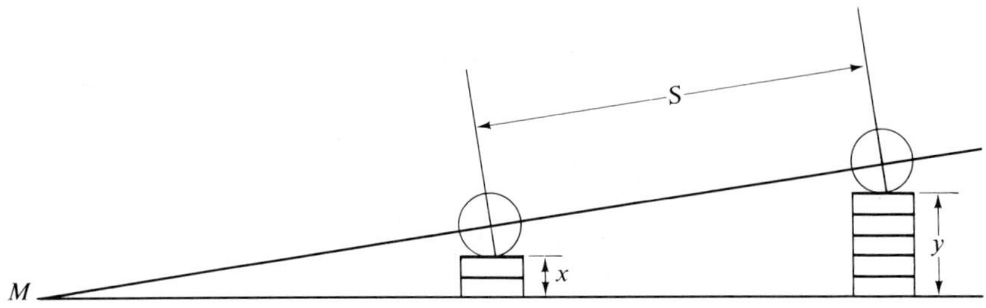

9) Compute the height of the gauge blocks in Problem 8 if $x = 1.20$ in. and $S = 10$ in. Angle M is 30°.

10) Use trigonometric functions to find the dimensions x and y on the diagram below.

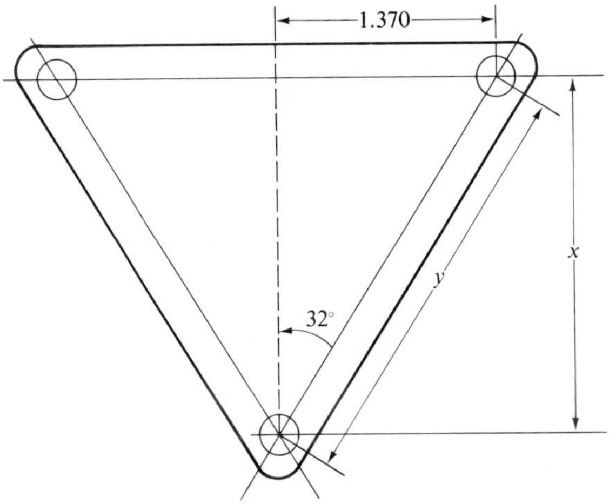

11) Compute the angle x on the tapered cylinder.

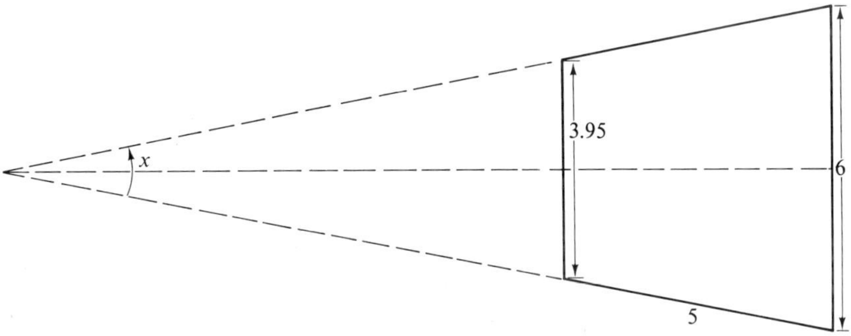

12) Determine the angle x for the lathe center shown below to the nearest degree.

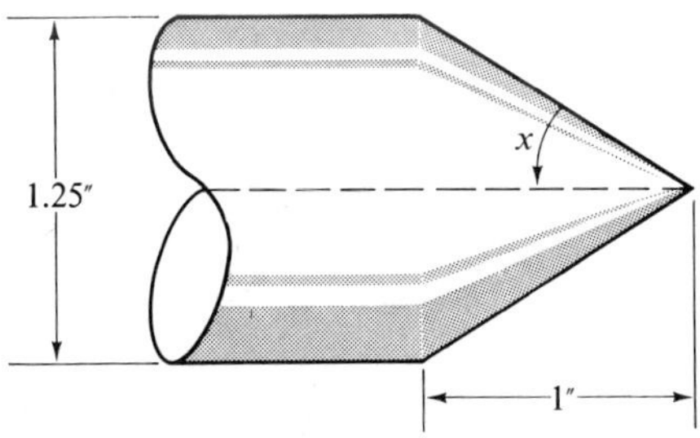

13) When light is reflected from a plane surface, the angle of incidence is equal to the angle of reflection. How far up on wall D would a light ray strike, if its source is located as in the diagram below?

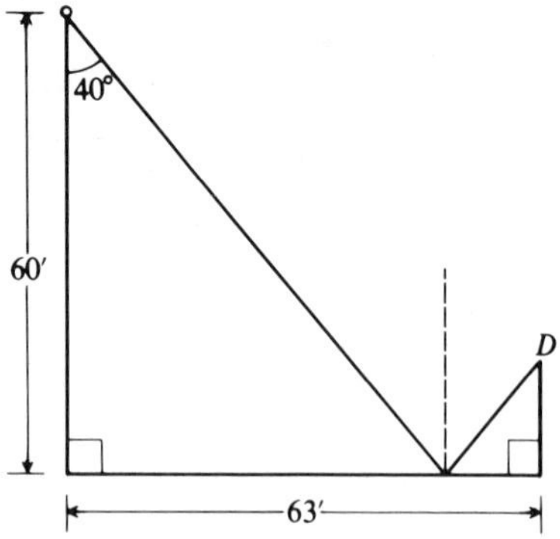

14) Find the length of chord AB shown in the diagram below. The central angle is 110° and the radius is 16″. (Hint: the line perpendicular to AB from the center O bisects angle AOB).

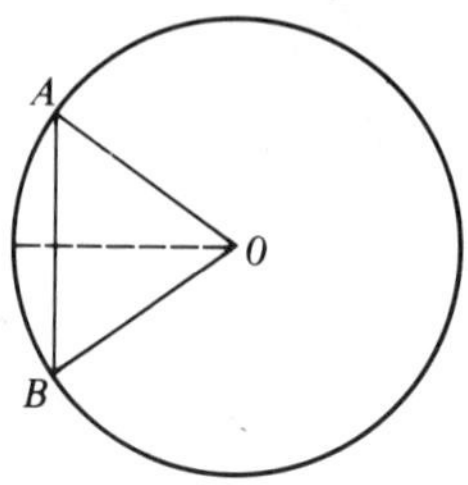

15) A flange is to have five equally spaced holes drilled as shown below. How far apart will the centers of two adjacent holes be located? The inner radius of the flange is 4″ and the outer radius is 6″.

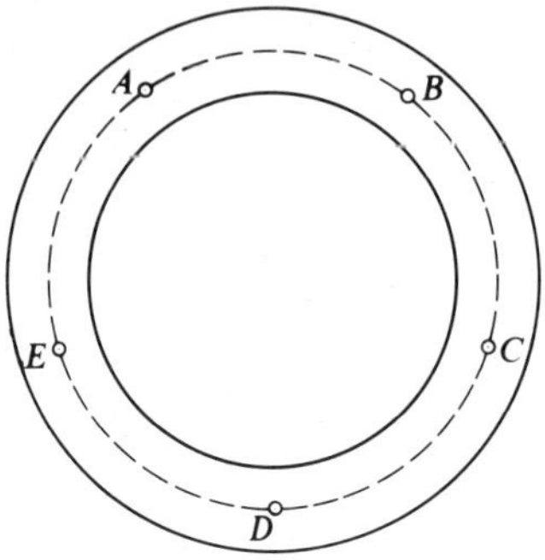

16) The end view of the roof of a building is shown below. Find angle AOB, the angle at which the rafters would be cut to butt the center beam.

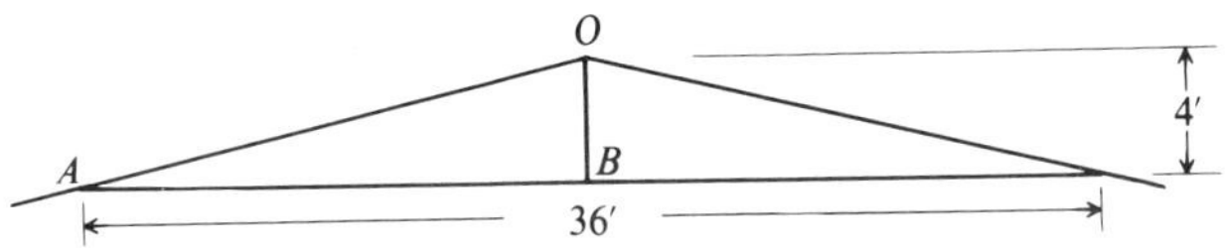

17) A right triangle called an *impedance triangle* is used to analyze circuits involving alternating current. The angle θ shown in the accompanying diagram is called the *phase angle*. If $M = 2213$ and $I = 5890$, find R and θ. (θ is the Greek letter theta.)

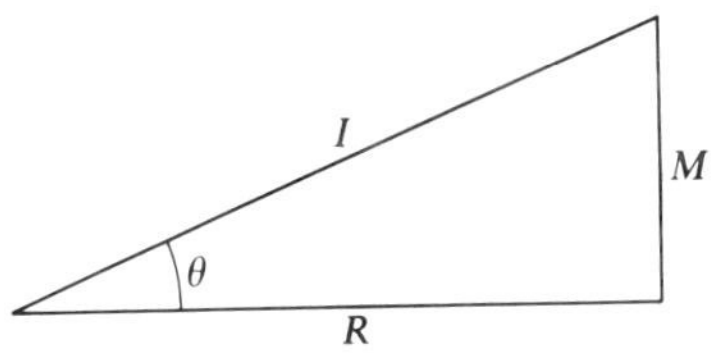

18) To measure the height of a vertical cliff, a surveyor measured a horizontal distance of 100 feet from its base and then found the angle of elevation of its top to be 63°. Determine the height of the cliff.

10.8 Coordinates Revisited

At this point the reader may wonder how he could solve a triangle that is not a right triangle. Such a question is quite valid, since problems involving nonright triangles arise frequently in everyday application of trigonometry to technology.

The basis for extending trigonometry to solutions of nonright triangles lies in the rectangular coordinate system discussed previously in the chapter on algebra. The following exercises provide a review of the basics of that system and are included for that purpose.

Exercise 62

1) Locate the quadrant in which each of the following points would be found.

a) (3,2)	b) (−3,2)	c) (2,4)
d) (−2,−4)	e) (6,−2)	f) (3,0)
g) (−3,6)	h) $\left(\frac{1}{2},\frac{3}{8}\right)$	i) (3,4)
j) $\left(-\frac{4}{3},2\right)$	k) (.5,−3.5)	l) (15,−12)
m) (−7,−12)	n) (4.5,−2)	o) (11,11)

2) For each part in Problem 1, plot the point.

3) For each part in Problem 1, use the Pythagorean theorem to find the distance of each point from the origin. The following example may be helpful.

Example: (2,4)

$$c^2 = a^2 + b^2 = (4)^2 + (2)^2$$
$$= 16 + 4 = 20.$$
$$c = \sqrt{20} \text{ or } 2\sqrt{5}.$$

The hypotenuse of the resulting triangle is always considered to be a positive quantity.

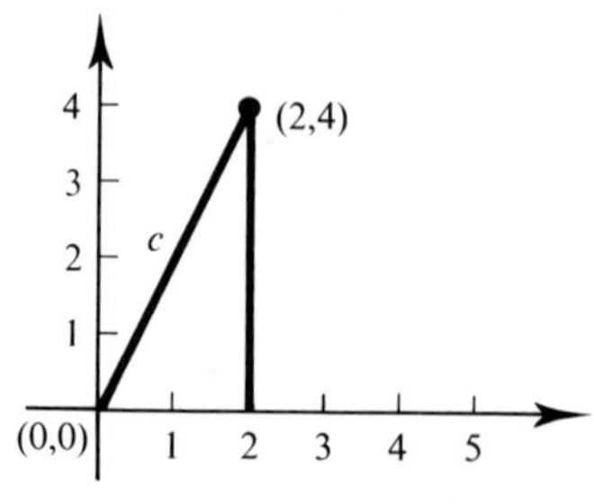

4) In each part of Problem 1, connect the point to the origin and, by using an appropriate trigonometric ratio, find the angle this line makes with the positive x axis to the nearest degree.

10.9 Standard Position of an Angle

An angle is considered to be in standard position in a rectangular coordinate system when its vertex is at the origin and its initial side is on the positive x axis. The measure of the angle is then determined by the amount of counterclockwise rotation between the initial and terminal sides (Figure 10.6).

By placing the angle in this position we could determine the coordinates of a point on the terminal side and, by dropping a perpendicular from this point to the initial side, we could determine the measure of the angle as well as the six trigonometric functions of the angle.

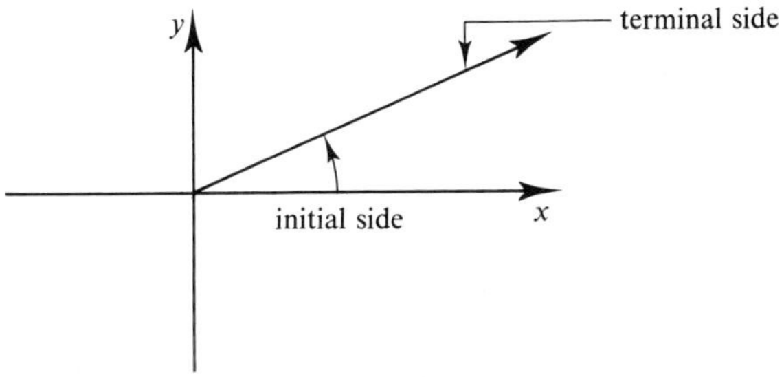

Figure 10.6

Example 1. Find the angle in standard position whose terminal side passes through (1,2).

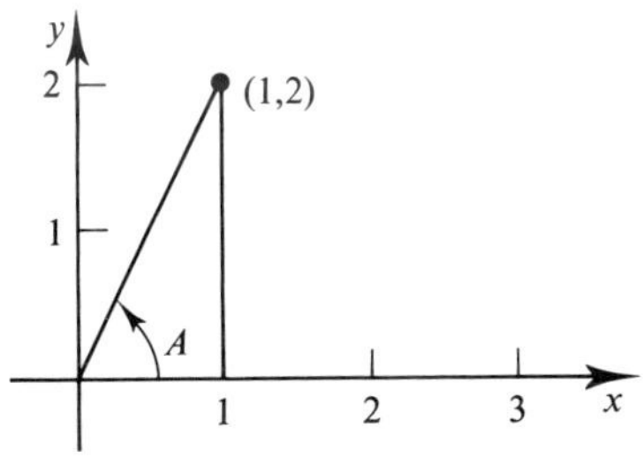

$\tan A = 2/1 = 2$. Using Table III, A is found to be approximately 63.5°, since 2 lies about midway between the entries $\tan 63° = 1.9626$ and $\tan 64° = 2.0503$.

Example 2. Find the six trigonometric ratios for an angle in standard position whose terminal side passes through $(x,1)$ and is 2 units long. (x is greater than 0.)

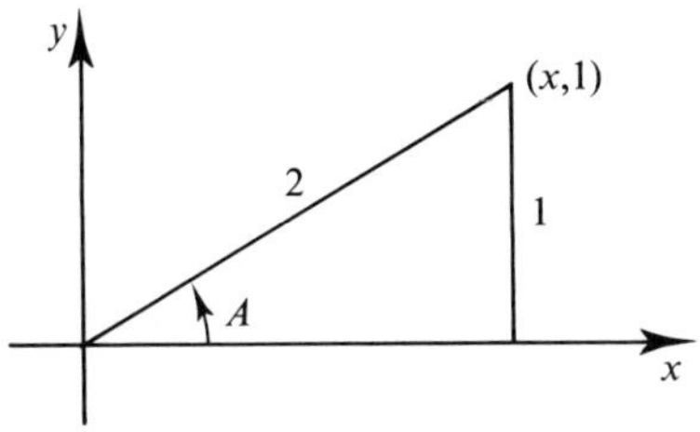

To determine the ratios we need to know both the adjacent and the opposite sides. Since the y coordinate of the point at the end of the terminal side is 1, the opposite side is one unit in length. We may now use the Pythagorean theorem to determine the adjacent side.

$$\begin{aligned} 2^2 &= 1^2 + x^2, \\ 4 &= 1 + x^2, \\ 3 &= x^2, \\ \sqrt{3} &= x. \end{aligned}$$

From this we have

$$\begin{aligned} \sin A &= \frac{1}{2}, \\ \cos A &= \frac{\sqrt{3}}{2}, \\ \tan A &= \frac{1}{\sqrt{3}}, \\ \cot A &= \sqrt{3}, \\ \sec A &= \frac{2}{\sqrt{3}}, \\ \csc A &= 2. \end{aligned}$$

Using $\sin A = 1/2$ or .5000 it is easy to determine that the measure of $\angle A$ is 30°.

10.10 Reference Angles

We now consider the question of how to determine the trigonometric ratios for obtuse angles. To answer this question we consider an angle, call it A, in standard position with the terminal side falling in quadrant II (Figure 10.7). Choosing any point (x,y) on the terminal side, we note that the x coordinate of this point is a negative number while the y coordinate is positive. For an acute angle whose terminal side is in the first quadrant, both the x and y coordinates of a reference point on the terminal side are positive. It is this difference that allows us to differentiate between acute and obtuse angles.

Notice that it is impossible to drop a perpendicular from a point on the terminal side of an obtuse angle to its initial side to form a right triangle. (Recall that the trigonometric ratios apply to right triangles.)

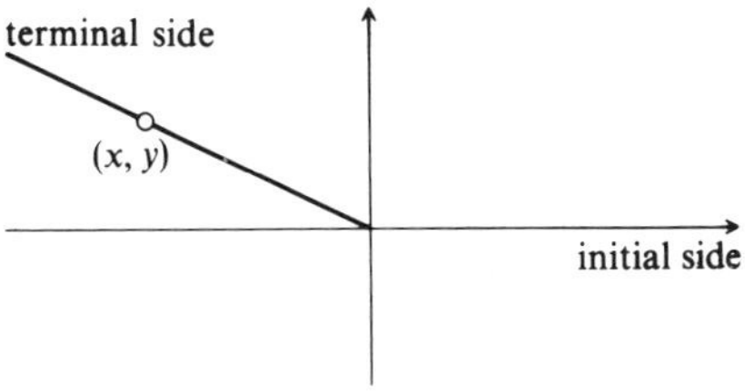

Figure 10.7

This problem is easily resolved by dropping a perpendicular to the negative extension of the initial side instead, forming a right triangle there (Figure 10.8).

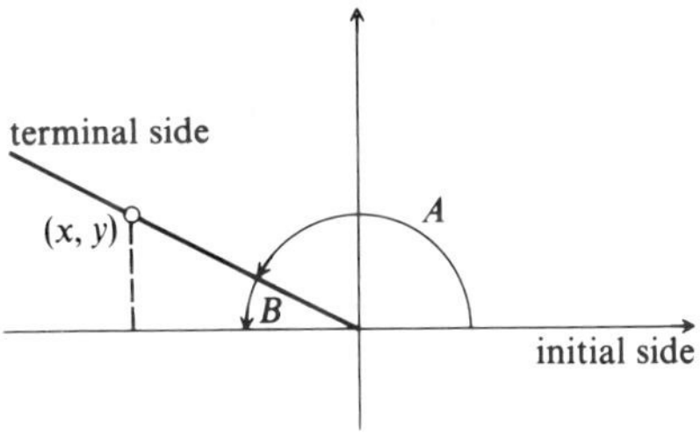

Figure 10.8

Using this triangle we could determine the six trigonometric ratios for $\angle B$ as well as the measure of $\angle B$. Cos B, tan B, sec B, cot B are all negative since the adjacent side is represented as a negative quantity.

$$\cos B = \frac{x \text{ (negative)}}{c \text{ (positive)}}$$
$$\tan B = \frac{y \text{ (positive)}}{x \text{ (negative)}}$$
$$\sec B = \frac{c \text{ (positive)}}{x \text{ (negative)}}$$
$$\cot B = \frac{x \text{ (negative)}}{y \text{ (positive)}}.$$

Sin B and csc B are positive quantities, since these ratios involve the hypotenuse (always positive) and the opposite side y which is positive in the second quadrant. B is known as the *reference angle* and is related to angle A in such a manner that $\angle A + \angle B = 180°$. Recall from the chapter on geometry that one complete rotation is 360° and $\angle A + \angle B$ form one half of a complete rotation. After finding the acute angle B as we did in Example 1 of the previous section, we determine

$$\angle A = 180° - \angle B.$$

Remember that the reference angle is the smallest angle measured between the terminal side and the x axis.

A similar discussion to the one above is used for angles between 180° and 360° or for that matter, for any angle regardless of its measure. This has been deferred to a later section.

Example 1. Find sin A, cos A, and tan A if A is 123°.

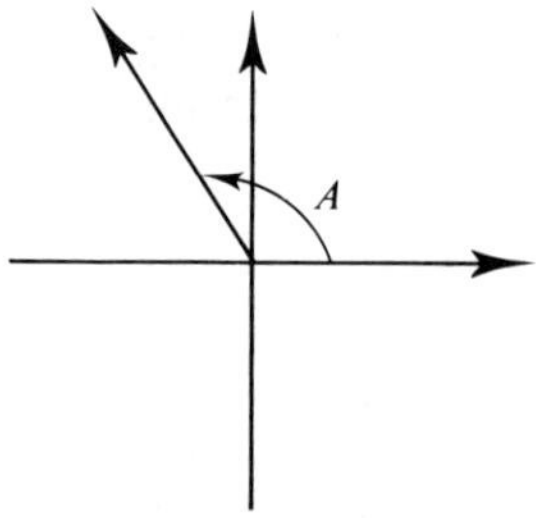

Here the reference angle is $180° - 123° = 57°$. The terminal side of the angle is in quadrant II so that sin A is positive while cos A and tan A are negative. Using Table III

$$\begin{aligned} \sin A &= \sin 57° = .8387, \\ \cos A &= -\cos 57° = -.5446, \\ \tan A &= -\tan 57° = -1.5399. \end{aligned}$$

Example 2. If $\sin A = .2756$ and tan A is negative, find $\angle A$. Since sin A is positive and tan A is negative, the angle must have its terminal side in quadrant II. From Table III we determine the reference angle to be approximately 16°, so that

$$\angle A = 180° - 16° = 164°.$$

Exercise 63

1) Determine whether the terminal side of angle A is in quadrant I or II for each of the following.
   a) sin A positive and tan A positive
   b) sin A positive and sec A negative
   c) cos A negative and cot A negative
   d) tan A positive and sec A positive
   e) cot A negative and sin A positive
   f) sin A positive and csc A positive

2) Which of the following angles are obtuse?
   a) $\sin A$ positive and $\tan A$ positive
   b) $\sec A$ negative and $\sin A$ positive
   c) $\sin A$ positive and $\tan A$ negative

3) Suppose the terminal side of an angle in standard position passes through the points below. Find the six trigonometric functions of each angle. Leave answers in fractional form.
   a) $(1,5)$ b) $(1,1)$ c) $(-1,3)$
   d) $(-2,5)$ e) $(-1,\sqrt{3})$ f) $(2,\sqrt{3})$

4) Find angle A to the nearest degree under the following conditions. Angle A is in either quadrant I or quadrant II.
   a) $\sin A = .4226$ b) $\tan A = .6494$
   c) $\cos A = -.3256$ d) $\tan A = -.2565$
   e) $\tan A = 1.9723$ f) $\tan A = 3.2154$
   g) $\sin A = .8750$; $\tan A$ is negative
   h) $\sin A = .9425$; $\cos A$ is positive
   i) $\cos A = -.4071$; $\sin A$ is positive
   j) $\tan A = 57.2900$; $\sin A$ is positive

5) Use Table III to find $\sin A$, $\cos A$, and $\tan A$ for each of the following angles:
   a) $62°$ b) $37°$ c) $122°$
   d) $1540°$ e) $30°$ f) $162°$
   g) $176°$ h) $0°$ i) $90°$

6) Can you use your tables to find $\csc 0°$? Why?

7) Can you use your tables to find $\sec 90°$? Why?

8) Standing at one point the angle of elevation of the top of a tower is 35°. From a point 20 feet closer to the tower, the angle of elevation is 50°. Find the height of the tower.

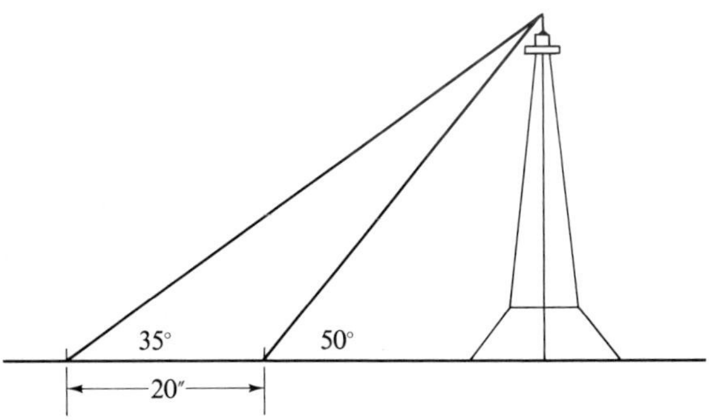

9) An air duct for heating is to be made by cutting a duct with square cross-section, then turning and joining the sections as in the diagram below. Find the cross-sectional area of the join.

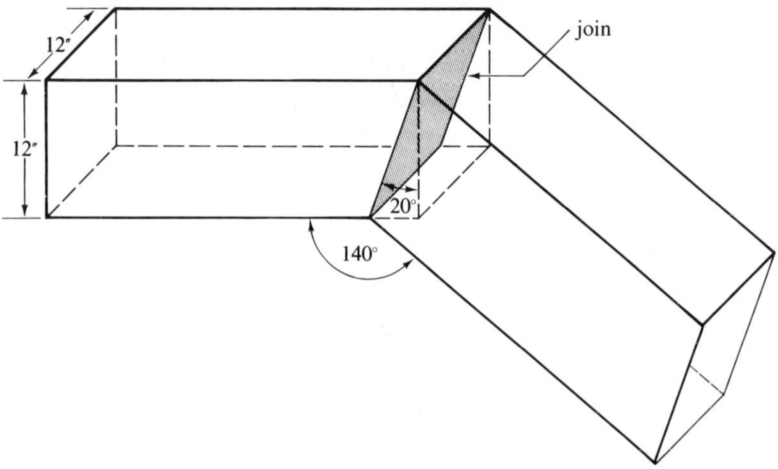

10) Compute distances x and y in the diagram below. Write answers to the nearest .1 inch. (Hint: compute distance d and angle T.)

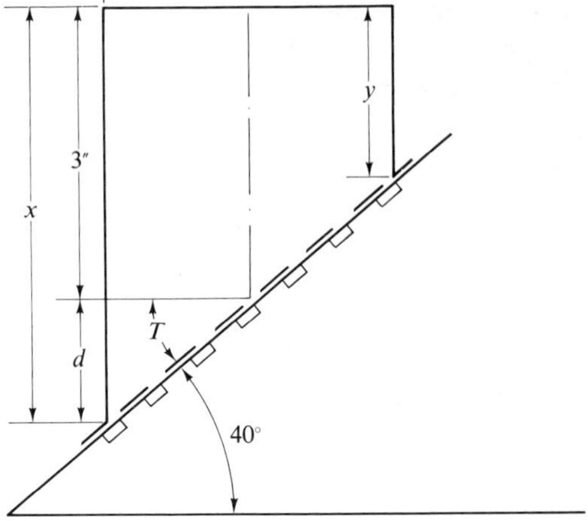

10.11 The Sine Law

Many methods for solving triangles which do not contain right angles exist, but only two of the more important ones will be covered here. These are the sine law and the cosine law. Each of these laws has special applications and will be treated separately.

The *sine law* states: *the ratio between any angle of a triangle and the side opposite that angle is a constant, or*

$$\frac{a}{\sin A} = \frac{b}{\sin B} = \frac{c}{\sin C}.$$

To demonstrate this law consider any triangle placed in such a manner on a coordinate system that its largest angle is at the origin in standard position (Figure 10.9).

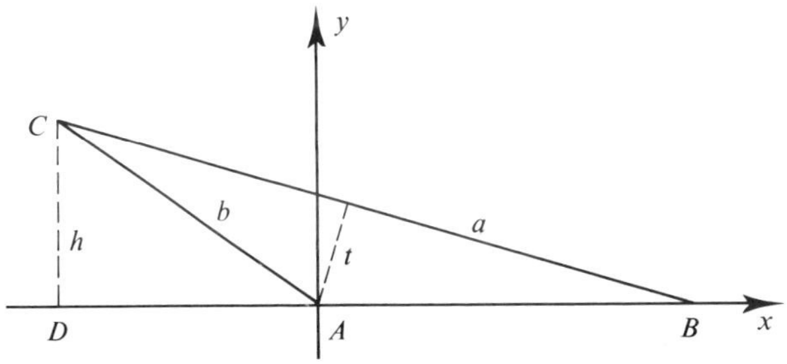

Figure 10.9

Drop a perpendicular from C to the horizontal line to the left of the origin and denote its length by h. From the diagram and right triangle BCD we have

$$\sin B = \frac{h}{a},$$
$$a \sin B = h.$$

From right triangle CAD we have

$$\sin A = \frac{h}{b},$$
$$b \sin A = h.$$

Both of these quantities are positive since the sine of an angle is positive for angles in quadrants I and II. We can now equate the two values of h, from which we obtain

$$a \sin B = b \sin A.$$

Dividing by sin A sin B yields

$$\frac{a}{\sin A} = \frac{b}{\sin B}.$$

We now construct a perpendicular from A to side BC and label the intersection E. AE, whose length we shall call t, is a side of two new right triangles, AEC and AEB. From right triangle AEC we have

$$\sin C = \frac{t}{b},$$
$$b \sin C = t.$$

From right triangle AEB we have

$$\sin B = \frac{t}{c},$$
$$c \sin B = t.$$

We now equate the two values of t, obtaining

$$b \sin C = c \sin B.$$

Dividing by sin B sin C yields

$$\frac{b}{\sin B} = \frac{c}{\sin C}.$$

Combining this result with

$$\frac{a}{\sin A} = \frac{b}{\sin B},$$

we obtain

$$\frac{a}{\sin A} = \frac{b}{\sin B} = \frac{c}{\sin C}.$$

The sine law can be used to solve triangles when certain parts are given, as the following examples illustrate.

Example 1. In triangle ABC, $\angle A = 40°$, $\angle B = 30°$, and $c = 5$ in. Find the other parts of the triangle.

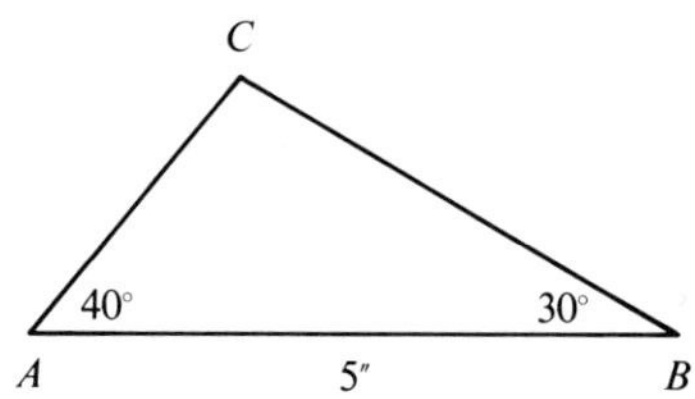

Since all triangles contain 180° we have

$$\angle C = 180° - (\angle A + \angle B) = 180° - 70° = 110°.$$

From the sine law we have

$$\frac{c}{\sin C} = \frac{a}{\sin A},$$
$$\frac{5}{\sin 110°} = \frac{a}{\sin 40°},$$
$$a = \frac{5 \sin 40°}{\sin 110°}$$
$$= \frac{5(.6428)}{.9397}$$
$$= 3.4202 \text{ inches.}$$

(Recall that $\sin 110° = \sin 70°$, where 70° is the reference angle.) Also

$$\frac{c}{\sin C} = \frac{b}{\sin B},$$
$$\frac{5}{\sin 110°} = \frac{b}{\sin 30°},$$
$$b = \frac{5 \sin 30°}{\sin 110°}$$
$$= \frac{5(.5)}{.9397}$$
$$= 2.6604 \text{ inches.}$$

Example 2. In triangle *ABC*, $\angle A = 26°$, $b = 4.21$ inches, and $a = 3$ inches. Solve for the other parts of the triangle.

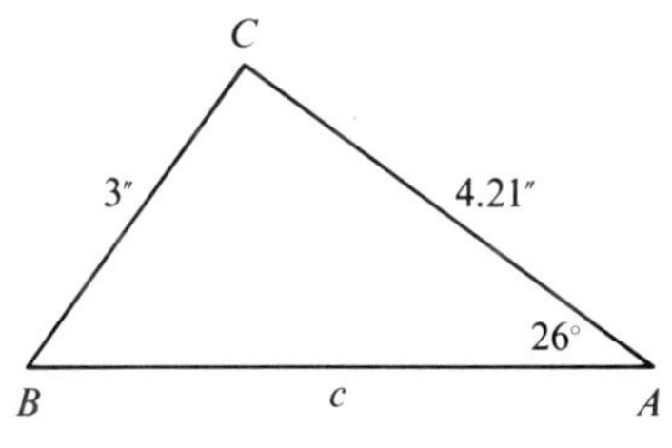

From the sine law we have

$$\frac{a}{\sin A} = \frac{b}{\sin B},$$
$$\frac{3}{\sin 26°} = \frac{4.21}{\sin B},$$
$$\sin B = \frac{4.21 \sin 26°}{3}$$
$$= \frac{4.21(.4384)}{3} = .6152.$$

If $\sin B = .6152$, $B = 38°$ to the nearest degree. Knowing $\angle A$ and $\angle B$ we determine $\angle C$, since

$$\begin{aligned}\angle C &= 180° - (\angle A + \angle B)\\ &= 180° - (38° + 26°) = 116°.\end{aligned}$$

Also,

$$\frac{a}{\sin A} = \frac{c}{\sin C},$$

$$\frac{3}{\sin 26°} = \frac{c}{\sin 116°},$$

$$c = \frac{3 \sin 116°}{\sin 26°}$$

$$= \frac{3(.8988)}{.4384} = 6.1505.$$

Example 3. In triangle ABC, $\angle A = 60°$, $b = 4$, and $c = 3$. Find the remaining parts of the triangle.

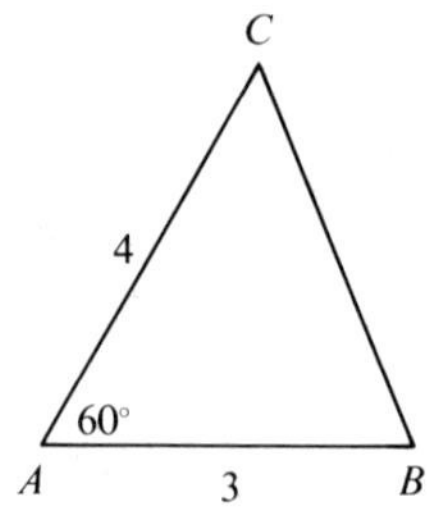

From the sine law we have

$$\frac{a}{\sin A} = \frac{b}{\sin B} \quad \text{and} \quad \frac{a}{\sin A} = \frac{c}{\sin C},$$

or

$$\frac{a}{\sin 60°} = \frac{4}{\sin B} \quad \text{and} \quad \frac{a}{\sin 60°} = \frac{3}{\sin C}.$$

Not enough information is given to allow us to solve for the remaining parts by using the sine law. To solve a triangle, one of the known sides must be opposite a known angle.

In all of the preceding examples the answers are only approximate, since the values taken from Table III are approximations.

Exercise 64

Solve the following triangles using the sine law:

1) $\angle B = 53°$, $\angle C = 42°$, $c = 7.4$.

2) $\angle C = 46°$, $\angle A = 73°$, $b = 236.7$.

3) $\angle A = 21°$, $\angle B = 48°$, $b = 8.78$.

4) $\angle C = 105°$, $\angle B = 32°$, $b = 36$.

5) $\angle B = 63°$, $\angle C = 18°$, $a = 0.62$.

6) $\angle A = 56°$, $\angle C = 15°$, $b = 4826$.

7) $a = 5$, $\angle A = 20°$, $\angle B = 75°$.

8) $a = 32$, $\angle A = 26°$, $\angle C = 81°$.

9) $a = 58.4$, $\angle B = 37°$, $\angle C = 100°$.

10) $b = 1.9$, $\angle A = 111°$, $\angle B = 5°$.

11) $b = 0.42$, $\angle A = 35°$, $\angle C = 91°$.

12 $b = 0.88$, $\angle B = 63°$, $\angle C = 34°$.

13) $c = 13.6$, $\angle A = 30°$, $\angle B = 72°$.

14) $c = 1.4$, $\angle A = 135°$, $\angle C = 34°$.

15) $\angle A = 30°$, $\angle B = 45°$, $a = 20$.

16) A and B are two points on opposite sides of a river. From A the line AC is laid off parallel to the bank so that $AC = 250$ feet and the angles $CAB = 120°$ and $ACB = 48°$ are measured. Find AB.

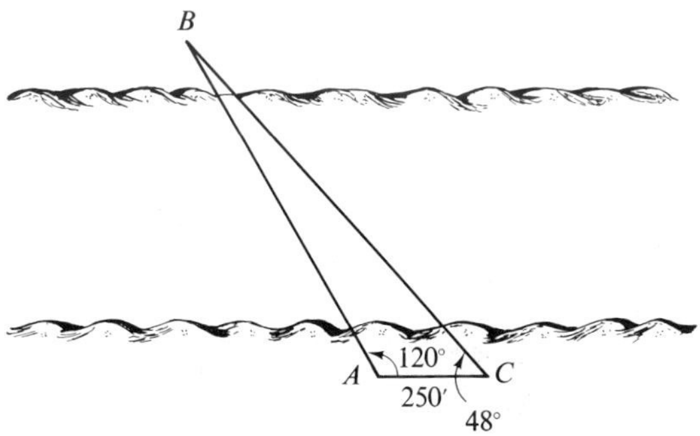

17) Two fire observation posts are 30 miles apart and one is due north of the other. A ranger at one tower observes a fire to be S 60° E and the ranger at the other tower observes the fire to be N 40° E. Find the distance of the fire from both observation posts.

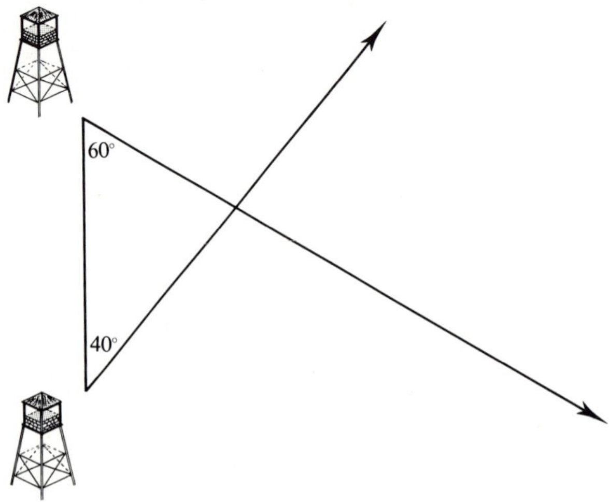

18) A city lot is in the shape of a right trapezoid (a trapezoid containing a right angle) as shown in the diagram below. If the width is 60 feet and the two parallel sides are 70 feet and 120 feet, find the length of the fourth side and the angle it makes with the two parallel sides. Find the area of the lot.

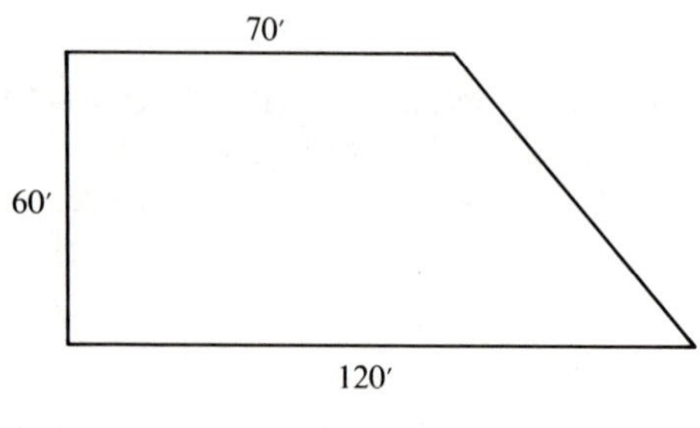

19) An iron wedge for splitting wood is designed with a base of $2\frac{1}{2}$ inches and a vertex angle of 17°. Find the length x of the edges.

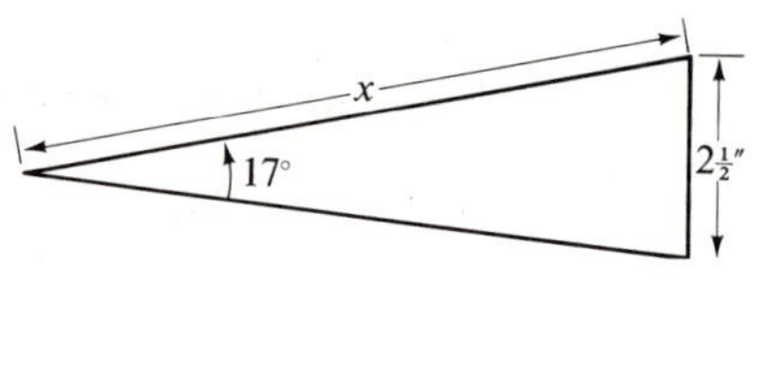

20) Find the area of a parallelogram if the angle between sides 8 inches and 12 inches long is 48°.

21) A bolt is to be machined so that the head is in the shape of a regular octagon. If the octagon can be inscribed in a circle with a $\frac{1}{2}$ inch radius, how long is the length of one side of the bolt head?

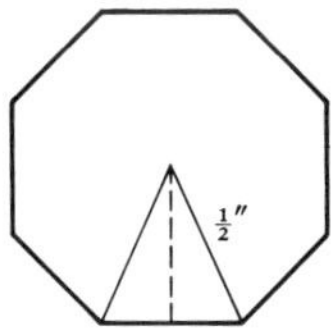

22) Solve Problem 21 if the bolt is to have a head with 6 sides, with 7 sides, with 5 sides, with 10 sides.

10.12 An Important Identity

In an earlier section, in right triangle ABC as shown in Figure 10.10, the sine of angle A was defined as a/c, and the cosine of angle A was defined

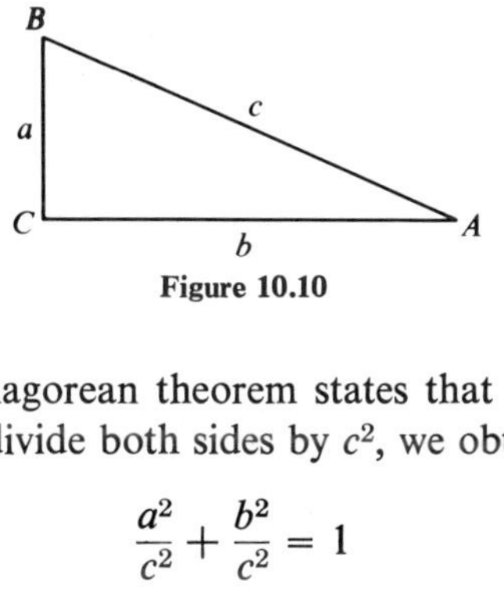

Figure 10.10

as b/c. Also, the Pythagorean theorem states that $a^2 + b^2 = c^2$. In this latter equation, if we divide both sides by c^2, we obtain

$$\frac{a^2}{c^2} + \frac{b^2}{c^2} = 1$$

which can be written as

$$\left(\frac{a}{c}\right)^2 + \left(\frac{b}{c}\right)^2 = 1.$$

Since $\sin A = a/c$ and $\cos A = b/c$, by substitution, the equation becomes $(\sin A)^2 + (\cos A)^2 = 1$ which is usually written as $\sin^2 A + \cos^2 A = 1$.

This statement is often referred to as the fundamental Pythagorean identity and states that the sine squared of any angle plus the cosine squared of the same angle is equal to 1.

Example 1. $\sin 45° = \frac{\sqrt{2}}{2}$, $\cos 45° = \frac{\sqrt{2}}{2}$

$$\sin^2 45° + \cos^2 45° = \left(\frac{\sqrt{2}}{2}\right)^2 + \left(\frac{\sqrt{2}}{2}\right)^2$$
$$= \frac{2}{4} + \frac{2}{4} = \frac{4}{4} = 1.$$

Example 2. $\sin 60° = \frac{\sqrt{3}}{2}$, $\cos 60° = \frac{1}{2}$

$$\sin^2 60° + \cos^2 60° = \left(\frac{\sqrt{3}}{2}\right)^2 + \left(\frac{1}{2}\right)^2$$
$$= \frac{3}{4} + \frac{1}{4} = 1.$$

This result is important in developing the result of the next section.

10.13 The Cosine Law

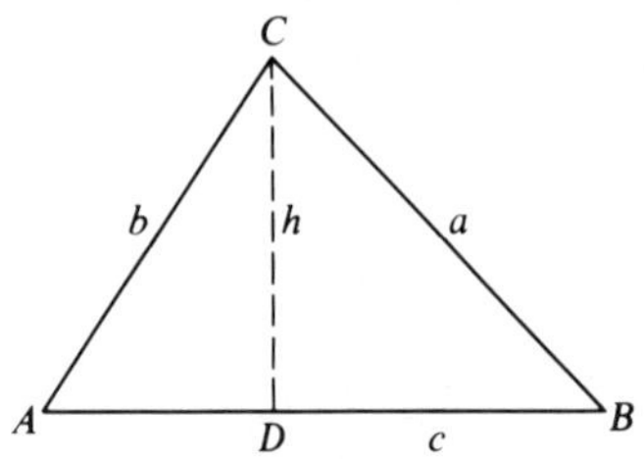

Figure 10.11

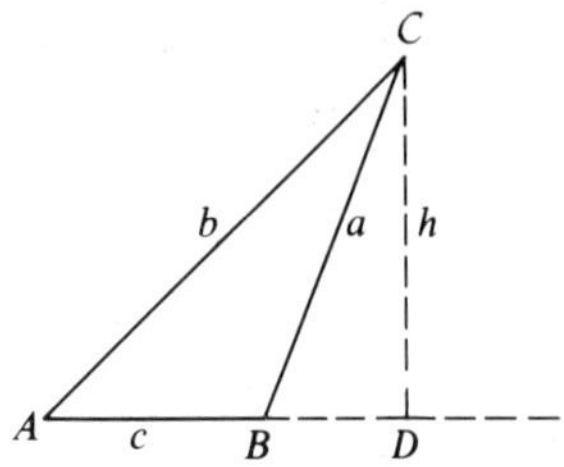

Figure 10.12

Consider the triangles illustrated in Figure 10.11 and Figure 10.12. In either triangle, by the Pythagorean theorem, $b^2 = h^2 + (AD)^2$. In Figure 10.11 $h = b \sin A = a \sin B$, and $DB = a \cos B$. Also $AD = AB - DB$. Since $AB = c$ and $DB = a \cos B$, $AD = c - a \cos B$,

$$\begin{aligned} b^2 &= a^2 \sin^2 B + (c - a \cos B)^2 \\ &= a^2 \sin^2 B + c^2 - 2ac \cos B + a^2 \cos^2 B \\ &= c^2 - 2ac \cos B + a^2 (\sin^2 B + \cos^2 B) \\ &= c^2 + a^2 - 2ac \cos B. \end{aligned}$$

The resulting equation

$$b^2 = a^2 + c^2 - 2ac \cos B$$

is one form of the *Cosine Law*. By a different use of labeling and angles the following alternate forms of the law can be written.

$$c^2 = a^2 + b^2 - 2ab \cos C$$
$$a^2 = b^2 + c^2 - 2bc \cos A$$

Stated in words, the cosine law says:

The square of any side of a triangle is equal to the sum of the squares of the other two sides decreased by twice the product of these two sides times the cosine of the angle included between the two sides.

Let us return to Figure 10.12 for a moment. In this figure, $h = a \sin \angle CBD = a \sin (180 - B) = a \sin B$, and $BD = a \cos \angle CBD = a \cos (180 - B) = -a \cos B$.
Then, again substituting the values into $b^2 = h^2 + (AD)^2$, where $AD = AB + BD = c - a \cos B$,

$$\begin{aligned} b^2 &= a^2 \sin^2 B + (c - a \cos B)^2 \\ &= a^2 \sin^2 B + c^2 - 2ac \cos B + a^2 \cos^2 B \\ &= c^2 - 2ac \cos B + a^2 (\sin^2 B + \cos^2 B) \\ &= a^2 + c^2 - 2ac \cos B. \end{aligned}$$

We can thus see that the law of cosines holds true for any triangle regardless of whether the triangle is acute or obtus

The cosine law gives us a method for solving triangles not solvable by use of the law of sines, that is, triangles involving three known sides or triangles involving two known sides and the angle included between these two sides.

Example 1. In triangle ABC, $\angle A = 60°$, $b = 4$, and $c = 3$. Find the remaining parts of the triangle.

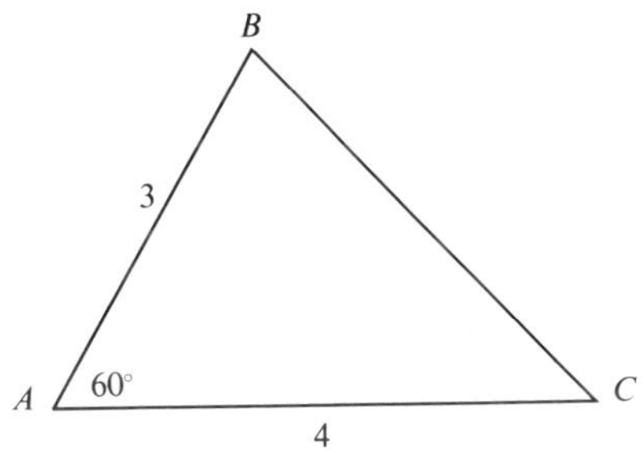

From the cosine law we have

$$\begin{aligned} a^2 &= b^2 + c^2 - 2bc (\cos A) \\ &= 4^2 + 3^2 - 2(4)(3) \cos 60° \\ &= 16 + 9 - 24(.500) \\ &= 25 - 12 = 13, \\ a &= \sqrt{13} = 3.606. \end{aligned}$$

To find the remaining parts we now use the sine law.

$$\frac{a}{\sin A} = \frac{b}{\sin B}.$$

$$\frac{3.606}{\sin 60^\circ} = \frac{4}{\sin B}.$$

$$\sin B = \frac{4 \sin 60^\circ}{3.606}$$

$$= \frac{4(.8660)}{3.606} = .9606$$

$$\angle B = 74^\circ.$$

Finally, to find $\angle C$, we have

$$\begin{aligned} \angle C &= 180^\circ - (\angle A + \angle B) \\ &= 180^\circ - (60^\circ + 74^\circ) = 46^\circ. \end{aligned}$$

Example 2. Solve for side c if $a = 132$, $b = 224$, and $\angle C = 29^\circ$. From the cosine law we have

$$\begin{aligned} c^2 &= a^2 + b^2 - 2ab\,(\cos C) \\ &= (132)^2 + (224)^2 - 2(132)(224) \cos 29^\circ \\ &= 17424 + 50176 - (59136)(.8746) \\ &= 17424 + 50176 - 51720 \\ &= 15880. \\ c &= 126. \end{aligned}$$

Example 3. Solve the triangle having $a = 5$ in., $b = 6$ in., $\angle C = 45^\circ$. From the law of cosines,

$$\begin{aligned} c^2 &= a^2 + b^2 - 2ab \cos C \\ c^2 &= 25 + 36 - 2(5)(6)(.7071) \\ &= 61 - 60(.7071) = 61 - 42.426 \\ &= 18.574 \\ c &= \sqrt{18.574} \\ &= 4.31 \text{ in. to the nearest hundredth inch.} \end{aligned}$$

Using the sine law,

$$\frac{\sin 45^\circ}{4.31} = \frac{\sin B}{6}.$$

$$\sin B = \frac{6(\sin 45^\circ)}{4.31} = \frac{6(.7071)}{4.31} = .9844.$$

From Table III, $\angle B = 80^\circ$ to the nearest degree. Therefore, $\angle A = 180^\circ - (B + C) = 180^\circ - (45^\circ + 80^\circ) = 55^\circ$. This gives us all the necessary values to complete the solution to the triangle.

When three sides of a triangle are given, it is useful to write the cosine law in one of the following forms

$$\cos A = \frac{b^2 + c^2 - a^2}{2bc}$$

$$\cos B = \frac{a^2 + c^2 - b^2}{2ac}$$

$$\cos C = \frac{a^2 + b^2 - c^2}{2ab}$$

These forms result from solving the previous equations for the cosine of the angle involved in each.

Example 4. Solve the triangle having $a = 3$, $b = 4$, $c = 5$.
By the cosine law,

$$\cos A = \frac{16 + 25 - 9}{2(4)(5)} = \frac{32}{40} = \frac{4}{5} = .533,$$

$\angle A = 37°$ to the nearest degree.

$$\cos B = \frac{9 + 25 - 16}{2(3)(5)} = \frac{18}{30} = \frac{3}{5} = .6,$$

$\angle B = 53°$ to the nearest degree.

$$\angle C = 180° - (53° + 37°) = 180° - 90°,$$

$\angle C = 90°$ which completes the solution.

Example 5. Solve for $\angle C$ if $a = 4$, $b = 2$, and $c = 5$.
From the cosine law we have

$$\begin{aligned}\cos C &= \frac{a^2 + b^2 - c^2}{2ab} \\ &= \frac{16 + 4 - 25}{2 \cdot 4 \cdot 2} \\ &= \frac{-5}{16} \\ &= -.3125.\end{aligned}$$

Since $\cos C$ is negative the angle is obtuse. We use the tables to find the reference angle to be approximately 72°. From this $\angle C = 180° - 72° = 108°$.

Example 6. Brace x needs to be constructed between two adjoining beams as shown. Find the length of the brace.

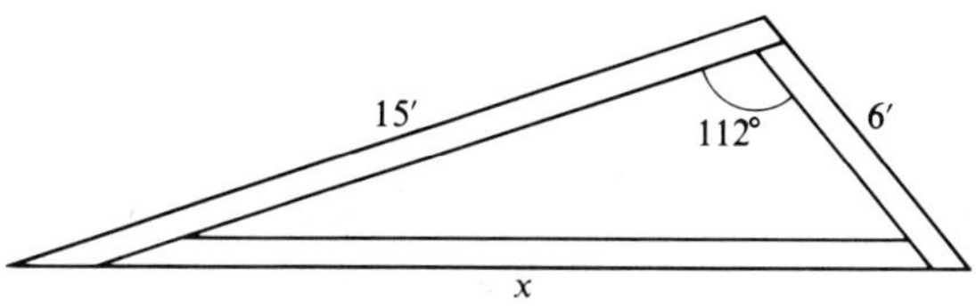

By the cosine law,

$$\begin{aligned} x^2 &= (15)^2 + (6)^2 - 2(6)(15)\cos 112^\circ \\ &= 225 + 36 - 180(-.3746) \\ &= 261 + 67.43 \\ &= 328.43 \\ x &= 18.12 \text{ to the nearest hundredth inch.} \end{aligned}$$

Exercise 65

1) Using $\sin^2 A + \cos^2 A = 1$, $0 < A < 90°$, find $\sin A$ if $\cos A =$
   a) 3/5 b) 5/13,
   c) .5 d) 3/7

2) Find $\cos A$ if $\sin A =$
   a) 1/8 b) $\sqrt{3}/2$
   c) 12/13 d) .7

Solve the following triangles:

3) $\angle A = 26°$, $b = 15$, $c = 19$

4) $\angle A = 36°$, $\angle B = 47°$, $c = 20$

5) $\angle C = 105°$, $a = 3.7$, $b = 6.5$

6) $a = 15$, $b = 18.3$, $c = 21.7$

7) A surveyor wishes to find the elevation of a hill by measuring an angle from the base of the hill to the top of the hill to be 40°. Moving 100 yards away from the hill, a sighting with the top of the hill yielded an angle of 32°. Find the height of the hill. (h in the diagram.)

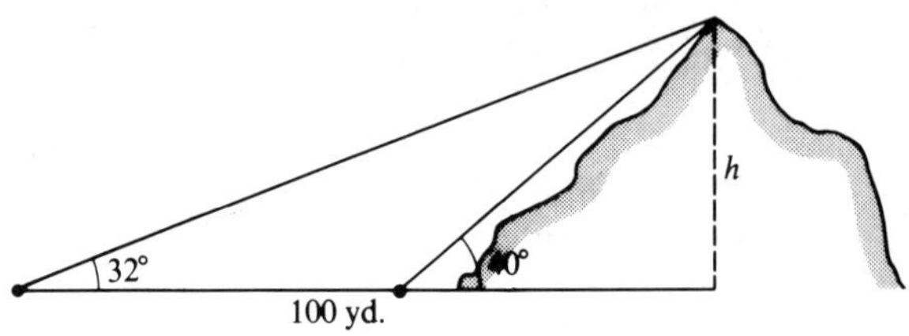

8) A city lot is the shape of a triangle with sides of 205 ft., 196 ft., and 287 ft. Find the size of each angle in the lot.

10.14 Trigonometric Functions of Any Angle

In an earlier section the measure of an angle was defined to be the amount of counterclockwise rotation between the initial and the terminal sides. Such an angle is considered to be a positive angle. If the rotation is in a clockwise direction, the angle is considered to be negative. Figure 10.13 shows this relationship.

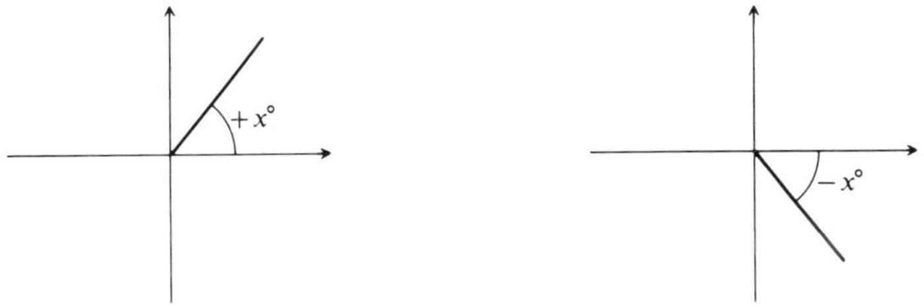

Figure 10.13

We also limited our discussion of angles in those sections to angles that were between 0° and 180° inclusive. Such a restriction is neither necessary nor desirable as can be seen in the following discussion and examples.

An angle whose terminal side falls in the third quadrant has a measure that lies between 180° and 270° if generated in a positive direction or whose measure lies between $-90°$ and $-180°$ if generated in a negative direction. Figure 10.14 shows this relationship for a 225° angle.

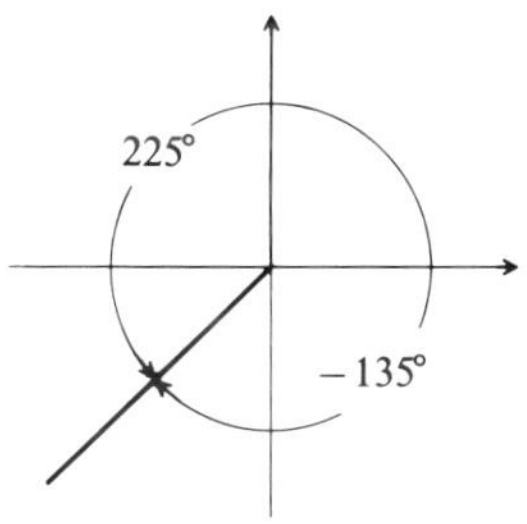

Figure 10.14

Several examples of angles greater than 180° are shown in the following figures. Note that some of the angles are greater than 360° and that the amount of rotation in either a positive or negative direction is indicated by the curved arrows.

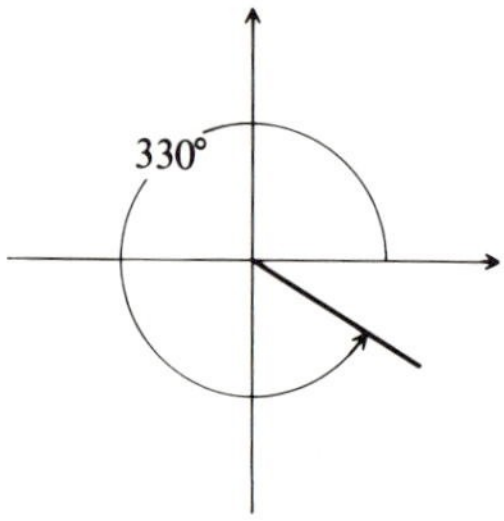

Figure 10.15

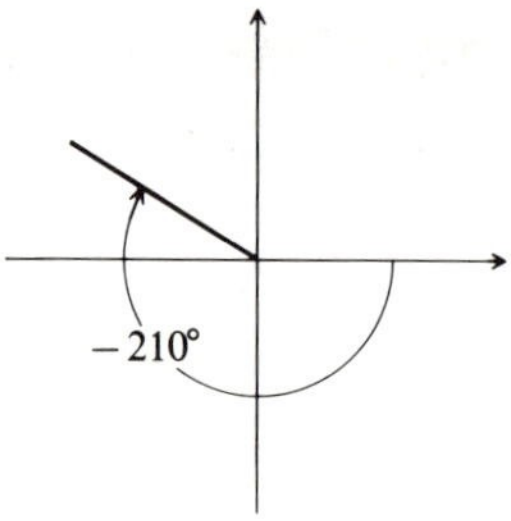

Figure 10.16

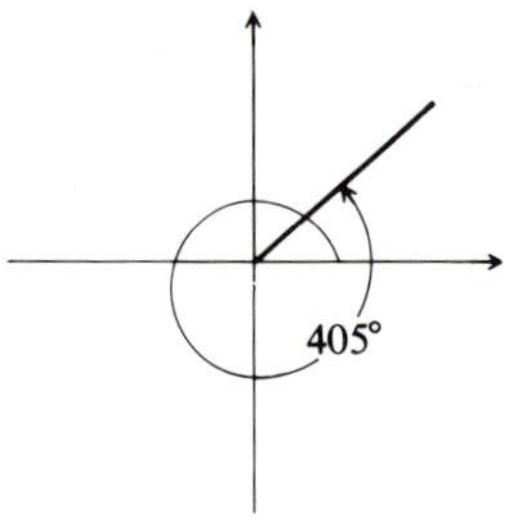

Figure 10.17

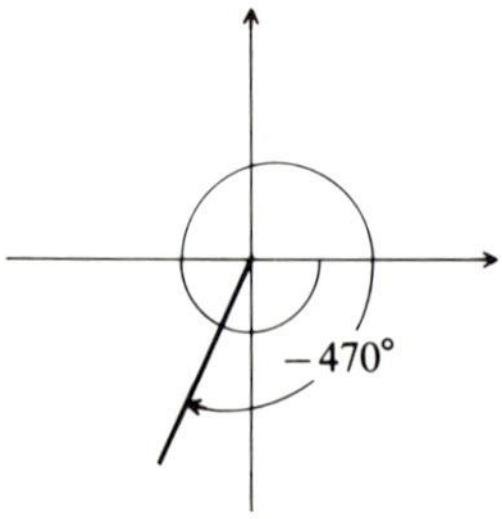

Figure 10.18

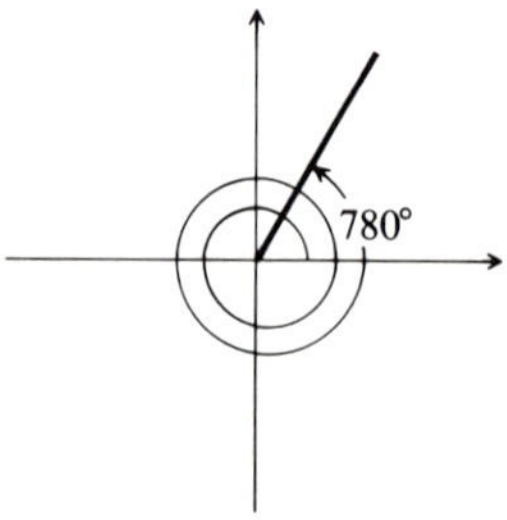

Figure 10.19

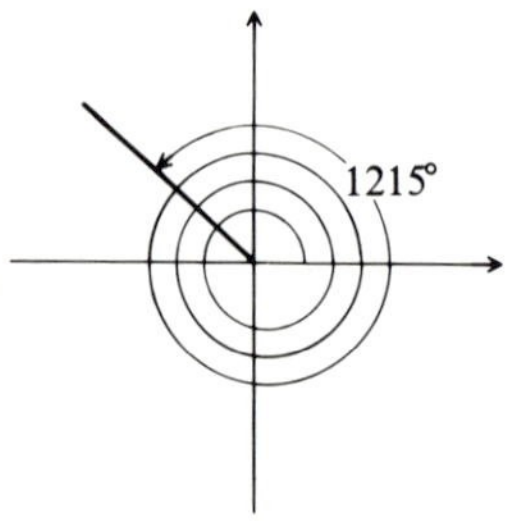

Figure 10.20

Figure 10.18 shows an angle in which the initial side has passed through one complete rotation, 360°, plus 110° all in the negative direction. In Figure 10.20, the initial side has passed through three complete rotations $3 \cdot 360° = 1080°$ plus 135°.

To find the trigonometric functions of angles greater than 180° or of angles generated in a negative direction, we follow the same procedures that we used for angles between 90° and 180°; we drop a perpendicular to the x axis and work with the reference angle formed by the terminal side and the x axis.

Example 1. What is the reference angle for 295°?

We subtract 295° from 360° to find the reference angle 65°.

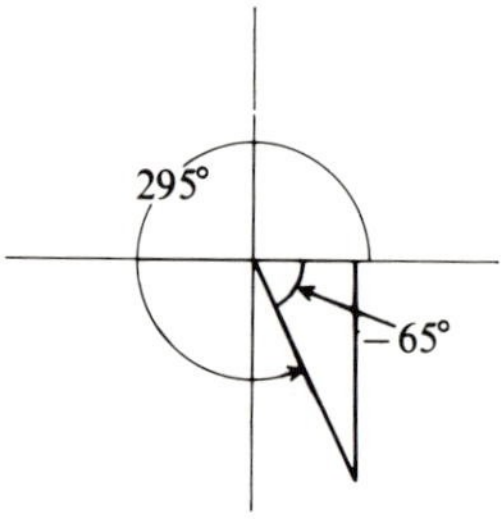

Example 2. Find the reference angle for $-196°$.
This angle would fall 16° into the second quadrant and the reference angle would contain 16°.

Example 3. Find the reference angle for 1363°.
Three complete revolutions of the terminal side would be 1080°. Therefore, we have $1363° - 1080° = 283°$. This angle falls in the fourth quadrant, and the reference angle is $360° - 283° = 77°$.

To find the trigonometric functions of the reference angle we use the same procedure we used before. However, we now must note the quadrant in which the angle lies so that the proper algebraic sign can be determined.

Example 4. Find cos 230°.
This angle lies in quadrant III and the reference angle is 50°. Since $\cos A = \text{adj}/\text{hyp}$ and since the adjacent side is negative in the third quadrant $\cos A$ will be negative also. From Table III $\cos 50° = .6428$. Therefore, $\cos 230° = -\cos 50° = -.6428$.

Example 5. Find tan $(-135°)$.
$-135°$ lies in quadrant III with a related angle of 45°. Tan $A = \text{opp}/\text{adj}$ and both the opposite and adjacent sides are negative in this quadrant. Thus $\tan(-135°)$ will be positive. Tan $(-135°) = \tan 45° = 1$.

Example 6. Find sin 390°.
390° lies in quadrant I and the reference angle is 30°. Sin $390° = \sin 30° = 1/2$.

As a final topic, we examine the procedure for finding an angle if its trigonometric function is known. The following examples illustrate this procedure. In each example we have imposed the arbitrary restriction $0° \leq A < 360°$.

Example 7. Given $\sin A = \sqrt{3}/2$, find A.
We know that $\sin 60° = \sqrt{3}/2$ and therefore that the reference angle is 60°. However, $\sin A$ is positive in quadrants I and II. Thus $A = 60°$ or $A = 180° - 60° = 120°$.

Example 8. Given $\tan A = -2.065$, find A.
From the table, $\tan 64° = 2.065$. Tan A is negative if the terminal side of A is in quadrants II or IV. Thus $A = 180° - 64°$ or $A = 360° - 64°$. $A = 116°$ or $A = 296°$.

Exercise 66

1) Find the reference angle for each of the following angles.

a) $220°$	b) $-120°$	c) $321°$
d) $456°$	e) $-229°$	f) $576°$
g) $-720°$	h) $391°$	i) $227°$
j) $-930°$	k) $-170°$	l) $-215°$
m) $318°$	n) $227°$	

2) Find sin A, cos A and tan A for each of the angles in Problem 1.

3) In which quadrants is the sin A positive?

4) In which quadrants is cos A positive?

5) In which quadrants is tan A positive?

6) If tan A is positive and sin A is negative, in what quadrant does angle A lie?

7) Determine what quadrant or quadrants contain the terminal side of angle A.
   a) sin A positive and tan A positive
   b) sin A positive and tan A negative
   c) cos A negative and cot A positive
   d) tan A positive and cot A positive
   e) sin A negative and cos A positive
   f) tan A negative and csc A positive
   g) cot A positive and sec A positive

8) Find angle A in each of the following. $0 \leq A < 360°$.

a) $\sin A = .3746$	b) $\tan A = .5095$
c) $\sec A = 1.051$	d) $\tan A = -1.428$
e) $\cos A = .9135$	f) $\tan A = .9325$
g) $\csc A = 1.662$	h) $\cos A = -.7071$

10.15 Trigonometric Identities

In Section 10.2 we discussed the reciprocal relationship existing between certain of the trigonometric functions. In Section 10.12 we developed the fact that the sine squared of an angle plus the cosine squared of the same angle was equal to one, called the fundamental Pythagorean identity. In this section and succeeding sections we will look at other useful relationships between functions. The relationships between various trigonometric functions are called *trigonometric identities*. These will be true for all angles except those causing a denominator to take on a value of zero.

10.16 Quotient Identities

From the definitions, in right triangle ABC as shown in Figure 10.21, $\sin A = a/c$ and $\cos A = b/c$.

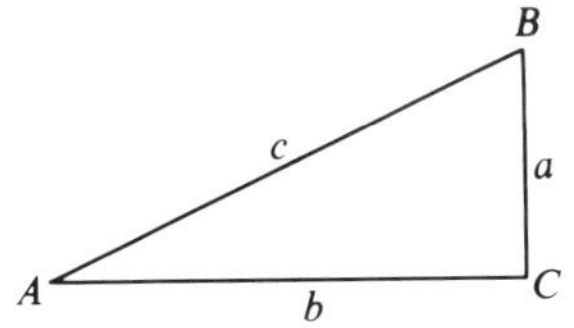

Figure 10.21

Dividing $\sin A$ by $\cos A$,

$$\frac{\frac{a}{c}}{\frac{b}{c}} = \frac{a}{c} \cdot \frac{c}{b} = \frac{a}{b}$$

which is $\tan A$.

Therefore, we obtain the quotient relationship

$$\frac{\sin A}{\cos A} = \tan A \text{ for any angle } A.$$

Also, using the same method we obtain

$$\frac{\cos A}{\sin A} = \cot A$$

Example 1. If $\sin x = 3/5$ and $\cos x = 4/5$, find $\tan x$ and $\cot x$.

$$\tan x = \frac{\sin x}{\cos x} = \frac{\frac{3}{5}}{\frac{4}{5}} = \frac{3}{5} \cdot \frac{5}{4} = \frac{3}{4}.$$

$$\cot x = \frac{\cos x}{\sin x} = \frac{\frac{4}{5}}{\frac{3}{5}} = \frac{4}{3}.$$

We could have obtained the second result by

$$\cot x = \frac{1}{\tan x} = \frac{1}{\frac{3}{4}} = \frac{4}{3}.$$

10.17 Other Pythagorean Identities

Using the quotient identities just established, the reciprocal identities and the fundamental Pythagorean identity, we can obtain some further identities. If we divide both sides of $\sin^2 x + \cos^2 x = 1$ by $\cos^2 x$,

we obtain
$$\frac{\sin^2 x}{\cos^2 x} + \frac{\cos^2 x}{\cos^2 x} = \frac{1}{\cos^2 x}$$

$$\left(\frac{\sin x}{\cos x}\right)^2 + 1 = \left(\frac{1}{\cos x}\right)^2$$

Since $\dfrac{\sin x}{\cos x} = \tan x$ and $\dfrac{1}{\cos x} = \sec x$, this becomes

$$\tan^2 x + 1 = \sec^2 x.$$

Also, dividing both sides of $\sin^2 x + \cos^2 x = 1$ by $\sin x$, we obtain

$$\frac{\sin^2 x}{\sin^2 x} + \frac{\cos^2 x}{\sin^2 x} = \frac{1}{\sin^2 x}$$

$$1 + \left(\frac{\cos x}{\sin x}\right)^2 = \left(\frac{1}{\sin x}\right)^2$$

or

$$1 + \cot^2 x = \csc^2 x.$$

The identities

$$\sin^2 x + \cos^2 x = 1$$
$$\sin^2 x = 1 - \cos^2 x$$
$$\cos^2 x = 1 - \sin^2 x$$
$$\tan^2 x + 1 = \sec^2 x$$
$$\tan^2 x = \sec^2 x - 1$$
$$1 + \cot^2 x = \csc^2 x$$
$$\cot^2 x = \csc^2 x - 1$$

are derived either directly or indirectly from the Pythagorean theorem and are called the *Pythagorean identities.*

Example 1. If $\tan x = \frac{5}{12}$, $0 < x < 90°$, find the other functions of x.

Since $\tan^2 x + 1 = \sec^2 x$,

$$\left(\frac{5}{12}\right)^2 + 1 = \sec^2 x,$$

$$\frac{25}{144} + \frac{144}{144} = \frac{169}{144} = \sec^2 x,$$

$$\sec x = \pm\sqrt{\frac{169}{144}} = \pm\frac{13}{12}.$$

Since $0 < x < 90°$, $\sec x = \frac{13}{12}$.

$$\cos x = \frac{1}{\sec x} = \frac{1}{13/12} = 12/13.$$

$$\cot x = \frac{1}{\tan x} = \frac{1}{5/12} = 12/5.$$

Also, since

$$\frac{\sin x}{\cos x} = \tan x, \sin x = \tan x \cdot \cos x = \frac{5}{12} \cdot \frac{12}{13} = \frac{5}{13}.$$

Since $\csc x = \frac{1}{\sin x}$,

$$\csc x = \frac{1}{5/13} = \frac{13}{5}.$$

Example 2. Simplify the expression $\frac{\cos x}{\sin x} + \frac{\sin x}{\cos x}$.

$$\frac{\cos x}{\sin x} \cdot \frac{\cos x}{\cos x} + \frac{\sin x}{\cos x} \cdot \frac{\sin x}{\sin x}$$

$$= \frac{\cos^2 x}{\sin x \cos x} + \frac{\sin^2 x}{\sin x \cos x}$$

$$= \frac{\cos^2 x + \sin^2 x}{\sin x \cos x} = \frac{1}{\sin x \cos x} = \csc x \cdot \sec x.$$

Example 3. Simplify the expression

$$3 \cos^2 x + \frac{3}{\csc^2 x}$$

$$3 \cos^2 x + \frac{3}{\csc^2 x} = 3 \cos^2 x + 3 \sin^2 x$$

$$= 3(\cos^2 x + \sin^2 x) = 3 \cdot 1 = 3.$$

10.18 Identities for Negative Angles

From the discussion on functions of any angle in Section 10.14, a counterclockwise rotation is called a positive angle and a clockwise rotation is considered negative. Therefore, if $\angle A$ terminates in quadrant I, $-\angle A$ terminates in quadrant IV, and if $\angle A$ terminates in quadrant II, $-\angle A$ terminates in quadrant III. In either case, the reference angle is the same size, but because of position the functions will vary in algebraic sign. Specifically, if $0 \leq A \leq 90°$, sin A is positive. Since $270° \leq -\angle A \leq 360°$, sin $(-A)$ is negative. That is,

$$\sin A = -\sin(-A) \quad \text{or} \quad -\sin A = \sin(-A).$$

Since the cosine of an angle is positive if the angle terminates in either quadrant I or IV and negative if the angle terminates in either quadrant II or III,

$$\cos A = \cos(-A).$$

Using these identities with the reciprocal and quotient identities developed previously, we can obtain the following further identities.

$$\tan A = -\tan(-A) \quad \text{or} \quad -\tan A = \tan(-A)$$
$$\cot A = -\cot(-A) \quad \text{or} \quad -\cot A = \cot(-A)$$
$$\sec A = \sec(-A)$$
$$\csc A = -\csc(-A) \quad \text{or} \quad -\csc A = \csc(-A)$$

Example 1. Find cos $(-30°)$.

$$\cos(-30°) = \cos 30° = \frac{\sqrt{3}}{2}.$$

Example 2. Find sin $(315°)$.

$$\sin 315° = \sin(-45°) = -\sin 45° = \frac{-\sqrt{2}}{2}.$$

Exercise 67

Using the identities developed in the preceding section, simplify the following.

1) $5\sin^2 x + 5\cos^2 x$
2) $\sin x \cdot \cot x \cdot \sec x$
3) $\dfrac{\cos x}{\tan x} + \sin^2 x \cdot \csc x$
4) $\tan x \cdot \cot x + \sin^2 x + \cos^2 x$
5) $\csc^2 x + \cot^2 x$
6) $\sin^3 x \cdot \cos x + \cos^3 \sin x$

Show that the following statements are true.

7) $\dfrac{1 + \tan^2 x}{\tan^2 x} = \csc^2 x$
8) $\dfrac{\tan x}{\cot x} + 1 = \sec^2 x$
9) $\dfrac{1 + \cos x}{\sin x} = \dfrac{\sin x}{1 - \cos x}$
10) $(\tan x + \cot x)^2 = \sec^2 x + \csc^2 x$
11) $3\sin^2 x + 3\cos^2 x - \cot^2 x = \dfrac{3\sin^2 x - \cos^2 x}{\cos^2 x}$
12) $(\tan x + \cot x)^2 + \csc^2 x = \sec^2 x$

Find the following without tables.

13) $\tan(-315°)$
14) $-\cos(-240°)$
15) $\sec(-390°)$
16) $-\csc(-45°) \cdot \cos(-45°)$
17) $\tan(-120°)$
18) $\cot(-270°)$
19) $\sin(-30°) \cdot \cos(-30°)$
20) $\sin^2(-15°) + \cos^2(-15°)$

10.19 Sum and Difference Identities

If points A, B, C, D, are points on a unit circle, that is, OA, OB, OC and OD are each one unit long, and if angles α* and β are located as shown in Figure 10.22.

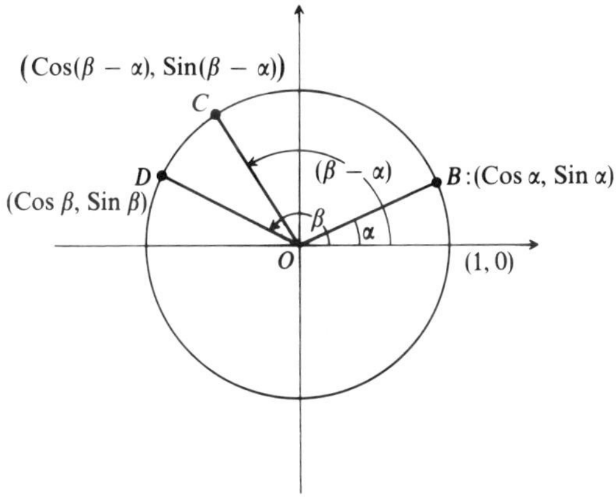

Figure 10.22

Angle DOB represents $(\beta - \alpha)$. Also, if angle DOC is the same measure as α, then angle COA also represents $(\beta - \alpha)$. Point D has coordinates $(\cos\beta, \sin\beta)$, point C has coordinates $(\cos(\beta - \alpha), \sin(\beta - \alpha))$, point B has coordinates $(\cos\alpha, \sin\alpha)$, point A has coordinates (1,0). Since angle DOB is the same as angle COA, and since all radii of the circle are equal, chord DB must equal chord CA. By the distance formula,

$$(DB)^2 = (\cos\beta - \cos\alpha)^2 + (\sin\beta - \sin\alpha)^2,$$
$$(CA)^2 = (\cos(\beta - \alpha) - 1)^2 + (\sin(\beta - \alpha) - 0)^2$$

Since $(DB)^2 = (CA)^2$

$$(\cos\beta - \cos\alpha)^2 + (\sin\beta - \sin\alpha)^2 = (\cos(\beta - \alpha) - 1)^2 + (\sin(\beta - \alpha))^2,$$
$$\cos^2\beta - 2\cos\beta\cos\alpha + \cos^2\alpha + \sin^2\beta - 2\sin\beta\sin\alpha + \sin^2\alpha = \cos^2(\beta - \alpha) - 2\cos(\beta - \alpha) + 1 + \sin^2(\beta - \alpha).$$

* Greek letters are often used to designate angles. For this reason, Greek letters will be used in the section to familiarize the reader with their use. The complete Greek alphabet is found on the end papers.

By grouping, this becomes

$$(\cos^2 \beta + \sin^2 \beta) + (\cos^2 \alpha + \sin^2 \alpha) - 2(\cos \beta \cos \alpha + \sin \beta \sin \alpha)$$
$$= [\cos^2 (\beta - \alpha) + \sin^2 (\beta - \alpha)] + 1 - 2 \cos (\beta - \alpha)$$

$$1 + 1 - 2(\cos \beta \sin \alpha + \sin \beta \sin \alpha)$$
$$= 1 + 1 - 2 \cos (\beta - \alpha)$$

$$2 - 2(\cos \beta \cos \alpha + \sin \beta \sin \alpha)$$
$$= 2 - 2 \cos (\beta - \alpha).$$

Subtracting 2 from both sides,

$$-2(\cos \beta \cos \alpha + \sin \beta \sin \alpha) = -2 \cos (\beta - \alpha),$$

and dividing by -2 we obtain

$$\cos (\beta - \alpha) = \cos \beta \cos \alpha + \sin \beta \sin \alpha.$$

Example 1. Find cos 15° without tables. Since cos 15° can be written as cos(45° − 30°), using the formula just derived, this becomes

$$\begin{aligned} \cos 15^\circ &= \cos 45^\circ \cos 30^\circ + \sin 45^\circ \sin 30^\circ \\ &= \frac{\sqrt{2}}{2} \cdot \frac{\sqrt{3}}{2} + \frac{\sqrt{2}}{2} \cdot \frac{1}{2} \\ &= \frac{\sqrt{6}}{4} + \frac{\sqrt{2}}{4} \\ &= \frac{1}{4}(\sqrt{6} + \sqrt{2}). \end{aligned}$$

Example 2. If $\cos \theta = 3/5$ and $\cos \phi = 5/13$, find $\cos (\theta - \phi)$. Since $\cos \theta = 3/5$, $\sin \theta = 4/5$, and since $\cos \phi = 5/13$, $\sin \phi = 12/13$; therefore,

$$\cos (\theta - \phi) = \frac{3}{5} \cdot \frac{5}{13} + \frac{4}{5} \cdot \frac{12}{13} = \frac{15}{65} + \frac{48}{65} = \frac{63}{65}.$$

Note: In Example 2, if we desired to know the size of $(\theta - \phi)$, we would convert 63/65 to decimal form as .9692 and find $(\theta - \phi) = 14^\circ$ to the nearest degree.

The identity for $\cos (\beta - \alpha)$ can be used to develop further identities for functions of sums and differences of angles. For instance,

$$\begin{aligned} \cos (\beta + \alpha) &= \cos [\beta - (-\alpha)] \\ &= \cos \beta \cos (-\alpha) + \sin \beta \sin (-\alpha) \\ &= \cos \beta \cos \alpha - \sin \beta \sin \alpha. \end{aligned}$$

Also,

$$\begin{aligned}\cos(90^\circ - \alpha) &= \cos 90^\circ \cos\alpha + \sin 90^\circ \sin\alpha \\ &= 0\cdot\cos\alpha + 1\cdot\sin\alpha \\ &= \sin\alpha\end{aligned}$$

Using this identity,

$$\cos\alpha = \cos[90^\circ - (90^\circ - \alpha)] = \sin(90^\circ - \alpha),$$

and

$$\begin{aligned}\cos(90^\circ + \alpha) &= \cos 90^\circ \cos\alpha - \sin 90^\circ \sin\alpha \\ &= 0\cdot\cos\alpha - 1\cdot\sin\alpha \\ &= -\sin\alpha.\end{aligned}$$

Now consider

$$\begin{aligned}\sin(\alpha - \beta) &= \cos[90^\circ - (\alpha - \beta)] = \cos[(90^\circ - \alpha) + \beta] \\ &= \cos(90^\circ - \alpha)\cos\beta - \sin(90^\circ - \alpha)\sin\beta \\ &= \sin\alpha\cos\beta - \cos\alpha\sin\beta\end{aligned}$$

Also, since $(\alpha + \beta) = [\alpha - (-\beta)]$,

$$\begin{aligned}\sin(\alpha + \beta) &= \sin[\alpha - (-\beta)] \\ &= \sin\alpha\cos(-\beta) - \cos\alpha\sin(-\beta) \\ &= \sin\alpha\cos(\beta) + \cos\alpha\sin\beta.\end{aligned}$$

Example 3. Find cos 75°.

$$\begin{aligned}\cos 75^\circ &= \cos(45^\circ + 30^\circ) \\ &= \cos 45^\circ \cos 30^\circ - \sin 45^\circ \sin 30^\circ \\ &= \frac{\sqrt{2}}{2}\cdot\frac{\sqrt{3}}{2} - \frac{\sqrt{2}}{2}\cdot\frac{1}{2} \\ &= \frac{\sqrt{6}}{4} - \frac{\sqrt{2}}{4} = \frac{1}{4}(\sqrt{6} - \sqrt{2})\end{aligned}$$

Example 4. Find sin 105° without tables.

$$\begin{aligned}\text{Sin } 105^\circ &= \sin(60^\circ + 45^\circ) \\ &= \sin 60^\circ \cos 45^\circ + \cos 60^\circ \sin 45^\circ \\ &= \frac{\sqrt{3}}{2}\cdot\frac{\sqrt{2}}{2} + \frac{1}{2}\cdot\frac{\sqrt{2}}{2} \\ &= \frac{\sqrt{6}}{4} + \frac{\sqrt{2}}{4} = \frac{1}{4}(\sqrt{6} + \sqrt{2}).\end{aligned}$$

Note also that since $\sin 105^\circ = \cos(90^\circ - 105^\circ) = \cos(-15^\circ) = \cos 15^\circ$, therefore,

$$\cos 15^\circ = \frac{1}{4}(\sqrt{6} + \sqrt{2}).$$

Example 5. In the diagram given, find $\sin \gamma$.

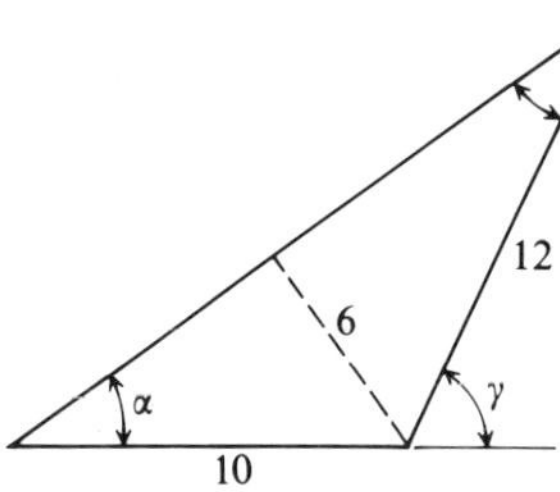

$$\gamma = (\alpha + \beta).$$

Therefore,

$$\sin \gamma = \sin (\alpha + \beta) = \sin \alpha \cos \beta + \cos \alpha \sin \beta.$$

Since

$$\sin \alpha = 6/10 = 3/5, \cos \alpha = 8/10 = 4/5$$

and

$$\sin \beta = 6/12 = 1/2;$$

thus

$$\cos \beta = \sqrt{3}/2,$$

and

$$\sin \gamma = \frac{3}{5} \cdot \frac{\sqrt{3}}{2} + \frac{4}{5} \cdot \frac{1}{2} = \frac{3\sqrt{3}}{10} + \frac{4}{10}$$

$$= \frac{3\sqrt{3} + 4}{10} = .9196 \text{ approx.}$$

10.20 Double Angle and Half Angle Identities

An identity for $\cos 2A$ can be derived, if $\cos 2A$ is written as $\cos (A + A)$.

$$\begin{aligned} \text{Cos } 2A &= \cos (A + A) = \cos A \cos A - \sin A \sin A \\ &= \cos^2 A - \sin^2 A. \end{aligned}$$

Since $\cos^2 A + \sin^2 A = 1$, this formula can also be written as

$$\cos 2A = 2 \cos^2 A - 1,$$

or

$$\cos 2A = 1 - 2 \sin^2 A.$$

Using the same method,

$$\begin{aligned} \sin 2A &= \sin (A + A) = \sin A \cos A + \cos A \sin A \\ &= 2 \sin A \cos A. \end{aligned}$$

Example 1. If $\sin A = 5/13$, find $\sin 2A$ and $\cos 2A$.
If $\sin A = 5/13$, $\cos A = 12/13$.
Therefore,

$$\cos 2A = \left(\frac{12}{13}\right)^2 - \left(\frac{5}{13}\right)^2 = \frac{144}{169} - \frac{25}{169} = \frac{119}{169}$$

$$\sin 2A = 2\left(\frac{5}{13}\right)\left(\frac{12}{13}\right) = \frac{120}{169}.$$

Again, the measure of $2A$ can be found by converting either value to decimal form and finding the angle in the table.

Since $\cos 2A = 2\cos^2 A - 1$,

$$\cos 2A + 1 = 2\cos^2 A$$

$$\cos^2 A = \frac{1}{2}(1 + \cos 2A)$$

$$\cos A = \pm\sqrt{\frac{1 + \cos 2A}{2}}.$$

Now, if $A = B/2$, then

$$\cos\frac{B}{2} = \pm\sqrt{\frac{1 + \cos B}{2}}.$$

That is, the cosine of one-half an angle is found by finding the square root of 1 plus the cosine of the angle all divided by 2. The sign is chosen depending on the quadrant in which one-half the angle lies.

On the other hand, if the formula $\cos 2A = 1 - 2\sin^2 A$ is used,

$$\cos 2A - 1 = -2\sin^2 A$$

$$\sin^2 A = \frac{1 - \cos 2A}{2},$$

$$\sin A = \pm\sqrt{\frac{1 - \cos 2A}{2}}.$$

If, as before, $\frac{B}{2} = A$,

$$\sin\frac{B}{2} = \pm\sqrt{\frac{1 - \cos B}{2}},$$

where again the sign is chosen depending on the size of $B/2$.

Example 2. Using the half-angle formula, find $\cos 15°$.

$$15° = \frac{30°}{2}.$$

$$\begin{aligned}\cos 15^\circ = \cos\frac{30^\circ}{2} &= +\sqrt{\frac{1+\cos 30^\circ}{2}}\\ &= \sqrt{\frac{1+\sqrt{3}/2}{2}}\\ &= \sqrt{\frac{2+\sqrt{3}}{4}}\\ &= \frac{1}{2}\sqrt{2+\sqrt{3}}.\end{aligned}$$

Example 3. If $\cos x = 3/7$, find $\sin x/2$.

$$\begin{aligned}\sin\frac{x}{2} &= \pm\sqrt{\frac{1-3/7}{2}}\\ &= \pm\sqrt{\frac{7-3}{14}} = \pm\sqrt{\frac{4}{14}} = \pm\sqrt{\frac{2}{7}} = \pm\frac{\sqrt{14}}{7}.\end{aligned}$$

Both signs must be used, since, if x is in quadrant four, $x/2$ is in quadrant four and the $\sin x/2$ would be negative. If x is in quandrant one, so is $x/2$ and $\sin x/2$ would be positive.

Example 4. Find $\cos 22\frac{1}{2}^\circ$.

$$\begin{aligned}\cos 22\tfrac{1}{2}^\circ = \cos\frac{45^\circ}{2} &= \sqrt{\frac{1+\cos 45^\circ}{2}}\\ &= \sqrt{\frac{1+\frac{\sqrt{2}}{2}}{2}} = \sqrt{\frac{2+\sqrt{2}}{4}}\\ &= \frac{1}{2}\sqrt{2+\sqrt{2}}.\end{aligned}$$

Exercise 68

Find the value of each of the following without tables.

1) $\sin 75^\circ$ 2) $\cos 105^\circ$ 3) $\sin 15^\circ$ 4) $\tan 22\frac{1}{2}^\circ$ 5) $\cos 135^\circ$

If $\sin\theta = 3/5$, find:

6) $\sin\frac{\theta}{2}$ 7) $\sin 2\theta$ 8) $\cos\theta$ 9) $\cos\frac{\theta}{2}$ 10) $\cos 2\theta$

If $\sin\theta = \frac{2}{3}$ and $\sin\phi = \frac{1}{4}$, $0 < \theta < 90$, $0 < \phi < 90$, find:

11) $\sin(\theta+\phi)$ 12) $\cos(\theta+\phi)$ 13) $\sin(\theta-\phi)$

14) $\cos(\theta-\phi)$ 15) $\sin 2\theta$ 16) $\cos\theta$ 17) $\sin 2\phi$

18) $\cos 2\phi$ 19) $\tan(\theta+\phi)$ 20) $\cot(\theta-\phi)$

Using the identities developed in the preceding section, derive the following identities.

21) $\tan(\theta + \phi)$ 22) $\cot(\theta + \phi)$ 23) $\tan 2\theta$

24) $\tan \frac{\theta}{2}$ 25) $\tan(\theta - \phi)$

10.21 Trigonometric Scales on the Slide Rule

The slide rule can be used to find the value of trigonometric ratios to three decimal places using the *S*, *T*, and *ST* scales. When these ratios are found, the rule may then be used to perform other calculations in trigonometric problems.

To find the sine of an angle between approximately 5.7° and 90°, we locate the desired angle on the *S* scale with the hairline and read the sine of the angle directly below the hairline on the *C* scale (or the D scale if the indices are lined up).

Example 1. Find sin 15°.

Locate 15° on the S scale with the hairline; read sin 15° = .259 on the C scale.

Example 2. Find sin 78°.

Locate 78° on the S scale with the hairline, then read sin 78° = .978 on the C scale.

Notice that since the sine of an angle is always less than or equal to 1, the decimal point will be in front of the first digit of the result.

If one has the sine of the angle and wishes to find the corresponding angle, the sine value is located on the C scale with the hairline and the angle is on the *S* scale directly below the hairline.

Example 3. If $\sin A = .435$, find A.

Locate .435 on the C scale with the hairline and read $A = 25.8°$ under the hairline on the C scale.

The operations discussed in this section for solving trigonometry problems involving the sine can be solved using the slide rule.

Example 4. If a 50-foot wire attached to the top of a pole is attached to the ground forming an angle of 57°, find the height h of the pole.

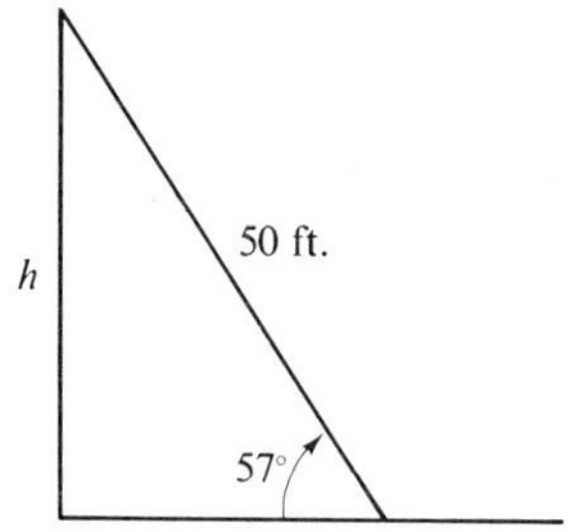

$$\sin 57° = \frac{h}{50},$$
$$h = 50 \sin 57°$$

To find sin 57°, locate 57° on the S scale with the hairline. Sin 57° is found on D, if the indices are lined up, as .839. To find the product

$$L = 50 \sin 57° = 50(.839),$$

move the right index of C under the hairline. Move the hairline to 50 on C. The product, 42.0, is now under the hairline on D. Therefore, $h = 42.0$ ft.

Example 5. Solve for sin A.

$$\frac{\sin 45°}{4.5} = \frac{\sin A}{6.2},$$
$$\sin A = \frac{6.2(\sin 45°)}{4.5}.$$

Locate 6.2 with the hairline on D. Slide 4.5 directly under the hairline. Move the hairline to 45° on S. Align the C and D indices, and read $A = 77°$ on S below the hairline.

The T scale is used in much the same manner to find the tangent of an angle. Notice however that the T scale reads from approximately 5.7° to 45° reading from left to right and from 45° to approximately 84.3°, reading from right to left. If the angle is between 5.7° and 45°, it is located on the T scale with the hairline and the tangent of the angle is located on the C scale under the hairline. If, however, the angle is between 45° and 84.3°, the angle is found on the T scale reading from right to left, and the tangent of the angle is read on the Cl scale.

Some rules have a folded or double T scale so that all angles are found on the appropriate T scale and the tangent of these angles is read directly on the C scale. If this is the case, it is suggested that the reader refer to the instruction manual accompanying the slide rule for a complete explanation of the procedure.

Example 6. Find tan 27°.

Locate 27° on the T scale with the hairline. Read tan 27° = .508 below the hairline on C. Note that since the tangent of any angle less than 65° is less than one, the decimal is in front of the first digit.

Example 7. Find tan 53°.

Locate 53° on the T scale with the hairline (reading from right to left). Read tan 53° = 1.325 below the hairline on Cl. Since the tangent of any angle greater than 45° is more than 1, the decimal is after the first digit of the result.

The sine and the tangent of any very small angle (from .57° to 5.7°) are almost equal to three decimal places. Since this is the limit of accuracy obtainable on the slide rule, both the sines and tangents of small angles can be found from the same scale, the ST, on the slide rule. Since both the sine and the tangent of small angles are very near zero, there is always one zero between the decimal and the first digit of the result.

Example 8. Find sin 4.2°.

Locate 4.2° on the ST scale. Read sin 4.2° = .0732 below the hairline on the C scale.

Example 9. Find tan .63°.

Locate .63° on the ST scale with the hairline. Read tan .63° = .0110 below the hairline on the C scale.

To find the tangent of an angle greater than 84°, the ST scale must also be used. This is done by finding the complement of the angle (90 − angle) and reading the tangent of the angle on the CI scale. Care must be taken in locating the decimal point, however, since the tangents of angles near 90° become very large.

Example 10. Find tan 87.4°.

$$90 - 87.4° = 2.6°$$

Locate 2.6 on the ST scale and read 220 on the CI scale. Since tan 2.6° = .0454, tan 87.4° = 1/tan 2.6° = 1/.0454 = 22.0.

In order to consider finding the cosine of an angle on the slide rule we must first recall the relationship between the sine and cosine from the trigonometry section. Since $\sin x = \cos (90 - x)$ and the sine (S) scale is graduated from 5.7° to 90°, directly below the S scale reading from right to left is the complement of each angle found on the S scale. On some rules these are renumbered in red or in some way to designate that they are read from right to left. Therefore, to find the cosine of an angle, locate the angle on the under side of the S scale reading from right to left and read the cosine of the angle on the C scale.

If the rule is not specially designated for finding cosines, one need merely find the sine of the complement of the angle.

Example 11. Find cos 29.5°.
Locate 29.5° on the under side of the S scale. (29.5° is the complement of 60.5°). Read cos 29.5° = .871 on the C scale.

10.22 Reciprocal Functions on the Slide Rule

It is very easy to locate the reciprocal functions on the slide rule by using the CI and DI scales. For instance, since $\csc x = 1/\sin x$, if an angle is located on the S scale reading from left to right, the sine of the angle is read from the C scale. The cosecant of the angle therefore can be read directly under the hairline on the CI scale, (or DI scale if the indices are aligned).

In the same manner the secant of an angle can be found by locating the angle on the S scale, reading from right to left, and the cosine of the angle is read on the C scale. The secant of the angle is the reciprocal of the cosine and is therefore found under the hairline on the CI scale.

The cotangent of an angle can also be found by using the reciprocal scales. Notice, however, that the cotangent and the tangent are related not only as reciprocals but also as complements. That is, $\tan x = 1/\cot x$ and $\tan x = \cot (90 - x)$. Therefore, when the $\tan x$ is found on the C scale (when x is between 5.7° and 45°), $\cot x$ is found under the hairline on CI scale. When the $\tan x$ is found on the CI scale (when x is between 45° and 84°), $\cot x$ is found on the C scale.

Example 1. Find sec 41.3°.
Locate 41.3° on S scale reading from right to left. Cos 41.3° is on C, and sec 41.3 = 1.33 is found on the CI scale.

Example 2. Find csc 28.4°.
Locate 28.4° on S scale reading from left to right. Sin x is on the C scale; thus csc x is found under the hairline on the CI scale. Csc 28.4° = 2.10.

Example 3. Find cot 32.8°.
Locate 32.8° on the T scale. Since tan 32.8° is found on the C scale, cot 32.8° is on CI. Cot 32.8° = 1.55.

Example 4. Find cot 67.9°.
Locate 67.9° on the T scale reading from right to left. Since tan 67.9° is located on the CI scale, cot 67.9° is located on the C scale under the hairline. Cot 67.9° = .406.

Exercise 69

Find the value of the following.

1) sin 41°	2) tan 34.8°	3) sin 75.2°
4) cos 29.4°	5) tan 16.3°	6) tan 3.2°
7) sin 1.7°	8) cos 87°	9) tan 23.5°
10) cos .8°		

Find the value of angle x in each of the following.

11) $\tan x = .614$	12) $\tan x = 6.20$	13) $\cos x = .0742$
14) $\sin x = .314$	15) $\tan x = .866$	16) $\sin x = .866$
17) $\cos x = .866$	18) $\cos x = .702$	19) $\sin x = .500$
20) $\tan x = 12.25$		

Find the value of the following reciprocal functions.

21) csc 49°	22) sec 57°	23) cot 19.3°
24) csc 26.7°	25) sec 74.2°	26) csc 59.7°
27) cot 45°	28) cot 67.3°	29) csc 16.7°
30) cot 28.8°		

Find the value of angle x in each of the following.

31) $\csc x = 1.85$ 32) $\sec x = 2.71$ 33) $\cot x = 4.52$

34) $\csc x = 3.92$ 35) $\sec x = 1.085$ 36) $\cot x = .875$

37) $\csc x = 1.84$ 38) $\cot x = .324$ 39) $\cot x = .0452$

40) $\csc x = 1.045$

Solve the following using the slide rule. Problems 41 – 45 relate to triangle ABC.

41) $b = 2.4$, $\angle A = 13°$, $\angle C = 110°$; find a

42) $b = 8.5$, $c = 6.3$, $B = 42.5°$; find a

43) $c = .84$, $A = 54°$, $C = 90°$; find b

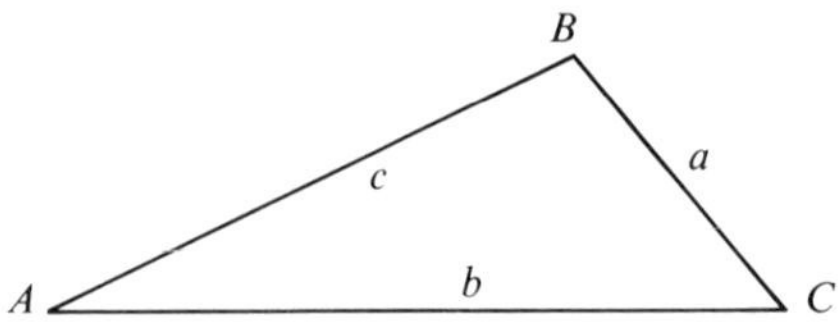

44) $a = 12$, $A = 30°$, $B = 80°$; find b, c, C

45) $a = 540$, $c = 600$, $C = 30°$; find A, B, b

46) Two diagonals of a parallelogram are 5 and 6 and they form an angle of 45°. Find the length of each side.

47) The rectangular components of a vector $\vec{r}$ are 12 and 7.5 as shown in the figure. Find the magnitude (length of $\vec{r}$) and the direction (value of θ) of the vector.

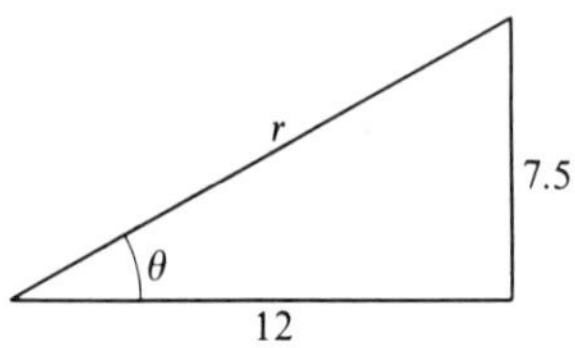

48) Find the angle made by the diagonal of a cube with the diagonal of a face of the cube drawn from the same vertex.

49) If in a circle a chord of 25.3 inches subtends a central angle of 43°, what is the radius of the circle?

50) A car is traveling at a rate of 60 ft. per sec. up a grade inclined 7° with the horizontal. How far must the car travel for a rise of 350 ft.?

Review Exercise Chapter 10

Simplify each of the following.

1) $\sin(90° + x)$

2) $-\tan(x - 90°)$

3) $\cot 270° + \tan 45°$

4) $\sin 4a$

5) $\cos^2 2a + \sin^2 2a$

6) $\cot 3x$

7) $\cos \frac{x}{2}$

8) $\cos 3a - \sin 3a$

Prove the following identities true.

9) $\cos^4 x - \sin^4 x = \cos 2x$

10) $\sec^2 x \csc^2 x = \sec^2 x + \csc^2 x$

11) $\sqrt{\frac{1 + \cos x}{1 - \cos x}} = \csc x + \cot x$

12) $\frac{2 \tan x}{1 + \tan^2 x} = \sin 2x$

Evaluate the following without tables.

13) $\tan(-120°)$

14) $\cot(75°)$

15) $\sin(-15°)$

16) $\csc(105°)$

17) $\sec(-195°)$

18) $-\sin(-150°)$

Perform the following operations on the slide rule.

19) $\frac{7 \cos 27°}{15.2}$

20) $100 \tan 2.6°$

21) $\frac{\sin 15.3°}{\cos 15.3°}$

22) $\frac{19.5 \cos 75.3°}{14.8}$

11 Radian Measure and Sine Wave Analysis

11.1 Radian Measure

In the chapter on geometry and in the previous chapters on trigonometry, the unit of measure used for angles was the degree. This unit is but one of two important units for measuring angles. The second unit is the *radian* which has the following definition. *A radian is the measure of the central angle of a circle which intercepts an arc equal in length to the radius.*

In Figure 11.1, angle AOB contains one radian since the length of arc AB is equal to the length of the radius of the circle.

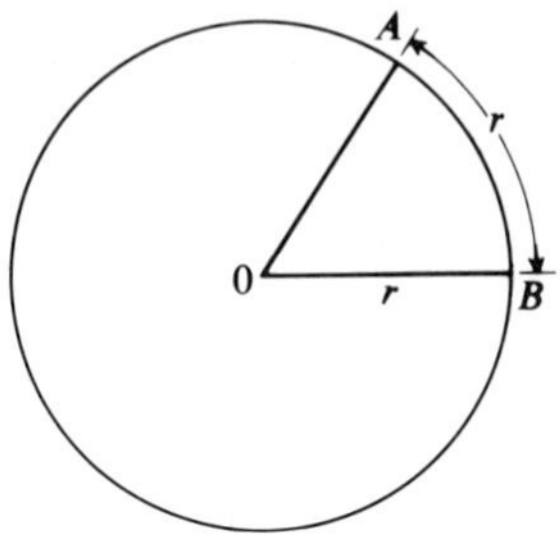

Figure 11.1

The next logical question is one which inquires about the relationship between the radian and the degree. To discover this relationship, we need to recall two important measures from the field of geometry.

I. One degree is $1/360$ of the total rotation about a point.

II. The circumference of a circle is given by the formula $C = 2\pi r$.

The first measure leads to the conclusion that a 360°-central angle would intercept the entire circumference of a circle. The second measure leads to the conclusion that the radius of any circle can be laid out along the circumference 2π times and that radian measure does not depend on the length of the radius. Figure 11.2 illustrates this fact.

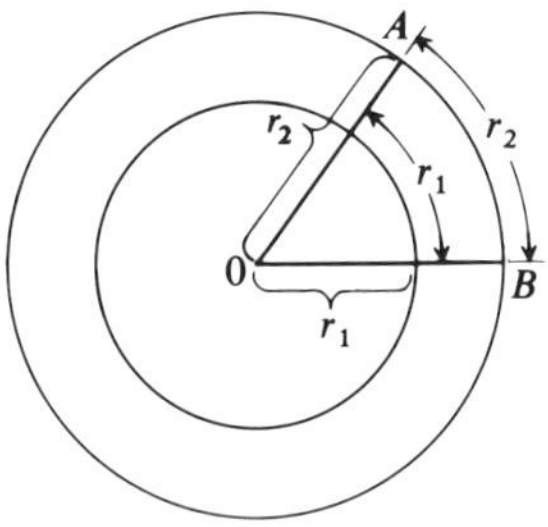

Figure 11.2

If we let $r_1 = 1$, we see that one complete rotation about a point would form a central angle of 2π radians, if given in radian measure, and 360°, if degree measure is used. This leads to the relationship

$$2\pi \text{ radians} = 360°,$$

or

$$\pi \text{ rad.} = 180°.$$

From this we see that

$$1° = \frac{\pi}{180} \text{ rad.}$$

and

$$1 \text{ rad.} = \frac{180°}{\pi} = 57.3° \text{ to the nearest tenth.}$$

Normally the unit measure radian is understood and not formally written. Thus

$$1° = \frac{\pi}{180} \text{ rad.}$$

is written

$$1° = \frac{\pi}{180}.$$

Similarly

$$1 \text{ rad.} = \frac{180°}{\pi}$$

is written

$$1 = \frac{180°}{\pi}.$$

11.2 Conversion of Degree and Radian Measure

Frequently it is desirable to change from degree measure to radian measure or from radian measure to degree measure. These conversions are easy to accomplish as the following examples illustrate.

Example 1. Convert 30° to radians.

$$1° = \frac{\pi}{180}.$$

Thus $30° = 30 \cdot 1° = 30 \cdot \frac{\pi}{180} = \frac{\pi}{6}$, and $30° = \frac{\pi}{6}$.

Example 2. Convert 115° to radians.

$$115° = 115 \cdot 1°.$$

Thus $115° = 115 \cdot \frac{\pi}{180} = \frac{23\pi}{36}$, and $115° = \frac{23\pi}{36}$.

Example 3. Convert $-720°$ to radians.

$$-720° = -720 \cdot 1°,$$
$$-720° = -720 \cdot \frac{\pi}{180} = -4\pi$$
$$-720° = -4\pi.$$

Example 4. Convert $\pi/6$ to degree measure.

$$\pi = 180°,$$

and thus
$$\frac{\pi}{6} = \frac{180°}{6} = 30°.$$

Example 5. Convert $\frac{4\pi}{3}$ to degree measure.

$$\frac{4\pi}{3} = \frac{4 \cdot 180°}{3} = 240°$$
$$\frac{4\pi}{3} = 240°.$$

Example 6. Convert -7π to degree measure.

$$-7\pi = -7 \cdot 180° = -1260°$$
$$-7\pi = -1260°.$$

Example 7. Convert 2.3 to degree measure.

$$1 \text{ radian} = 57.3°$$
$$2.3 = 2.3(57.3°).$$
$$2.3 = 131.79°.$$

The trigonometric functions are frequently expressed in terms of radian measure and can be handled in one of two ways. For example, $\sin \pi/3$ could be evaluated by changing $\pi/3$ to 60° and then evaluating as $\sin 60° = \sqrt{3}/2$.

Some tables contain a column headed *radians*. To use such a table we would change $\pi/3$ to $3.1416/3 = 1.0472$. We then have $\sin 1.0472$ which would be evaluated by reading the entry under the column headed *sin* that is opposite 1.0472 in the column headed *radians*.

11.3 Arc Length and Radians

The length of an arc intercepted by a central angle of a circle can be computed directly by expressing the central angle in terms of radians and multiplying by the length of the radius. To see this, we set up the proportion

$$\frac{2\pi r}{2\pi} = \frac{s}{\theta}.$$

This proportion states that the ratio between the circumference of a circle and the central angle that intercepts it is the same as the ratio of the length of any arc s and the central angle that intercepts it. If we reduce this proportion we have

$$r = \frac{s}{\theta}, \qquad \text{or} \qquad s = r\theta.$$

From this we see that *the length of an arc (s) of any circle is equal to the product of the radius of the circle and the radian measure of the central angle that intercepts the arc.*

Example 1. How long would the arc of a circle of radius 3 in. be if the intercepting central angle contains 60° as shown in the following figure?

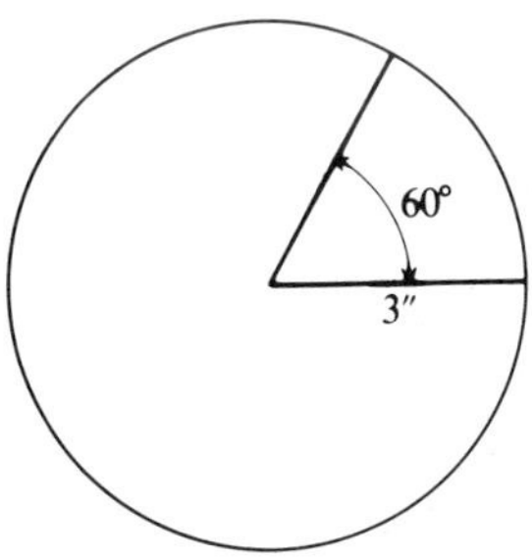

Expressing 60° in radians we have $60^\circ = \pi/3$. Thus

$$s = \frac{\pi}{3}(3) = \pi \text{ inches.}$$

This answer is clearly correct since the circumference of this circle is $2\pi r$ or 6π units long. A central angle of 60° would intercept 1/6 of the total circumference which is π units.

Example 2. Find the length of the arc intercepted by a 78° central angle in a circle with radius 10 in.

$$78^\circ = 78 \cdot \frac{\pi}{180}$$

$$s = \frac{78\pi}{180} \cdot 10$$

$$s = \frac{13\pi}{3} = 13.61 \text{ in.}$$

11.4 Area of a Circular Sector

In the last section we developed a formula for the length of an arc of a circle by using the central angle measured in radians. In a similar manner it is possible to develop a formula for finding the area of a sector of a circle. Since a central angle of 2π radians sweeps through the entire circle we can set up the proportion.

$$\frac{\pi r^2}{2\pi} = \frac{A}{\theta}.$$

This proportion states that the ratio between the area of a circle and the central angle that sweeps out the area is the same as the ratio between a circular sector and the central angle that sweeps out its area. If we reduce the proportion we have

$$\frac{r^2}{2} = \frac{A}{\theta}$$

or

$$A = \frac{1}{2}r^2\theta.$$

Stated as a rule we have the following. *The area of a circular sector is equal to one-half the product of the square of the radius and the radian measure of the central angle that sweeps out the area.*

Example 1. Find the area of a sector of a circle of radius 6 in. if the central angle intercepting the sector is 30°.

$$30° = \frac{\pi}{6}.$$

Thus

$$A = \frac{1}{2}\,6^2\,\frac{\pi}{6} = 3\pi \text{ sq. in.}$$

Example 2. Find the area of a sector of a circle of radius 3 ft. if the central angle intercepting the sector is 125°.

$$125° = \frac{25\pi}{36}.$$

$$A = \frac{1}{2}\,3^2 \cdot \frac{25\pi}{36} = \frac{25\pi}{8} \text{ sq. ft.}$$

11.5 Radian Measure on the Slide Rule

In the trigonometry section we discussed the concept of radian measure as a measure of angles. This is putting angle measurement in terms of real numbers rather than degrees. The conversion given at that point from one system to another was

$$\frac{\theta \text{ (degrees)}}{180} = \frac{\theta \text{ (radians)}}{\pi}.$$

One radian is therefore approximately 57.3°.

For angles less than 5.7° the radian equivalent is (to slide rule accuracy) the value of the sine or tangent of the angle. Therefore, the conversion can be read directly from the ST scale by putting .0 in front of the digits read from the C scale.

Example 1. 2.56° = .0447 radians.

We can discuss further conversions by use of the same example. For instance, multiplying both sides of the equation by 10 we get 25.6° = .447 radians. The same result is obtained by setting up the ratio 25.6/57.3 = .447. Using this procedure in reverse, we can find the value of .256° as .00447 radians. In this manner, most radian-to-degree and degree-to-radian conversions can be accomplished by use of the ST scale and the C scale.

Example 2. Convert 49.3° to radians.
Locate 4.93° on the ST scale. Read 4.93° = .0862 on the C scale. Multiplying both sides by 10, we obtain 49.3° = .862 radians.

Example 3. Convert .396° to radians.
Locate 3.96° on the ST scale. Read 3.96° = .0692 radians. Therefore, .396° = .00692 radians.

Exercise 70

Convert each of the following to degree measure.

1) $\pi/2$	2) $\pi/6$	3) $2\pi/3$	4) $3\pi/2$
5) $5\pi/2$	6) $3\pi/4$	7) $7\pi/8$	8) $-5\pi/4$
9) $-\pi/18$	10) 6π	11) $11\pi/3$	12) $7\pi/4$

Convert each of the following to radian measure in terms of π.

13) 30°	14) 60°	15) 45°
16) 135°	17) 720°	18) 225°
19) −45°	20) −90°	21) 300°

Change each of the following to degree measure.

22) 3	23) 5.4	24) 3/4
25) −.7	26) −2.5	27) −4.6

Evaluate each of the following.

28) $\sin \frac{\pi}{3}$	29) $\cos \frac{\pi}{4}$	30) $\tan \frac{5\pi}{3}$
31) $\sin \frac{-3\pi}{2}$	32) $\cot \left(\frac{-5\pi}{3}\right)$	33) $\sin \frac{5\pi}{3}$
34) $\cos \frac{13\pi}{36}$	35) $\sin 2.4$	

Find the following angles to the nearest degree. Also express the angles in terms of radians $0 \leq A < 360°$.

36) $\sin A = .2419$	37) $\cos A = .6820$
38) $\tan A = 1.963$	39) $\sin A = -.9135$
40) $\csc A = -1.179$	41) $\cot A = 4.705$

Convert the following angles to radians.

42) 58.7°	43) .896°	44) .245°
45) 7.56°	46) 87.8°	

11.6 Linear and Angular Velocity

Radian measure has numerous applications, but we will consider only a few that relate directly to technical applications at the elementary levels. One of the more important applications is that of *linear* and *angular velocity*.

One of the easier ways to gain insight into these topics is to consider a point moving along the circumference of a circle with constant speed. The distance that the point moves in a given period of time determines its velocity. For example, suppose the point is located on the tire tread of a car, and the wheel is turning at a constant speed of 5 revolutions per second. The car moves forward at a rate of 5 times the circumference of the tire each second. If the radius of the wheel, tire included, is 1.2 ft., the circumference is $C = \pi d = (3.14)2.4 = 7.536$ ft.

The car is moving at $5 \cdot (7.536)$ or 37.68 ft. per sec. This motion is along a line and is called *linear velocity.*

If we let s represent the distance traveled by the moving point, t represent the time it takes to travel this distance, and v represent the linear velocity, then

$$\frac{s}{t} = v \quad \text{or} \quad s = v \cdot t.$$

If the radius swings through θ angular units measured in radians in the given period of time, then the angular velocity is given by

$$\omega^* = \frac{\theta}{t}.$$

Angular velocity can be directly related to linear velocity through the formula

$$s = r\theta.$$

If we divide both sides by t, we have

$$\frac{s}{t} = \frac{r\theta}{t},$$

but

$$\frac{s}{t} = v \quad \text{and} \quad \frac{\theta}{t} = \omega$$

so that

$$v = r\omega.$$

In this formula v is the linear velocity, r the radius of the circle, and ω the angular velocity.

* Greek letter omega.

Example 1. Find the linear velocity of a point on the circumference of a 12-in. phonograph record being played at $33\frac{1}{3}$ R.P.M.

$$v = r\omega.$$

Each revolution generates a central angle of 2π radians. From this we see that $\omega = 33\frac{1}{3} \cdot 2$ rad./min. If $r = 6$, we have

$$\begin{aligned} v &= 6 \cdot 33\tfrac{1}{3} \cdot 2\pi \text{ in./min.} \\ &= 6 \cdot \frac{100}{3} \cdot 2\pi \\ &= 400\pi \text{ in./min.} \end{aligned}$$

Example 2. A pulley on a lathe rotates with an angular velocity of 3400π rad/min. If the radius of the pulley is 6 inches, find the linear velocity of a point on its rim.

$$\begin{aligned} v &= r\omega \\ v &= 6(3400\pi) \\ &= 20400\pi \text{ in./min.} \end{aligned}$$

Example 3. Ten minutes are required to run 140 ft. of film through a movie projector at constant speed. What is the diameter in inches of the film on each reel at the instant when one makes 7 R.P.M. and the other 10 R.P.M.?

$$\begin{aligned} v = \frac{s}{t} = \frac{140}{10} &= 14 \text{ ft./min.} \\ &= 168 \text{ in./min.} \end{aligned}$$

A point on the reel passes through 2π radians in one revolution. Thus one wheel has an angular velocity of 14π rad./min., while the other has an angular velocity of 20 rad./min. From the formula $v = r\omega$ we see

$$168 = r(14\pi) \quad \text{and} \quad 168 = r(20\pi)$$

$$r = \frac{84}{7\pi} \qquad\qquad r = \frac{42}{5\pi}$$

The diameter of the film is

$$D = \frac{168}{7\pi} \text{ in.} \quad \text{and} \quad D = \frac{84}{5\pi} \text{ in.}$$

$$= 7.64 \text{ in.} \qquad\qquad = 5.35 \text{ in.}$$

Exercise 71

1) Find the length of arc intercepted by a 150° central angle in a circle whose radius is four feet. Carry answer to two decimal places.

2) Find the length of arc through which a pendulum 4 ft. long swings if it swings through 45° of arc.

3) Find the radius of a circle in which a central angle of 2 radians intercepts an arc of 80 ft.

4) A curve is to be laid out along a new highway right-of-way. What radius should be used if the curve is to be an arc of a circle and if the road is to change its direction by 10° in a distance of 200 ft.?

5) Find the length of a belt around two wheels of radii 2 and 4 ft., if the distance between their centers is 8 ft. (Hint: The belt is perpendicular to the radius of each wheel at the point that it first makes contact.)

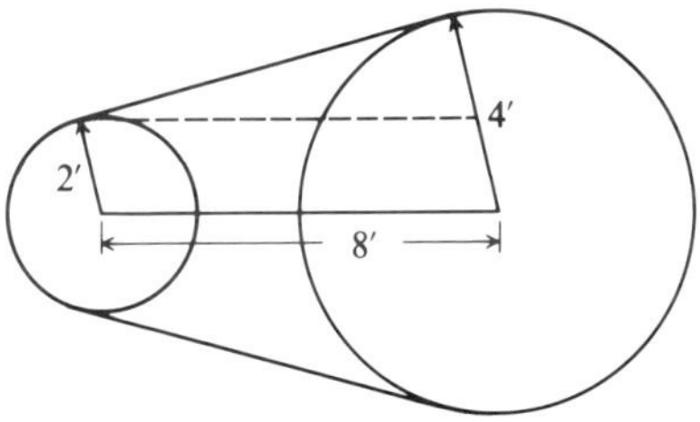

6) Find the measure of the central angle of a circle where the radius is 26 cm. and the intercepted arc is 34 cm.

7) Find the area of a sector of a circle of radius 3 ft., if the central angle is 35°.

8) Find the area of a sector of a circle of radius 750 mm., if the central angle is 120°.

9) Find the area of a sector of a circle, if the radius is 3 ft. and the intercepted arc is 2.8 ft.

10) The radius of the earth is approximately 3960 miles. Assuming the earth to be a sphere, how far is it from 10° N latitude to 13° N latitude?

11) Give the linear velocity of a point on the earth's equator ($D = 7920$ mi).

12) A 10-in. phonograph record plays at a constant speed of $33\frac{1}{3}$ R.P.M. What is the linear velocity of a point on its rim? What is the angular velocity of the same point?

13) What is the angular velocity of a point on the rim of a 6 in. pulley that is turning at 1760 R.P.M.? What is the angular velocity of the same point?

14) An artificial earth satellite has a circular orbit 370 miles high. If it passes over a point on the earth's surface once every 1.42 hours, what is its velocity?

15) A pulley 6 in. in diameter rotates at 1760 R.P.M. A second pulley 8 in. in diameter is driven by the first by use of a v-belt. What is the linear velocity of a point on the rim of each wheel?

16) What is the angular velocity of the 8 in. wheel in Problem 15?

17) The pulley on the crankshaft of a 1971 automobile has a diameter of 9 in.; the diameters of some other pulleys are as follows: alternator 4 in., waterpump 4.2 in., air conditioning compressor 6 in. If the engine of the car is turning at 3400 R.P.M., how many R.P.M.'s are each of the above-mentioned pulleys turning?

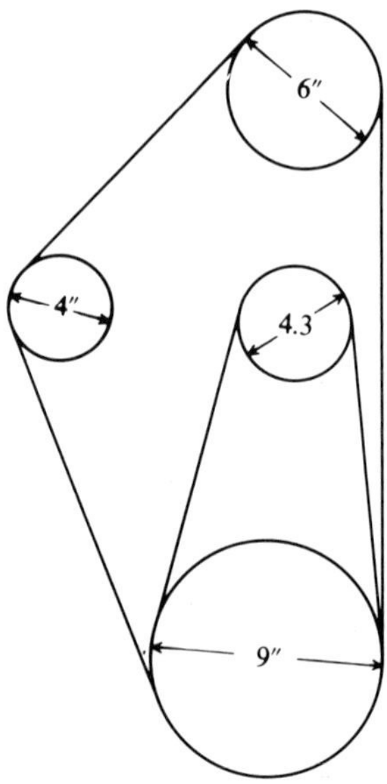

18) What is the linear velocity of a point on the rim of each pulley in Problem 17?

11.7 The Sine Wave

Many areas of technical mathematics involve the sine wave and analysis of its various properties. Such diverse topics as alternating current theory, sound wave transmission, spring vibrations, and radio and T.V. waves are but a few that are directly related. If we were to observe the motion of the piston in a piston engine, we would find that it describes a sine wave when it is considered with relationship to a change in time.

11.8 The Graph of the Sine Wave

Perhaps the best approach to understanding the sine wave and its properties is that of graphing. To this end we will sketch the graph of

$$y = \sin x$$

over the interval from 0° to 360°. Just as we did in previous sections on graphing, we will plot the x coordinate on the horizontal axis and the y coordinate on the vertical axis. Our only problem will be that of deciding where to plot the x coordinates, since they are given in degree measure. However, this problem is easily solved by relating degree measure to linear measure through conversion to radian measure. Recall that

$$2\pi = 360°,$$

and consequently

$$360° \approx 6.28.$$

We will now locate 360° approximately 8.28 units to the right of the origin and all other angles in proportion to this measure. See Figure 11.3.

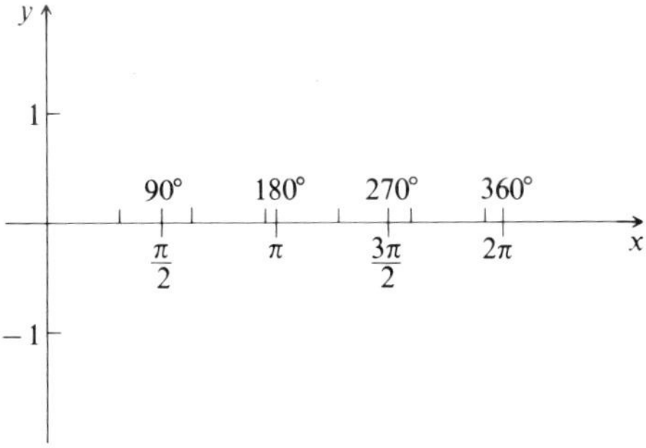

Figure 11.3

$$\pi \approx 3.14 \approx 180^\circ$$
$$\pi \approx 1.57 \approx 90^\circ$$

It should be pointed out that points on the x axis can be labeled in either radian measure or degree measure. One should also note that it is not necessary to let one unit be the same on each axis, although we will do so throughout the following sections.

With this interpretation in mind we return to the problem of sketching the graph of $y = \sin x$. To do this, we will let x assume the values 0°, 15°, 30°, 60°, $\cdots$ 360° and determine the corresponding values of y. The following table shows this relationship. The entries for sin x are taken from Table IV and rounded to two decimal places.

x	Sin x	Ordered Pair	x	Sin x	Ordered Pair
0	0	(0°, 0)	195°	−.26	(195°, −.26)
15°	.26	(15°, .26)	210°	−.5	(210°, −.5)
30°	.5	(30°, .5)	225°	−.71	(225°, −.71)
45°	.71	(45°, .71)	240°	−.87	(240°, −.87)
60°	.87	(60°, .87)	255°	−.97	(255°, −.97)
75°	.97	(75°, .97)	270°	−1	(270°, −1)
90°	1	(90°, 1)	285°	−.97	(285°, −.97)
105°	.97	(105°, .97)	300°	−.87	(300°, −.87)
120°	.87	(120°, .87)	315°	−.71	(315°, −.71)
135°	.71	(135°, .71)	330°	−.5	(330°, −.5)
150°	.5	(150°, .5)	345°	−.26	(345°, −.26)
165°	.26	(165°, .26)	360°	0	(360°, 0)
180°	0	(180°, 0)			

We now locate these ordered pairs on a rectangular coordinate system and join them with a smooth curve as in Figure 11.4.

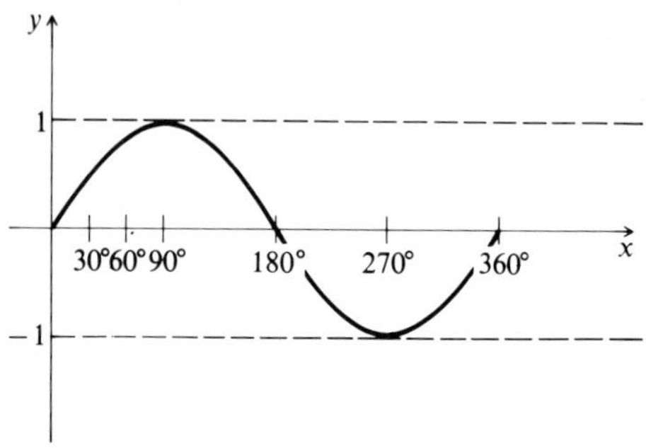

Figure 11.4

Notice that the maximum displacement from 0 for the curve is one. Also notice that if we were to continue sketching the graph for angles greater than 360° we would have a repetition of the segment that we have already sketched. The maximum displacement from zero is called the *amplitude* of the sine wave and the *period* is the wave length of the curve. In the case of the sine the period is 360° or 2π.

11.9 The Basic Sine Wave

The Basic Sine Wave is one that has an equation of the form $y = A \sin x$, where A is a positive number. In the last section we considered the graph of $y = \sin x$ which is a basic sine wave in which $A = 1$. We will now consider what effect increasing the value of A has on the sine wave. Suppose, for example, that A is set equal to 2 and the set of ordered pairs satisfying $y = 2 \sin x$ is determined. Each ordered pair satisfying this relationship would have a y coordinate which is two times the corresponding y coordinate for $y = \sin x$. The following table shows this comparison $0° \leq x \leq 180°$.

x	$y = \sin x$	$y = 2 \sin x$
0°	0	0
30°	$1/2 = .5$	1
60°	$\sqrt{3}/2 = .866$	$\sqrt{3} = 1.732$
90°	1	2
120°	$\sqrt{3}/2 = .866$	$\sqrt{3} = 1.732$
150°	$1/2 = .5$	1
180°	0	0

Figure 11.5 shows the graphs of each of these sine waves for comparison.

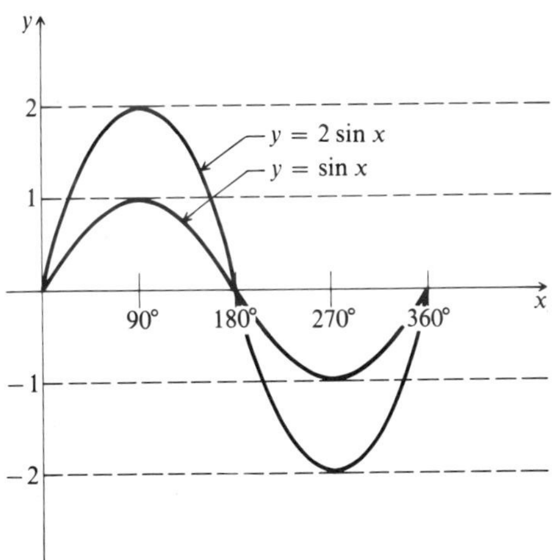

Figure 11.5

Notice that the amplitude of $y = 2 \sin x$ is twice the amplitude of $y = \sin x$ but that the period has not been altered. Thus the amplitude of $y = A \sin x$, in general, is always A times the amplitude of $y = \sin x$ or A.

We now inquire as to what would cause a change in the period of the sine wave. To answer this question we will consider the equation $y = \sin 2x$ in comparison with the basic sine wave $y = \sin x$. The following table provides the means of discovering the relationship. We restrict x to the interval $0 \leq x \leq 180°$ for this purpose with the realization that the points in the third and fourth quadrants have a y coordinate that is the additive inverse of those in quadrants one and two.

x	$y = \sin x$	$2x$	$y = \sin 2x$
0°	0	0°	0
15°	.26	30°	.5
30°	.5	60°	.87
45°	.71	90°	1
60°	.87	120°	.87
75°	.97	150°	.5
90°	1	180°	0
105°	.97	210°	−.5
120°	.87	240°	−.87
135°	.71	270°	−1
150°	.5	300°	−.87
165°	.26	330°	−.5
180°	0	360°	0

Figure 11.6 shows the graphs of the two sine waves for comparison. To sketch $y = \sin 2x$ obtain the ordered pairs $(x, \sin 2x)$ and not $(2x, \sin 2x)$ which is a common error.

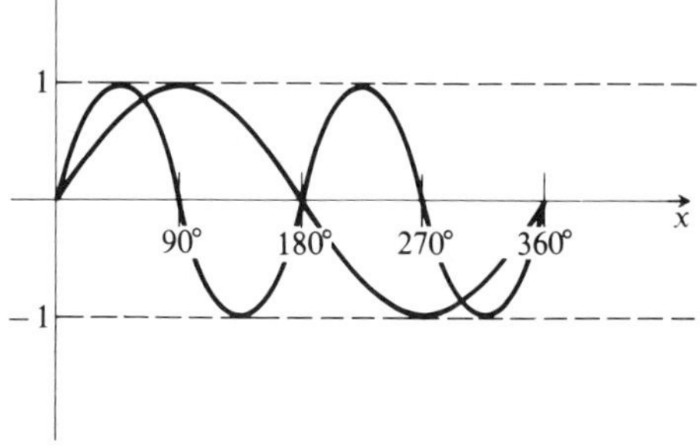

Figure 11.6

Notice that the amplitude is not affected but that the period is cut in half. The period of $y = \sin 2x$ is π.

If we stop to think for a few moments, we will realize that multiplying the angle by a factor of $B > 0$, such as $y = \sin Bx$, alters the period of the basic sine wave to a new period of $2\pi/B$. In general, the graph of $y = A \sin Bx$ has amplitude A and period $2\pi/B$.

Example 1. Sketch the graph of $y = 2 \sin 3x$.
The amplitude is 2 and the period is $2\pi/3$. Knowing the basic pattern of the sine wave allows us to sketch the graph without constructing a table of values. We must begin at the origin where $y = 0$ and pass through its basic cycle within $2\pi/3$ units or 120°.

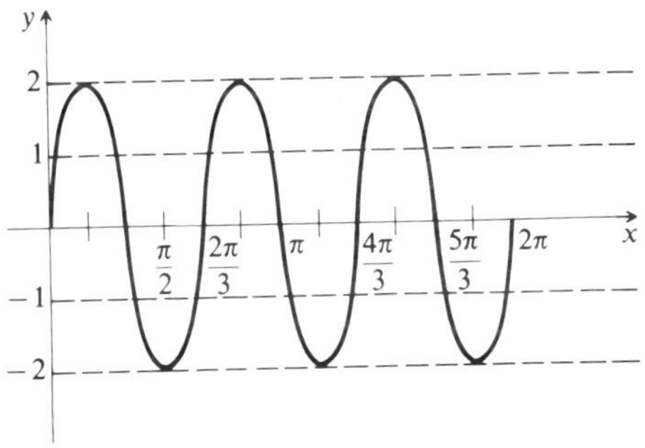

Example 2. Sketch the graph of $y = 4 \sin 1/2x$.
The amplitude is 4 and the period is $\dfrac{2\pi}{1/2} = 4\pi$.

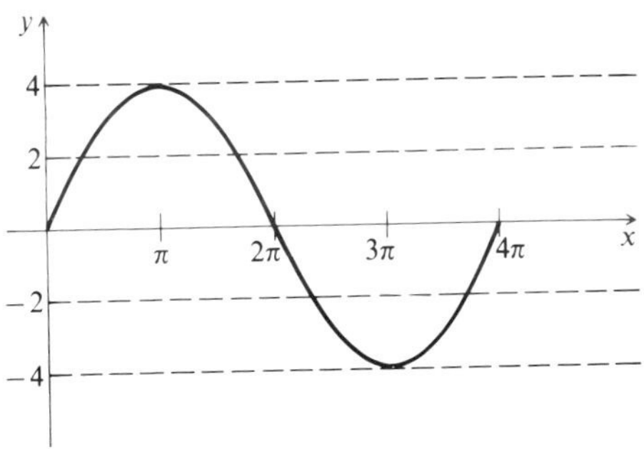

Exercise 72

State the period and amplitude of each of the following, and sketch the graphs $0 \le x < 2\pi$.

1) $y = \sin x$
2) $y = 3 \sin x$
3) $y = 4 \sin x$
4) $y = \sin 3x$
5) $y = \sin 2x$
6) $y = \sin 1/2x$
7) $y = \sin 1/3x$
8) $y = 3 \sin 2x$
9) $y = 3 \sin 3x$
10) $y = 1/2 \sin 3x$
11) $y = 2 \sin 2x$
12) $y = \sin \frac{\pi}{4}x$
13) $y = 7 \sin 3x$
14) $y = 5 \sin 4x$
15) $y = 2 \sin \frac{\pi}{2}x$

11.10 The Phase Angle of the Sine Wave

In Section 11.9 we defined the basic sine wave to be one of the form $y = A \sin x$. The period is 2π, and the wave passes through the origin for any value of A. If a sine wave does not pass through the origin, we say that it is *out of phase* with the basic sine wave. The number of degrees through which we need to shift the curve to the right or left along the x axis to cause it to pass through the origin is called the *phase angle*.

Equations of the form

$$y = A \sin(Bx + C) = A \sin B\left(x + \frac{C}{B}\right)$$

are out of phase with the basic sine wave as we shall soon discover. Here as before A is the amplitude and $2\pi/B$ is the period. C/B is the phase angle, and it is expressed either in radians or degrees.

To discover how the phase angle effects the basic sine wave we consider

$$y = \sin(x + 30°).$$

To do this we will proceed in the following manner. We first choose a value for x. We then add 30° to this value and compute $\sin(x + 30°)$ which gives us the y value. The following table was constructed using this procedure.

x	$x + 30°$	$y = \sin(x + 30°)$
$-30°$	$0°$	0
$0°$	$30°$	$1/2$
$30°$	$60°$	$\sqrt{3}/2$
$60°$	$90°$	1
$90°$	$120°$	$\sqrt{3}/2$
$120°$	$150°$	$1/2$
$150°$	$180°$	0
$180°$	$210°$	$-1/2$

Figure 11.7 shows the graph of $y = \sin(x + 30°)$.

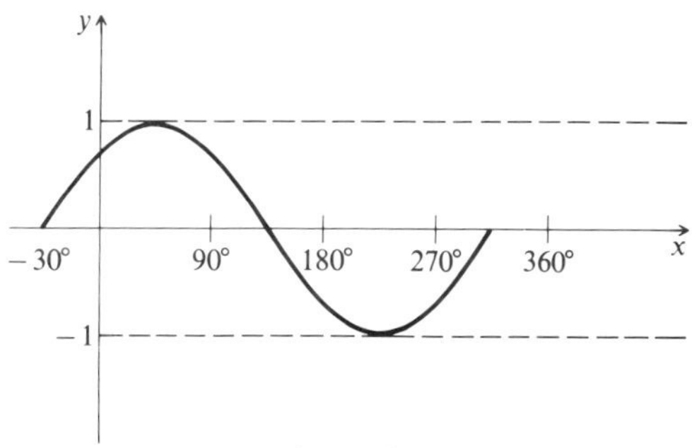

Figure 11.7

Here the essential feature is that the curve has been shifted 30° to the left. The period and amplitude are 2π and 1 respectively.

We now consider the case where the phase angle is negative. An example of this case is the equation $y = \sin(x - 45°)$. Using the procedures outlined above we construct the following table.

x	$x - 45°$	$\sin(x - 45°)$
$0°$	$-45°$	$-\sqrt{2}/2$
$45°$	$0°$	0
$90°$	$45°$	$\sqrt{2}/2$
$135°$	$90°$	1
$180°$	$135°$	$\sqrt{2}/2$
$225°$	$180°$	0

Figure 11.8 shows the graph of $y = \sin(x - 45°)$.

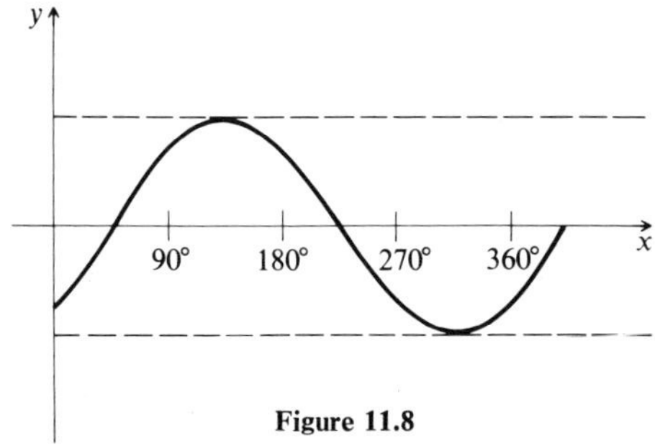

Figure 11.8

Once again the essential feature is the shift to the right of the basic sine wave. From the previous two examples we conclude that:

1) A positive phase angle causes a left shift of a basic sine wave.
2) A negative phase angle causes a right shift of a basic sine wave.

Example 1. Sketch the graph of $y = 2\sin(x + 90°)$.
The amplitude is 2. The period is 2π. The phase shift is 90° to the left. The graph is shown below.

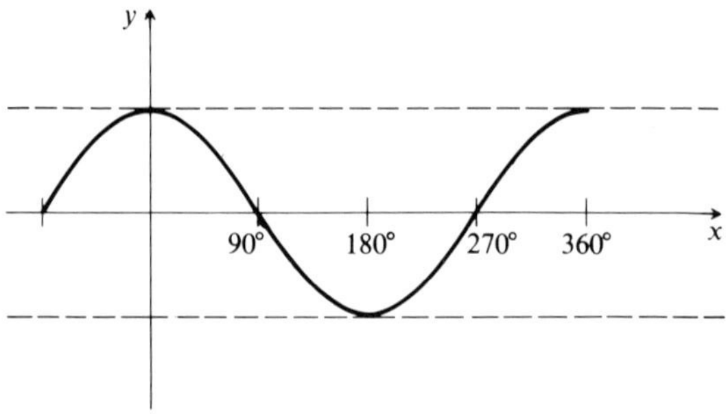

Example 2. Sketch the graph of $y = 2\sin(2x - 45°)$.
The amplitude is 2. The period is $2\pi/2 = \pi$. The phase shift is 22.5° to the right. The graph is shown below.

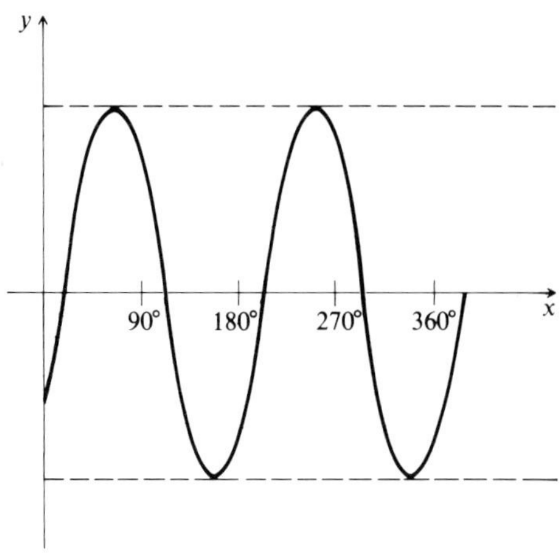

Example 3. Sketch $y = \cos x$ from $x = 0$ to $x = 180°$.

x	$y = \cos x$
0°	1
45°	$\sqrt{2}/2$
90°	0
135°	$-\sqrt{2}/2$
180°	-1

The graph is shown in the diagram by the solid line. The broken line is the graph of $y = \sin x$.

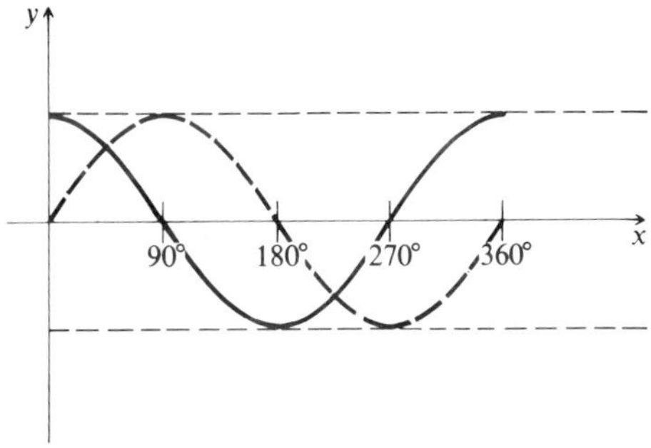

From the graph we observe that $y = \cos x$ is 90° out of phase with $y = \sin x$ or $\cos x = \sin(x + 90°)$.

Exercise 73

Sketch the graph of each of the following. Unless otherwise stated, $0° \leq x < 360°$.

1) $y = \sin(x + 30°)$
2) $y = \sin(x + 60°)$
3) $y = \sin(x - 45°)$
4) $y = \sin(x + 90°)$
5) $y = \sin(x - 180°)$
6) $y = 2\sin(2x + 60°)$
7) $y = 3\sin(x - 90°)$
8) $y = 4\sin(4x - 30°)$; $0° \leq x < 180°$
9) $y = 2\sin(2x - 90°)$
10) $y = 2\sin(x + 1)$

(For Problem 10, recall that unless otherwise indicated radian measure is used. Thus 1 represents radians.)

11) $y = 3\sin(2x + \pi)$
12) $y = -2\sin x$
13) What is the phase angle for Problem 12 with respect to the basic sine wave?
14) Find at least three examples of sine wave patterns from library research of technical journals.

Review Exercise Chapter 11

Convert the following from degrees to radians.

1) $85°$ 2) $455°$ 3) $-75°$ 4) $615°$

Convert the following from radians to degrees.

5) $\frac{5\pi}{12}$ 6) $\frac{7\pi}{18}$ 7) $\frac{11\pi}{6}$ 8) $\frac{14\pi}{3}$

Solve the following.

9) Find the area of the sector of a circle if the radius is 5 ft. and an intercepted arc is 6 ft.

10) If a point is on the circumference of an 8 ft. circle and moves with an angular velocity of 15π radians per second, find the linear velocity of the point.

Find the period and amplitude of the following and sketch. Designate the phase angle if out of phase.

11) $y = 3 \sin 2x$ 12) $y = \frac{3}{5} \sin \frac{\pi}{3}x$

13) $y = 1/2 \sin (4x + 6)$ 14) $y = 5 \sin (3x + \pi)$

15) $y = 2 \sin 2x + \sin \frac{x}{2}$

12
Vectors and Polar Coordinates

12.1 Vectors

When a quantity has both magnitude (size) and direction, it is called a *vector* quantity. If the quantity has magnitude only, it is called a *scalar* quantity. Some physical examples of vector quantities are force and velocity. Some examples of scaler quantities are weight and speed.

On a plane, a vector is the set of all line segments having a fixed length and a given direction. On the coordinate system in Figure 12.1, some representatives of vector v are given.

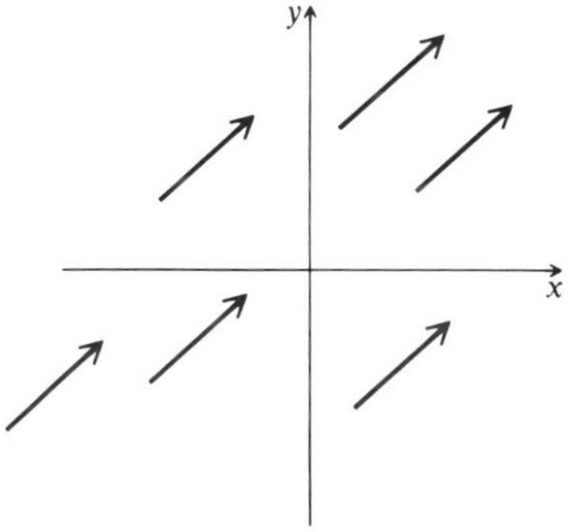

Figure 12.1

The notation $\vec{v}$ designates a vector quantity. Since all representatives of a given vector have the same magnitude and direction, we will choose only one *representative* of the vector. We are at liberty to choose the most convenient representative possible.

If $\vec{v}$ is a given vector, the $-\vec{v}$ is a vector having the same magnitude but the opposite direction. (See Figure 12.2).

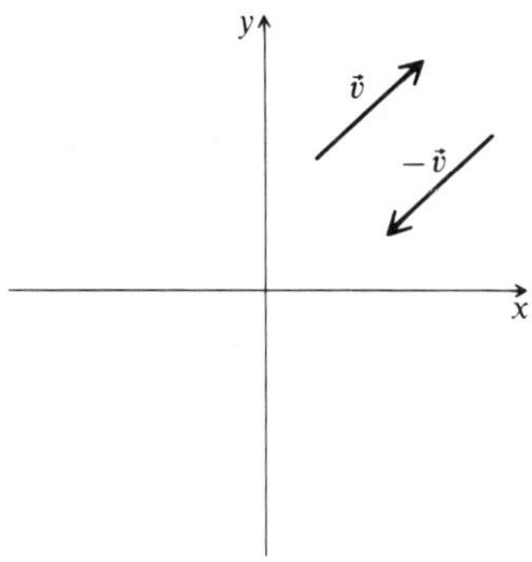

Figure 12.2

On a plane the direction of a vector is determined by the coordinates of the endpoints of a vector. That is, if vector $\overrightarrow{AB}$ (the vector from point A to point B) has endpoints of A: (x_1,y_1) and $B(x_2,y_2)$ as shown in Figure 12.3, the total y distance the vector covers is called the vertical component of the vector found by $(y_2 - y_1)$, and the total x distance the vector covers is called the horizontal component found by $(x_2 - x_1)$.

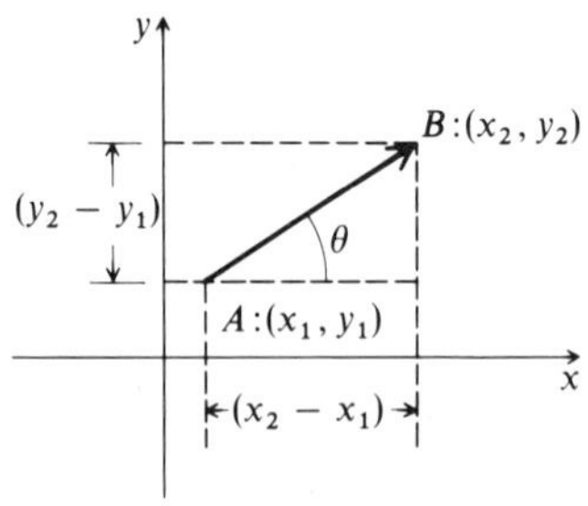

Figure 12.3

The direction angle of the vector is the angle the vector makes with a line parallel to the positive x axis. This angle, θ, is such that

$$\tan \theta = \frac{y_2 - y_1}{x_2 - x_1}.$$

If the coordinates of the endpoints are known, the magnitude of a vector can be determined by the distance formula. This length or magnitude is designated by $|\vec{v}|$. (See Figure 12.4).

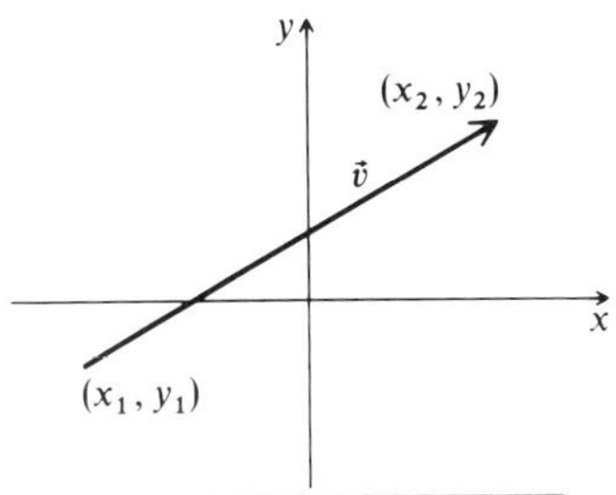

$$|\vec{v}| = \sqrt{(x_2 - x_1)^2 + (y_2 - y_1)^2}$$

Figure 12.4

If the representative for $\vec{v}$ is chosen with its tail (initial point) at the origin, then $|\vec{v}| = \sqrt{x^2 + y^2}$.

12.2 Scalar Multiplication

If k is scalar quantity and $\vec{u}$ is a vector, then $k\vec{u}$ is a vector in the same direction or in the opposite direction, if k is negative, and having a magnitude k times that of vector $\vec{u}$. Figure 12.5 illustrates scalar multiplication of a vector.

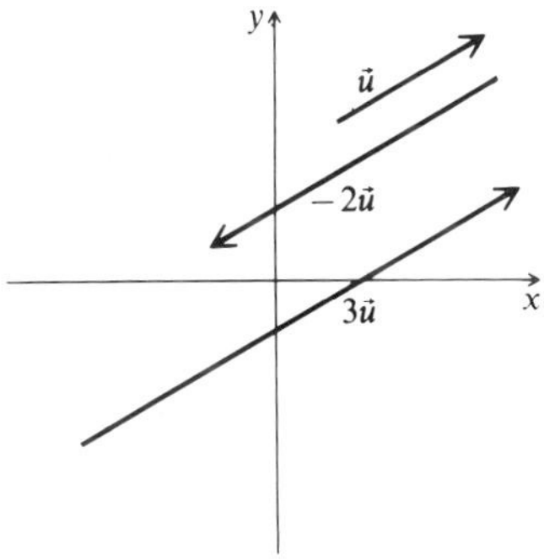

Figure 12.5

Since $|k|$ is the absolute value of k and $|\vec{u}|$ is the magnitude of vector $\vec{u}$ which is also a scalar, then the magnitude of ku is given by

$$|k|\,|\vec{u}| = |k\vec{u}|.$$

12.3 Vector Addition

If u and v are vectors, then the *sum* or *resultant* of $\vec{u}$ and $\vec{v}$, designated as $\vec{u} + \vec{v}$, is a vector whose x value is the sum of the x values of $\vec{u}$ and $\vec{v}$ and whose y value is the sum of the y values of $\vec{u}$ and $\vec{v}$. The direction of $\vec{u} + \vec{v}$ will be the angle formed by a representative of this sum with the positive x axis. If $\vec{u}$ and $\vec{v}$ are chosen with their tail at the origin as in Figure 12.6, then $\vec{r} = \vec{u} + \vec{v}$ will have endpoint coordinates of $(x_1 + x_2, y_1 + y_2)$ and $|\vec{r}| = \sqrt{(x_1 + x_2)^2 + (y_1 + y_2)^2}$. The direction of $\vec{r}$, given by angle θ can be determined by $\tan \theta = \dfrac{y_1 + y_2}{x_1 + x_2}$.

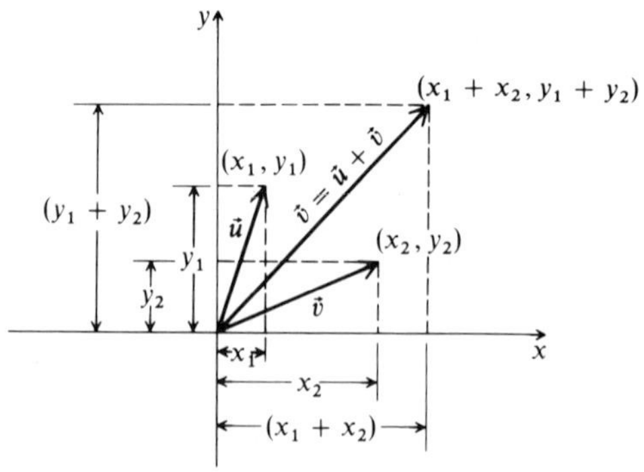

Figure 12.6

Example 1. If two vectors have initial points at the origin and terminal points at (3,2) and (7,12) respectively, find their sum.
Their sum is a vector with initial point at the origin and terminal point at $(3 + 7, 2 + 12) = (10,14)$.

One useful way of finding the sum of two vectors in a plane is by the *parallelogram method.* Using this method, a representative of each vector is chosen in such a way that the initial point of the second vector is at the terminal point of the first vector. The sum of the vectors is represented by the vector whose initial point is at the initial point of the first vector and whose terminal point is at the terminal point of the second vector. The resulting sum, often called the *resultant,* then becomes the diagonal of a parallelogram whose sides are the given vectors. The length of the resultant can be obtained by solving the triangle formed by the vectors. Figure 12.7 illustrates this procedure.

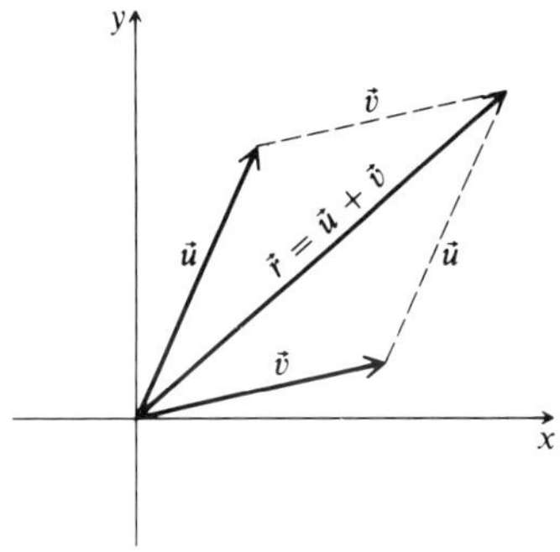

Figure 12.7

Also illustrated in Figure 12.7 is the fact that vector addition is commutative. That is, $\vec{u} + \vec{v} = \vec{r} = \vec{v} + \vec{u}$.

Example 2. If two forces f_1 and f_2 of magnitudes 50 lb. and 90 lb. respectively are acting on an object in the directions shown, find the magnitude and direction of the resultant force.

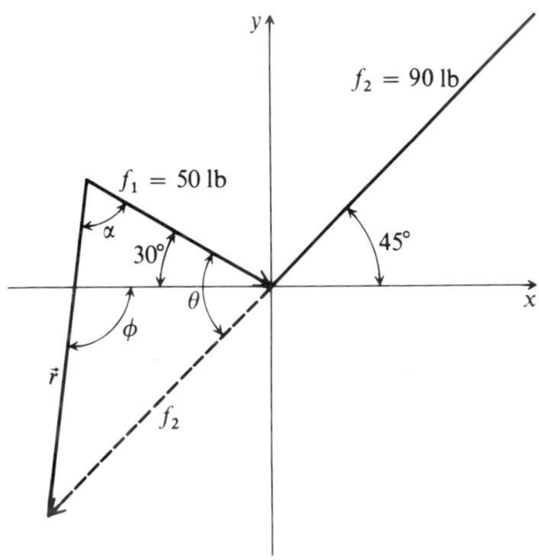

The angle θ between the vectors is $(30° + 45°) = 75°$. By using the law of cosines,

$$\begin{aligned} |\vec{r}|^2 &= (50)^2 + (90)^2 - 2(50)(90)\cos 75° \\ &= 2500 + 8100 - 9000(\cos 75°) \\ &= 10600 - 9000(.2588) \\ &= 8270.8 \\ |r| &= 90.94. \end{aligned}$$

Angle α can be found by the law of sines.

$$\frac{\sin \alpha}{90} = \frac{\sin 75}{r}$$

$$\sin \alpha = \frac{90 \sin 75}{r} = \frac{90(.9659)}{90.94} = .9559$$

$$\alpha = 72.9^\circ \text{ to the nearest tenth degree.}$$

$$\phi = \alpha + 30^\circ = 102.9^\circ.$$

Therefore, the resultant has a magnitude of 90.94 lbs. in a direction 102.9°. (The resultant is in a clockwise direction from the positive x axis.)

12.4 Vector Subtraction

The difference of two vectors $\vec{u}$ and $\vec{v}$ is the vector $\vec{d} = \vec{u} - \vec{v}$ which must be added to $\vec{v}$ to obtain $\vec{u}$. That is, $\vec{d} + \vec{v} = \vec{u}$ and $\vec{u}$ becomes the resultant of $\vec{d}$ and $\vec{v}$. (See Figure 12.8). Note also that $\vec{u} - \vec{v}$ can be thought of as $\vec{u} + (-\vec{v})$ and can therefore be determined by the difference of the horizontal and vertical components of the two vectors, if these values are known.

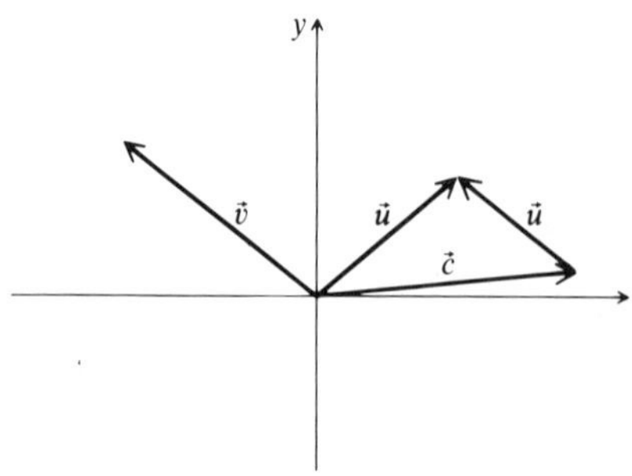

Figure 12.8

Example 1. If vectors $\vec{m}$ and $\vec{n}$ have their initial point at the origin and have terminal points at $(-3,2)$ and $(5,6)$ respectively, find $\vec{m} + \vec{n}$, $\vec{m} - \vec{n}$ and $\vec{n} - \vec{m}$.

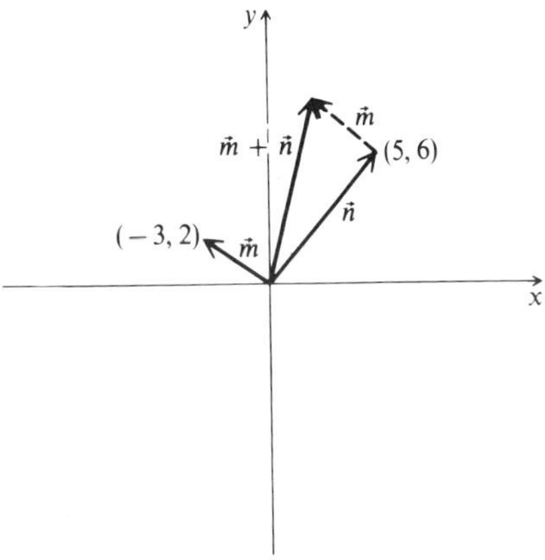

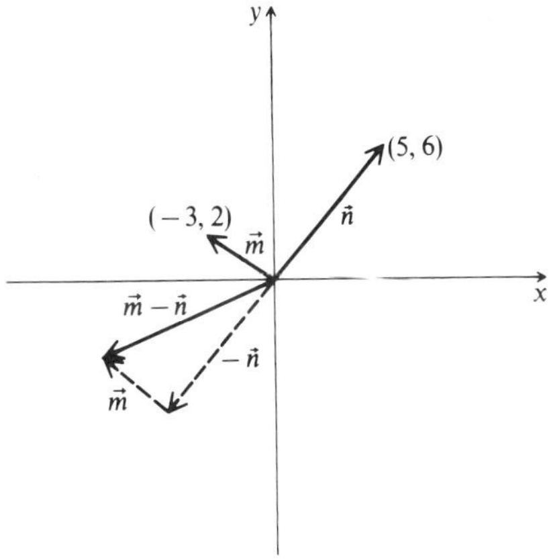

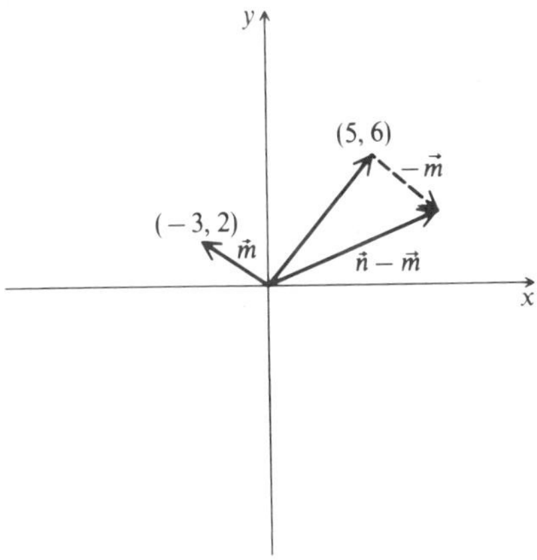

$\vec{m} + \vec{n}$ has a representative with initial point at the origin and terminal point at $(-3 + 5, 2 + 6) = (2,8)$. (See diagram (a).) $\vec{m} - \vec{n} = \vec{m} + (-\vec{n})$ and has a representative with initial point at the origin and terminal point at $(-3 - 5, 2 - 6) = (-8, -4)$. (See diagram (b).) $\vec{n} - \vec{m} = \vec{n} + (-\vec{m})$ and has a representative with initial point at the origin and terminal point at $(5 - (-3), 6 - 2) = (8,4)$. (See diagram (c).) Note that $\vec{n} - \vec{m} = -(\vec{m} - \vec{n})$.

12.5 Vectors in Terms of Unit Vectors

A *unit vector* is a vector one unit in length in a given direction. A unit vector in the direction of any given vector can be shown by multiplying the given vector by one over the magnitude (a scalar). This leaves the direction unchanged and makes the magnitude one unit. That is, if $\vec{v}$ is a vector with magnitude $|\vec{v}|$, then a unit vector along $\vec{v}$ is given by $\dfrac{\vec{v}}{|\vec{v}|}$.

Example 1. Given $\vec{v}$ with magnitude 5 in. A unit vector in the direction of $\vec{v}$ would be $\vec{v}/5$.

If a unit vector along the x axis is called $\vec{i}$ and a unit vector along the y axis is called $\vec{j}$, then a vector a units long along the x axis can be written as $a\vec{i}$ and a vector b units long along the y axis can be written as $b\vec{j}$. By using this analogy, any vector with initial point at the origin and terminal point at the point (a,b) is the result of $a\vec{i} + b\vec{j}$. This is illustrated in Figure 12.9.

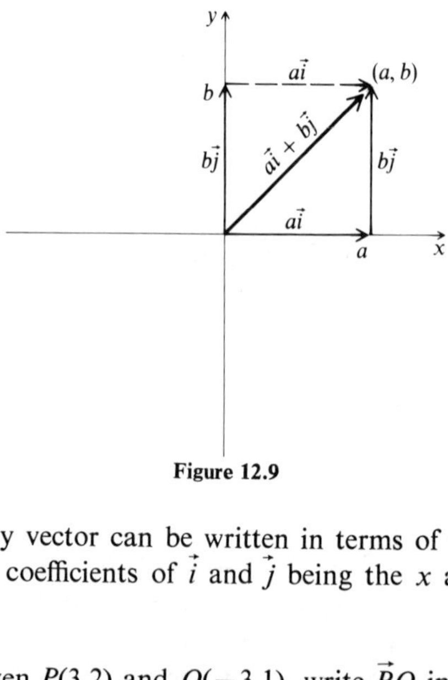

Figure 12.9

Therefore, any vector can be written in terms of unit vectors along x and y with the coefficients of $\vec{i}$ and $\vec{j}$ being the x and y components respectively.

Example 2. Given $P(3,2)$ and $Q(-3,1)$, write $\vec{PQ}$ in unit vector form. The x component of $\vec{PQ}$ is $3 - (-3) = 6$ and the y component is $(2 - 1) = 1$. Therefore, $\vec{PQ} = 6\vec{i} + \vec{j}$.

If vectors are given in unit vector form, the sum or difference of the two vectors is a vector whose components are the sums or differences of the x and y components respectively.

Example 3. Given $\vec{v} = i - j$ and $\vec{u} = 4i + 6j$, find $\vec{v} + \vec{u}$ and $\vec{v} - \vec{u}$.

$$\vec{v} + \vec{u} = (1 + 4)i + (-1 + 6)j = 5i + 5j$$
$$\vec{v} - \vec{u} = (1 - 4)i + (-1 - 6)j = -3i - 7j.$$

The magnitude of $\vec{v} = a\vec{i} + b\vec{j}$ is found as before by finding the length of the vector.

$$|\vec{v}| = \sqrt{a^2 + b^2}.$$

If we desired a unit vector in the direction of $\vec{v}$, we could write

$$\frac{\vec{v}}{|\vec{v}|} = \frac{ai + bj}{\sqrt{a^2 + b^2}} \quad \text{or} \quad \frac{a}{\sqrt{a^2 + b^2}}\, i + \frac{b}{\sqrt{a^2 + b^2}}\, j.$$

Exercise 74

Determine the magnitude of the vectors having the following endpoints.

1) (1,3); (5,6)
2) (7,−1); (−4,−3)
3) (−2,2); (−2,−2)
4) (7,4); (−3,−2)
5) (3,−5); (−11,7)

6)–10) Find the direction angle for Problems 1)–5).

Each of the following points represent the endpoint of a vector with initial point at the origin. In each problem, find $\vec{u} + \vec{v}$ and $\vec{u} - \vec{v}$.

11) $\vec{u}$: (3,−2); $\vec{v}$: (6,−5)
12) $\vec{u}$: (−7,−4); $\vec{v}$: (2,−3)
13) $\vec{u}$: (−9,−6); $\vec{v}$: (3,−5)
14) $\vec{u}$: (1,3); $\vec{v}$: (5,6)
15) $\vec{u}$: (7,−1); $\vec{v}$: (−4,−3)

Solve the following.

16) If an airplane is flying east at 150 mph and a wind is blowing from the northeast at 30 mph, what is the resulting speed and direction of the airplane? (Hint: northeast is 45° from the east direction.)

17) If a 50-lb. box is set on a plane inclined at 15°, what part of the force is acting in the direction of the plane? Assume no friction.

18) In Problem 17 if a friction force of 15 lbs. is acting up the plane, is the box sliding along the plane?

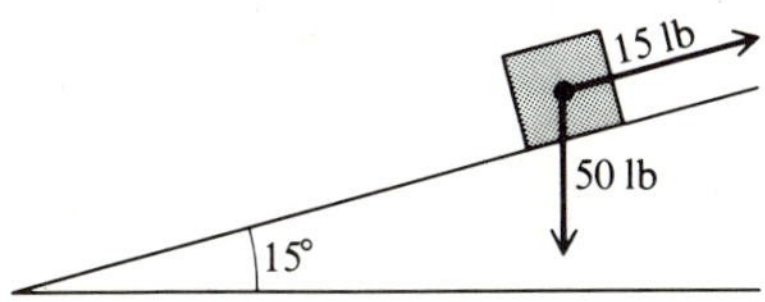

19) A ship travels 150 miles due west from port. If the ship then turns toward the north by 75 degrees and travels 110 miles, how far is the ship from port?

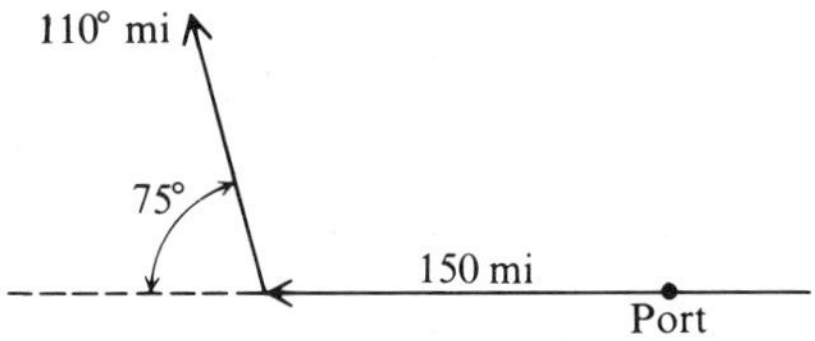

20) Two men are pushing on a box weighing 275 lbs. One man is pushing against the box with a force of 175 lbs. at an angle of 30° with the direction the box is moving. If the other man is pushing at an angle of 20° with the direction of motion, how much force must he exert to move the box.

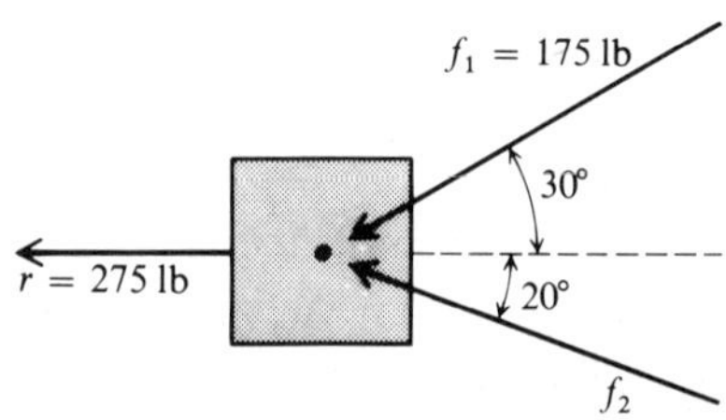

21)–25) Write a unit vector in the same direction as each vector in exercises 1)–5).

12.6 Polar Coordinates

In rectangular form, any point on a plane may be represented by an ordered pair. Let us now consider the point in a different context.

Any point having rectangular coordinates (a,b) can be considered as the terminal point of a vector with initial point at the origin. This vector has horizontal component a and vertical component b and magnitude of $\sqrt{a^2 + b^2}$. This vector will be referred to as $\vec{r}$. (See Figure 12.10). Every vector $\vec{r}$ forms an angle, θ with the positive x axis. Using this notation, any point in rectangular form can be written in terms of a directed distance $\vec{r}$ and an angle θ. (The ordered pair (r,θ) designates the endpoint of a vector r forming an angle θ with the positive x axis.) When points are given in this way, they are in *polar coordinates*. The direction from the origin to the given point is considered as a positive r, and the vector projected through the origin in the opposite direction is considered as negative r. The positive angle θ is measured by a counterclockwise rotation as shown in previous sections.

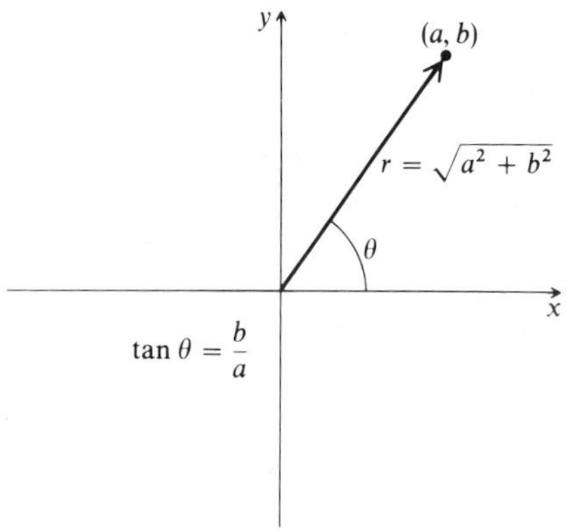

Figure 12.10

The polar representation of a point is not unique as is the rectangular representation. Figure 12.11 shows various ways of representing the point (r,θ) which is the polar form of point (a,b). Figure 12.11(a) shows both

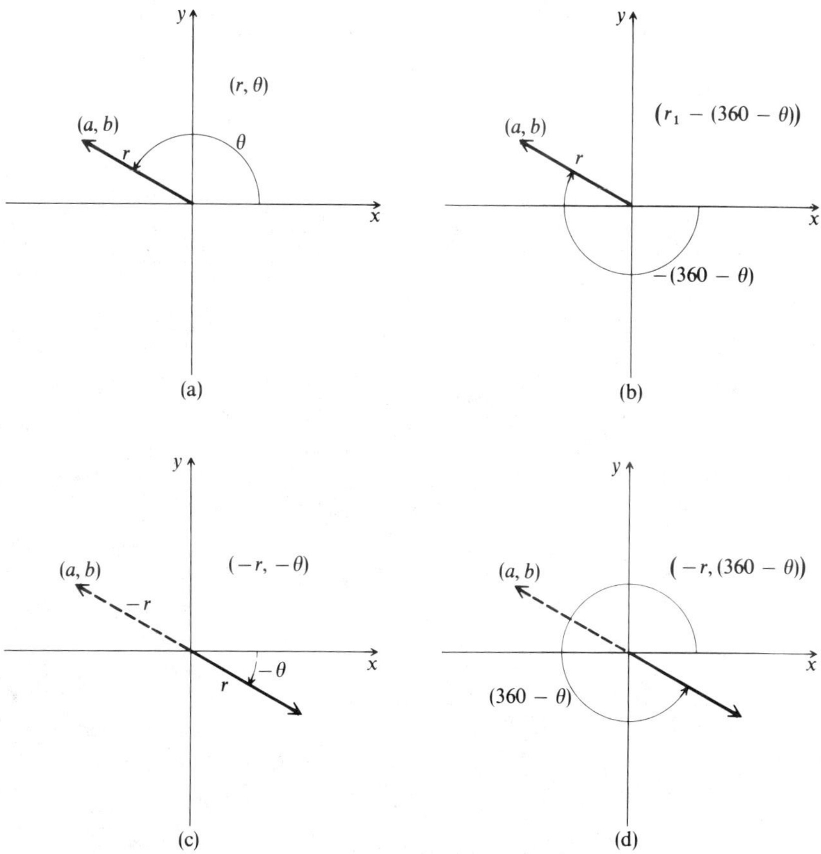

Figure 12.11

r and θ as positive; Figure 12.11(b) shows the angle as negative and r as positive; Figure 12.11(c) shows the point represented by both a negative vector and a negative angle and Figure 12.11(d) shows the point represented as a positive angle with a negative vector. Even these ways do not exhaust the possibilities. Since θ is a rotational angle, a rotation through further complete revolutions in each case shown will result in different polar coordinates for the point. In practice, however, the simplest positive or negative angle is normally used.

Example 1. Represent (5,30°) graphically in four different ways.

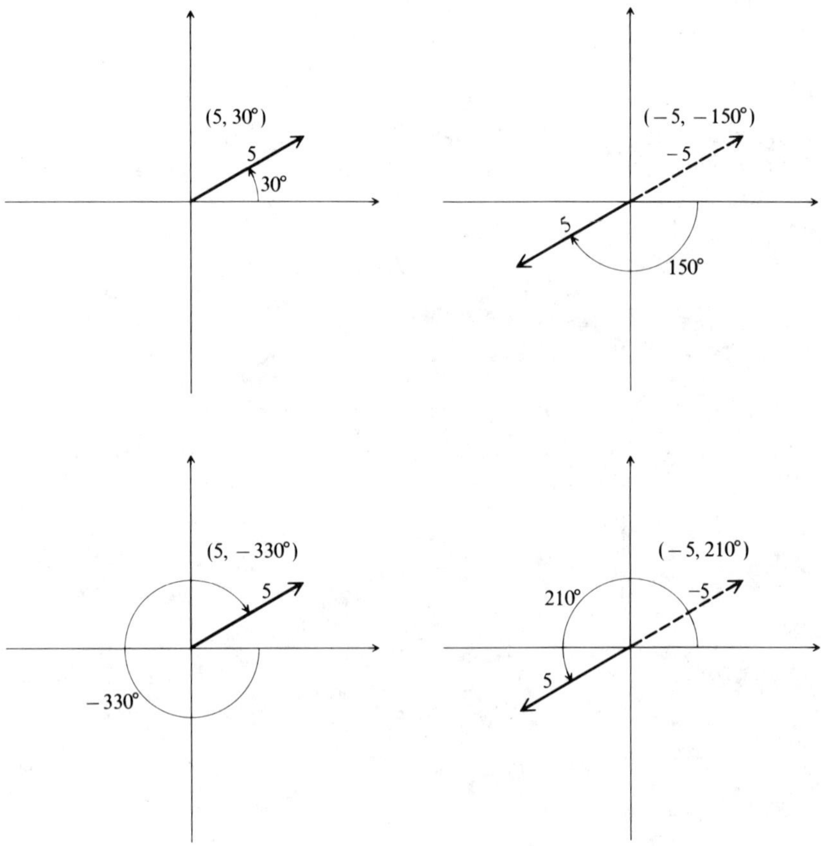

Example 2. Give three equivalent forms for the point (−3,−45°).

(3,135°); (−3,315°); (3,−225°).

All four forms are shown graphically below.

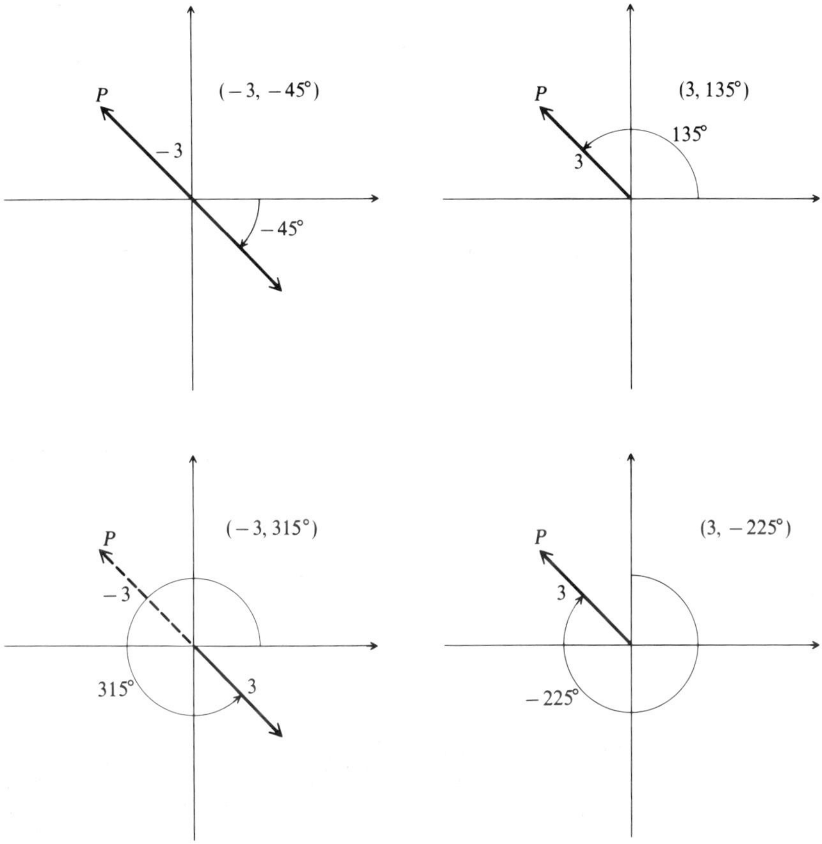

12.7 Graphing in Polar Coordinates

Polar coordinates involve only a directed distance from the origin and an angle formed with the positive horizontal axis. The origin is referred to as the *pole* and the positive horizontal axis is called the *polar axis*. A polar graph, therefore, is marked only in angles and distances from the pole shown as concentric circles. (See Figure 12.12). Each mark on the circumference represents 5° and each concentric circle one unit. A point can be represented on a polar graph as the endpoint of a vector forming a given angle with the polar axis.

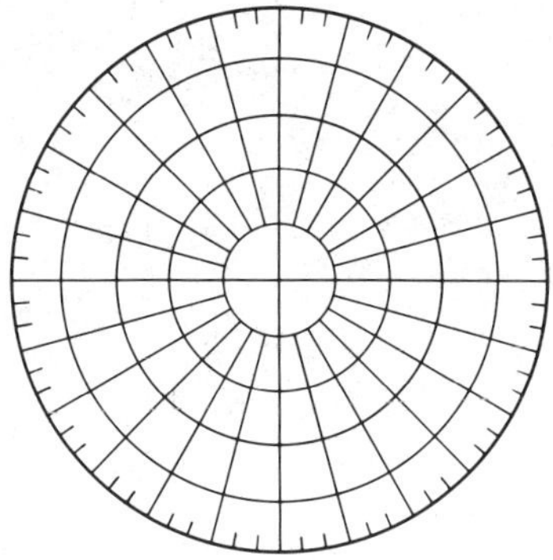

Figure 12.12

Example 1. Show the points (3,75°), (−2,15°), (5,210°) and (4,135°) on a polar graph.

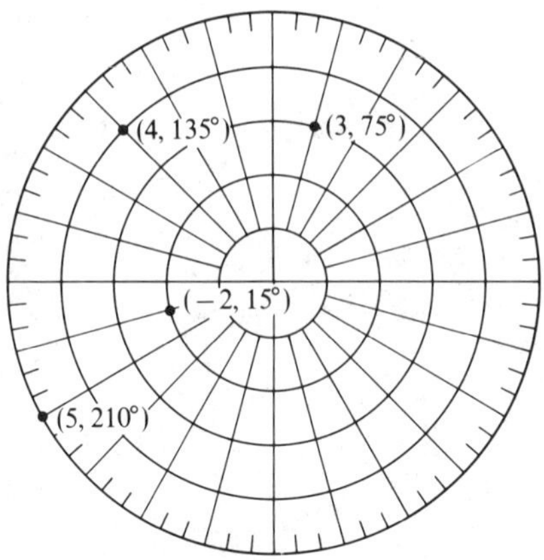

12.8 Equation in Polar Form

When equations are written in terms of the variables r and θ, they are said to be in polar form. It is often necessary, in using polar coordinates, to consider θ in radians (real numbers) so that both coordinates of each ordered pair (r,θ) will be in terms of real numbers. For instance, the equation $r = \theta$ is in polar form. For each real number r, there is a real number θ so that the angle θ must be in radian measure. The student must be familiar with angles both in terms of degrees and radians so that conversion from one form to another is a minor part of the computation.

Polar equations can be graphed by choosing a set of ordered pairs satisfying the equation and plotting each of these ordered pairs on a polar graph. Sufficient points must be taken to determine the general shape of the graph.

Example 1. Graph $r = \theta$.

r	0	.26	.52	.79	1.05	1.57	2.09	2.36	2.62
θ	0	$\pi/12$	$\pi/6$	$\pi/4$	$\pi/3$	$\pi/2$	$2\pi/3$	$3\pi/4$	$5\pi/6$

3.14	3.67	3.93	4.19	4.71	5.24	5.50	5.76	6.28
π	$7\pi/6$	$5\pi/4$	$4\pi/3$	$3\pi/2$	$5\pi/3$	$7\pi/4$	$11\pi/6$	2π

The graph of this equation is a spiral.

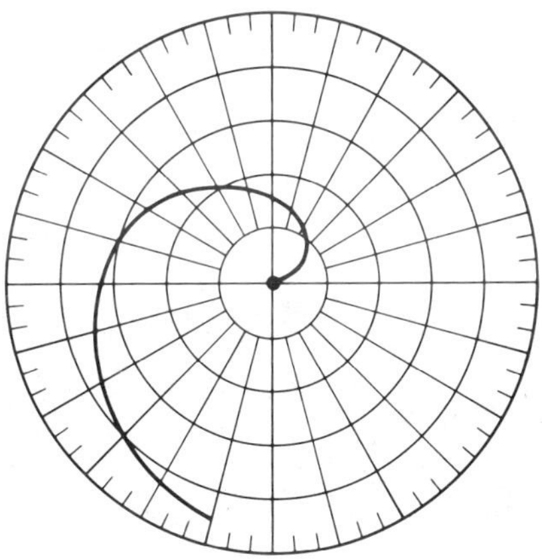

Example 2. Graph the equation $r = \sin\theta$.

r	0	.5	.707	.866	1	.866	.707	.5
θ	0	$\pi/6$	$\pi/4$	$\pi/3$	$\pi/2$	$2\pi/3$	$3\pi/4$	$5\pi/6$

0	−.5	−.707	−.866	−1	−.866	−.707	−.5	0
π	$7\pi/6$	$5\pi/4$	$4\pi/3$	$3\pi/2$	$5\pi/3$	$7\pi/4$	$11\pi/6$	2π

Notice that points in quadrants III and IV only repeat points already located. The graph describes the graph of a circle with center at $(1,\pi/2)$ having a diameter of 1.

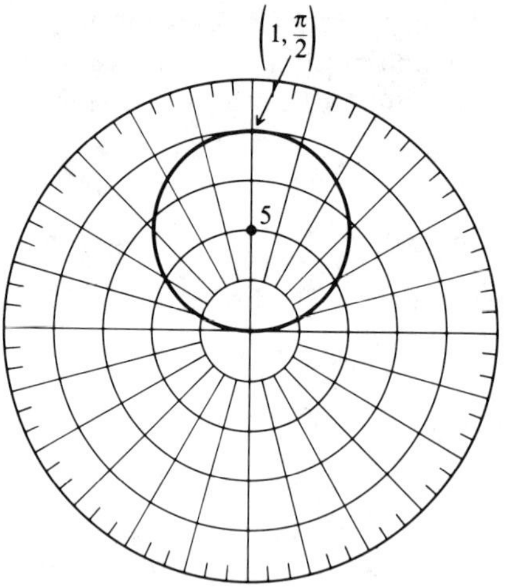

Exercise 75

Represent each of the following points in two different ways and show each representation graphically.

1) $(5,45^\circ)$ 2) $(-6,\pi/3)$ 3) $(\sqrt{3},-60^\circ)$

4) $(3,\pi/12)$ 5) $\left(-2,\frac{15\pi}{6}\right)$ 6) $(1,-720^\circ)$

7) $(-7,-570^\circ)$ 8) $\left(\frac{\sqrt{3}}{2},\frac{-23\pi}{6}\right)$

Graph the following curves in polar form.

9) $r = \cos\theta$ 10) $2r = \cos\frac{\theta}{2}$ 11) $3r = 4\theta$

12) $r = 2\sin\theta + \cos\theta$ 13) $r^2 = 4\cos 2\theta$ 14) $r = 3\cos 2\theta$

15) $r = 3\cos\theta$ 16) $r = 4\cos 3\theta$ 17) $r = \sin 3\theta$

18) $(r-2)^2 = 8\theta$ 19) $r^2 = 2\sin 2\theta$ 20) $r\theta = \sin\theta$

12.9 Converting from Rectangular to Polar Form

We have previously discussed a point as the endpoint of a vector $\vec{r}$ as a motivation for the system of polar coordinates. We can follow this concept of the rectangular coordinates of a point to convert the coordinates of the point into polar coordinates. Notice from Figure 12.12 that the rectangular coordinates of a point (a,b) determine the legs of a right triangle with θ, the angle formed with the positive x axis, as one of the acute angles of the triangle.

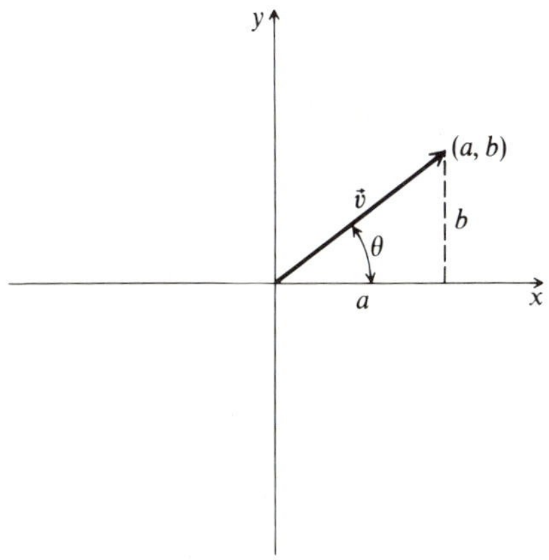

Figure 12.13

As previously indicated, $|\vec{r}| = \sqrt{a^2 + b^2}$ or $r = \pm\sqrt{a^2 + b^2}$, where the positive sign is used to designate the direction from the origin to (a,b) and the negative sign designates the opposite direction. We can also observe, however, that in the right triangle $\sin\theta = b/r$ and $\cos\theta = a/r$. This leads us to the important conversion equations

$$a = r\cos\theta$$
$$b = r\sin\theta.$$

That is, for any point (x,y) in rectangular form,

$$x = r\cos\theta$$
$$y = r\sin\theta$$
$$r = \sqrt{x^2 + y^2}.$$

Example 1. Write the point $(\sqrt{3},1)$ in polar coordinates.

$$r = \sqrt{x^2 + y^2} = \sqrt{3 + 1} = \sqrt{4} = 2.$$

$x = r\cos\theta$. Thus $\cos\theta = \dfrac{x}{r} = \dfrac{\sqrt{3}}{2}$. $\theta = \pi/6$. The point is therefore $(2,\pi/6)$ in polar coordinates, (or $(-2,7\pi/6)$, $(-2,-5\pi/6)$, $(2,-11\pi/6)$).

Example 2. Write the point $(-3,5)$ in polar coordinates.

$$r = \sqrt{9 + 25} = \sqrt{34}.$$

$\cos\theta = \dfrac{x}{r} = \dfrac{-3}{\sqrt{34}}$ and $\sin\theta = \dfrac{y}{r} = \dfrac{5}{\sqrt{34}}$ so $\tan\theta = \dfrac{y}{x} = \dfrac{5}{-3} = -1.666$. Thus $\theta = 59°$ approximately or 1.03 radians. The polar coordinates of the point are $(\sqrt{34},59°)$.

Example 3. Write $(5,-\pi/6)$ in rectangular form.

$$r = 5,\ \theta = -\pi/6.$$

$$x = r\cos\theta = 5\cos -\pi/6 = \frac{5\sqrt{3}}{2}.$$

$$y = r\sin\theta = 5\sin -\pi/6 = 5(-1/2) = -5/2.$$

The rectangular coordinates of the point are

$$\left(\frac{5\sqrt{3}}{2}, -5/2\right).$$

Equations can also be converted from one system to another by using the same conversion identities. The following examples are used to illustrate this procedure.

Example 4. Write $x^2 + y^2 = 16$ in polar form.

$$x = r\cos\theta,\ y = r\sin\theta$$

$$r^2\cos^2\theta + r^2\sin^2\theta = 16$$

$$r^2(\cos^2\theta + \sin^2\theta) = 16$$

$$r^2 = 16$$

$$r = 4 \quad \text{or} \quad r = -4$$

are polar forms of the equation.

Example 5. Write $3x + 2y = 1$ in polar form.

$$3r\cos\theta + 2r\sin\theta = 1.$$

$$r(3\cos\theta + 2\sin\theta) = 1.$$

$$r = \frac{1}{3\cos\theta + 2\sin\theta}.$$

Example 6. Write $r = \cos\theta - \sin\theta$ in rectangular form.

$$r = \sqrt{x^2 + y^2}$$

$$\cos\theta = \frac{x}{r},\ \sin\theta = \frac{y}{r}.$$

$$\sqrt{x^2 + y^2} = \frac{x}{\sqrt{x^2 + y^2}} - \frac{y}{\sqrt{x^2 + y^2}}$$

$$\sqrt{x^2 + y^2} = \frac{x - y}{\sqrt{x^2 + y^2}}$$

$$x^2 + y^2 = x - y$$

$$x^2 - x + y^2 + y = 0.$$

Exercise 76

Convert the following to rectangular form.

1) $(3,30°)$ 2) $(-2,\pi/12)$ 3) $(-1,-\pi/4)$ 4) $(-7,19°)$

5) $(3,-215°)$ 6) $(3/2,120°)$ 7) $(1/4,275°)$ 8) $(3/8,4\pi)$

9) $(5,16°)$ 10) $(-4,3\pi/2)$

Convert the following to polar form.

11) $\left(\frac{\sqrt{3}}{2},1\right)$ 12) $\left(\frac{\sqrt{2}}{2},\frac{\sqrt{2}}{2}\right)$ 13) $(5,5)$ 14) $(-\sqrt{3},-1)$

15) $(-6,-4)$ 16) $(8,-2)$ 17) $(-\sqrt{2},1)$ 18) $(\sqrt{3},3)$

19) $(15,3)$ 20) $(-7,-5)$

Write the following equations in rectangular form.

21) $r^2 = 3\cos^2\theta$ 22) $r = \frac{\theta}{2}$

23) $r = \cos\theta + \sin\theta$ 24) $r = a\sin\theta\cos^2\theta$

25) $r = 5$ 26) $r = a\cos\theta + b\sin\theta$

27) $r = 4(1 - \cos\theta)$ 28) $r = \csc\theta$

29) $r = 1 + 2\cos\theta$ 30) $r = \sin\theta$

Write the following equations in polar form.

31) $x^2 + y^2 = 25$ 32) $y = x^2 + 5$

33) $4x - 3y = 6$ 34) $y^2 = x + 5$

35) $x = 6$ 36) $y = 4$

37) $9x^2 + 4y^2 = 36$ 38) $xy = 9$

39) $x = y^2 + y$ 40) $x - 5y = 6$

12.10 Graphing Complex Numbers

A complex number has previously been defined as a number of the form $a + bi$, where a and b are real numbers and $i = \sqrt{-1}$. If bi is considered as a pure imaginary number, the complex numbers can be thought of as a combination of the set of real numbers and the set of pure imaginary numbers where every complex number has a real part and an imaginary part.

If on a coordinate graph we label the horizontal axis the real axis and label the vertical axis imaginary, we can represent a complex number $a + bi$ as a vector with initial point at the origin and terminal point at the point (a,b). (See Figure 12.14). The only difference between this representation and that of the vectors previously discussed is that we must always remember to designate the second coordinate as imaginary when writing the number as a sum.

The length of the vector, called the magnitude or absolute value of the complex number, is given as before by $\sqrt{a^2 + b^2}$.

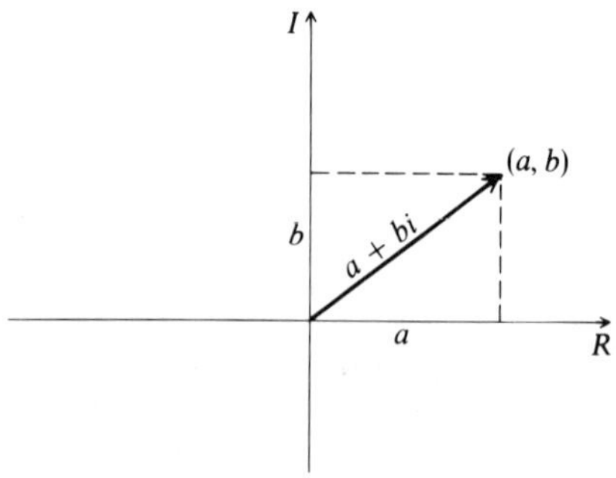

Figure 12.14

Example 1. Represent the complex numbers $5 + 2i$, $-3 - 7i$, $4 - i$ and $-2 + 2i$ graphically and find the magnitude of each.

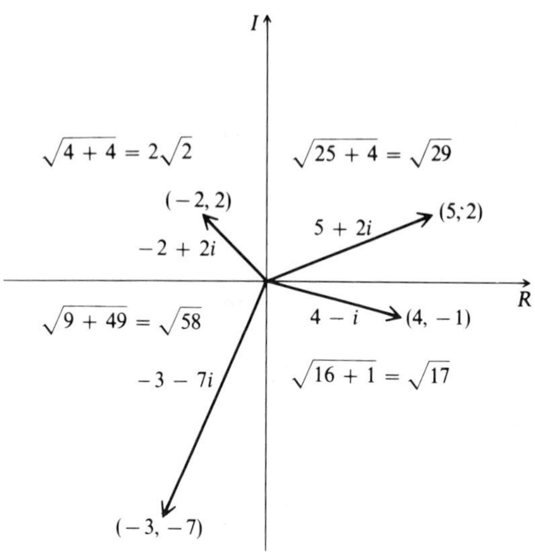

This vector representation of complex numbers leads immediately to a representation addition and subtraction of complex numbers. Graphically, these sums and differences will be the resultants of the representative vectors of the complex numbers using the method discussed in the section on addition of vectors.

12.11 The Trigonometric or Polar Form of Complex Numbers

If we consider the polar form of the point (a,b), we obtain $r = \sqrt{a^2 + b^2}$ and $\tan \theta = b/a$. If, however, this point represents the complex number $a + bi$, we can write the *trigonometric form* of the complex number using relationships previously discussed. The angle θ is called the *argument of the complex number* and r is called the *modulus*. (See Figure 12.15.)

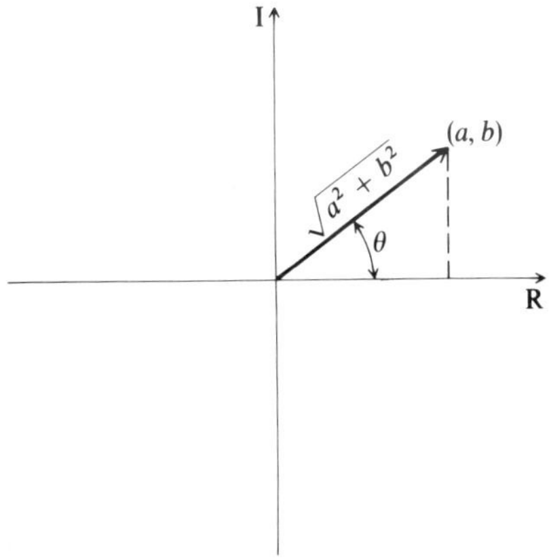

Figure 12.15

The $\sin \theta = b/r$ so that $b = r \sin \theta$ and $\cos \theta = a/r$ so that $a = r \cos \theta$. If these values are substituted for a and b in the complex number $a + bi$, we obtain $r \cos \theta + ir \sin \theta = r(\cos \theta + i \sin \theta)$, called the *trigonometric form of the complex number.*

Example 1. Write $1 + i\sqrt{3}$ trigonometric form.

$$r = \sqrt{1 + 3} = 2, \tan \theta = \frac{\sqrt{3}}{1}, \theta = \frac{\pi}{3},$$

$$1 + i\sqrt{3} = 2 \cos \frac{\pi}{3} + 2i \sin \frac{\pi}{3} = 2\left(\cos \frac{\pi}{3} + i \sin \frac{\pi}{3}\right).$$

The trigonometric form of a complex number involves one value for r and one value for θ. Therefore, a useful abbreviation is often used for the trigonometric form of complex numbers. The number $r(\cos\theta + i\sin\theta)$ can be abbreviated as $r \text{ cis } \theta$.

Example 2. Write $\sqrt{3} + i\sqrt{3}$ in trigonometric form.

$$r = \sqrt{3+3} = \sqrt{6} \qquad \tan\theta = \frac{\sqrt{3}}{\sqrt{3}} = 1 \qquad \theta = \frac{\pi}{4}$$

$$\sqrt{3} + i\sqrt{3} = \sqrt{6}\text{ cis }\frac{\pi}{4}.$$

Example 3. Write $5 \text{ cis } \frac{\pi}{6}$ in algebraic form.

$$5\text{ cis }\frac{\pi}{6} = 5\cos\frac{\pi}{6} + 5i\sin\frac{\pi}{6} = 5\left(\frac{\sqrt{3}}{2}\right) + 5i\left(\frac{1}{2}\right)$$

$$= \frac{5\sqrt{3}}{2} + i\frac{5}{2}.$$

Example 4. Write $3 \text{ cis } 22°$ in algebraic form.

$$3\cos 22° + 3i\sin 22° = 3(.9272) + i(3)(.3746)$$

$$= 2.7816 + 1.1238i$$

Exercise 77

Graph the following complex numbers.

1) $5 + 3i$ 2) $-3 + i$ 3) $12 - 5i$ 4) $-8 - 6i$

5) $-2 + 5i$ 6) $4 - i$ 7) $-7i$ 8) $3i$

9) 7 10) $-7 - 2i$

Put the following complex numbers in trigonometric form. Use tables where necessary and put arguments to the nearest degree.

11) $1 + i\sqrt{3}$ 12) $2 + 2i$ 13) $-4 - 4i$

14) $-3 + 3i$ 15) $-\sqrt{2} - i\sqrt{2}$ 16) 6

17) $-2i$ 18) $1 + i\sqrt{2}$ 19) $5 - 3i$

20) $-2 + 3i$

Put the following complex numbers in algebraic form.

21) $3(\cos 60° + i\sin 60°)$ 22) $-4(\cos 45° - i\sin 45°)$

23) $\sqrt{2}(\cos 225° + i\sin 225°)$ 24) $1(\cos 300° + i\sin 300°)$

25) 6 cis 30°

26) 2 cis 135°

27) −3 cis 315°

28) 5 cis 330°

29) 2 cis 22°

30) 9 cis 456°

12.12 The Product of Complex Numbers in Trigonometric Form

If we find the product of two complex numbers using the method of binomial multiplication,

$$\begin{aligned}(r_1 \text{ cis } \theta) \cdot (r_2 \text{ cis } \phi) &= (r_1 \cos\theta + r_1 i \sin\theta)(r_2 \cos\phi + r_2 i \sin\phi)\\ &= r_1 r_2 \cos\theta\cos\phi + r_1 r_2 i \sin\theta\cos\phi\\ &\qquad + r_1 r_2 i \cos\theta\sin\phi + r_1 r_2 i^2 \sin\theta\sin\phi\\ &= r_1 r_2(\cos\theta\cos\phi - \sin\theta\sin\phi)\\ &\qquad + i r_1 r_2(\cos\theta\sin\phi + \sin\theta\cos\phi)\\ &= r_1 r_2[\cos(\theta+\phi) + i\sin(\theta+\phi)].\end{aligned}$$

Therefore, *to multiply two complex numbers in trigonometric form, multiply their moduli and add the arguments.*

Example 1. Find the product of 5 cis 15° and 3 cis 20°.

$$5 \cdot 3 \text{ cis}(15° + 20°) = 15 \text{ cis } 35°.$$

In the same manner, the quotient of two complex numbers in trigonometric form can be found by *finding the quotient* of their moduli and the difference in their arguments. That is,

$$\frac{r_1(\cos\theta + i\sin\theta)}{r_2(\cos\phi + i\sin\phi)} = \frac{r_1}{r_2}\cdot\frac{\cos\theta + i\sin\theta}{\cos\phi + i\sin\phi}\cdot\frac{\cos(-\phi) + i\sin(-\phi)}{\cos(-\phi) + i\sin(-\phi)}$$

$$= \frac{r_1}{r_2}\,\frac{\cos\theta\cos(-\phi) + i\cos\theta\sin(-\phi) + i\sin\theta\cos(-\phi) - \sin\theta\sin(-\phi)}{\cos\phi\cos(-\phi) + i\sin\phi\cos(-\phi) + i\cos\phi\sin(-\phi) - \sin\phi\sin(-\phi)}$$

$$= \frac{r_1}{r_2}\,\frac{(\cos\theta\cos\phi + \sin\theta\sin\phi) + i(\sin\theta\cos\phi - \cos\theta\sin\phi)}{(\cos\phi\cos\phi + \sin\phi\sin\phi) + i(\sin\phi\cos\phi - \cos\phi\sin\phi)}$$

$$= \frac{r_1}{r_2}\,\frac{\cos(\theta-\phi) + i\sin(\theta-\phi)}{\cos 0 - i\sin 0}$$

$$= \frac{r_1}{r_2}[\cos(\theta-\phi) + i\sin(\theta-\phi)].$$

Example 2. Divide 10 cis 45° by 2 cis 15°.

$$\frac{10 \text{ cis } 45°}{2 \text{ cis } 15°} = \frac{10}{2} \text{ cis}(45° - 15°) = 5 \text{ cis } 30°.$$

The argument of a complex number is not unique since coterminal angles have the same functional values. That is, r cis θ can be written as r cis$(\theta + 2k\pi)$, where k is any integer.

12.13 Powers of Complex Numbers

An extension of the idea of the product of complex numbers can be made to the power of complex numbers. $(r \text{ cis } \theta)^n$ is the product of n factors of r cis θ. Therefore, the power of a complex number in trigonometric form is found by *taking the power of the modulus and the exponent times the argument*. That is, $[r(\cos\theta + i \sin\theta)]^n = r^n(\cos n\theta + i \sin n\theta)$.

Example 1. Find $(3 \text{ cis } 20°)^3$.

$$\begin{aligned} 3(\cos 20° + i \sin 20°) &= 3^3(\cos 3 \cdot 20° + i \sin 3 \cdot 20°) \\ &= 27 \text{ cis } 60°. \end{aligned}$$

Example 2. Find $(\sqrt{2} + i\sqrt{2})^4$.
It is often easier to convert $\sqrt{2} + i\sqrt{2}$ to trigonometric form and then find the power of the number.

$$\begin{aligned} (\sqrt{2} + i\sqrt{2})^4 &= [2(\cos 45° + i \sin 45°)]^4 \\ &= 2^4(\cos 4 \cdot 45° + i \sin 4 \cdot 45°) \\ &= 16(\cos 180° + i \sin 180°) \\ &= 16(-1) = -16. \end{aligned}$$

The statement $[r(\cos\theta + i \sin\theta)]^n = r^n(\cos n \cdot \theta + i \sin n \cdot \theta)$ is true for any *real number n* and is called *de Moivres theorem* after the man credited with its development.

12.14 Roots of Complex Numbers

Consider the equation $x^3 + 1 = 0$. Since this equation is of the third degree, one would expect three roots of the equation. Solving for x^3, $x^3 = -1$ which leads to a solution

$$x = \sqrt[3]{-1} = -1.$$

Other roots of this equation exist, and they can be found by using the trigonometric form of the number. That is, writing -1 in trigonometric form we have

$$1(\cos 180^\circ + i \sin 180^\circ).$$

Applying de Moivres theorem with $n = 1/3$,

$$\begin{aligned}[1(\cos 180^\circ + i \sin 180^\circ)]^{1/3} &= (1)^{1/3}[\cos 1/3(180) + i \sin 1/3(180)] \\ &= 1(\cos 60 + i \sin 60) \\ &= 1(1/2 + i\sqrt{3}/2) \\ &= 1/2 + i\sqrt{3}/2.\end{aligned}$$

One root of the desired equation is therefore $1/2 + i\sqrt{3}/2$. Also, $1(\cos 180^\circ + i \sin 180^\circ) = 1(\cos 540^\circ + i \sin 540^\circ)$. Applying de Moivres theorem to this second form of the given number we obtain

$$\begin{aligned}1\left(\cos\frac{540}{3} + i \sin\frac{540}{3}\right) &= \cos 180^\circ + i \sin 180^\circ \\ &= -1 \text{ as a second root.}\end{aligned}$$

Again, $1(\cos 180^\circ + i \sin 180^\circ) = 1(\cos 900^\circ + \text{i} \sin 900^\circ)$. By de Moivres theorem with $n = 1/3$, this becomes $1(\cos 300 + i \sin 300) = 1/2 - i\sqrt{3}/2$ as the third root of the equation. Hence the three cube roots of one are those given above. The reader should cube each of these numbers to convince himself that each of them cubed is in fact one.

If the three roots of the equation are shown graphically, they divide the graph into three parts with an equal angle between each part.

Exercise 78

Find the product of the following complex numbers by first putting the numbers in trigonometric form.

1) $(3 + 3i)(1 - i)$

2) $(-2 - i\sqrt{2})(-1 - i\sqrt{3})$

3) $(5 - i)(6 + 2i)$

4) $(-\sqrt{3} + i)(-1 + i)$

5) $(6 - i)(5 + 2i)$

Find the quotient of the following. Write answers in both trigonometric and algebraic form.

6) $\dfrac{3 \text{ cis } 400^\circ}{1 \text{ cis } 40^\circ}$

7) $\dfrac{15 \text{ cis } 25^\circ}{3 \text{ cis } 85^\circ}$

8) $\dfrac{7 \text{ cis } 212^\circ}{3 \text{ cis } 140^\circ}$

9) $\dfrac{3 - i\sqrt{3}}{\sqrt{2} - \sqrt{2}i}$

10) $\dfrac{i}{1 + i}$

11) $\dfrac{7}{-i}$

12) $\dfrac{6 - i}{6 + i}$

Use de Moivres theorem, where applicable, to find the following:

13) $(5 + 5i)^5$ 14) $(4 \text{ cis } 20°)^3$ 15) $(1 \text{ cis } 15°)^7$

16) $(-3 - 2i)^3$ 17) $(3 \text{ cis } 300°)^4$ 18) $(8 \text{ cis } 120°)^{13}$

Find all roots of the following equations.

19) $x^3 + 8 = 0$ 20) $x^2 = -1$ 21) $x^5 + 1 = 0$

22) $x^7 + 2 = 0$ 23) $x^4 = 16$ 24) $3x^3 = -81$

25) $\frac{x^5}{4} + 8 = 0$

Review Exercise Chapter 12

Find the magnitude and direction of the vectors having the following endpoints.

1) (2,3); (6,12) 2) (−3,−2); (5,7)

Find a unit vector having the direction of each of the following.

3) $5i + 6j$ 4) (3,5); (7,2)

Find the sum and difference of the following.

5) $6i + 2j;\ -2i + j$ 6) $-3i - j;\ 6i - j$

Represent each of the following points in two different ways.

7) (7,15°) 8) $(-5,\pi/4)$ 9) $\left(\frac{\sqrt{3}}{2}, -\frac{\pi}{6}\right)$

Graph the following in polar form.

10) $r^2 = \sin\theta$ 11) $r = \cos\frac{\theta}{2}$

12) $r = 2\theta$

Convert the following to polar form.

13) (1,1) 14) $\left(\frac{\sqrt{3}}{2}, \frac{1}{2}\right)$

15) $\left(\frac{\sqrt{2}}{2}, \frac{\sqrt{2}}{2}\right)$ 16) $x^2 + y^2 = 5$

17) $x = y - 5$ 18) $y = 3$

Convert the following to rectangular form.

19) $(-5,\pi/3)$ 20) $(-\sqrt{3}/2,-\pi/4)$

21) $(4/3,7\pi/12)$ 22) $r = 3\cos\theta$

23) $2r = 3\theta$ 24) $r = 2 + 3\sin\theta$

Put the following complex numbers into trigonometric form.

25) $2 - i\sqrt{3}$ 26) $-4i$

27) $2 - i$

Put the following complex numbers into algebraic form.

28) $3 \text{ cis } 60°$ 29) $-5 \text{ cis } \pi/6$

30) $2/3 \text{ cis } 45°$

Answers to Odd-Numbered Exercises

Chapter 1

Exercise 1

1) 120 3) 47 5) $42a + 3b + 8c$ 7) 15294 9) 1844783 11) 74
13) 84940 15) 1227563 17) 67 19) 1600 21) 81534 23) 278178772
25) 5685102 27) 72 29) 56980 31) 194.333 approx. 33) 71
35) 31 37) 1010.092 approx. 39) 29 41) 250 ft. 43) $4.80
45) 46956 47) $50.59 49) 9213 bushels

Exercise 2

1) 130 3) -32 5) 57 7) $2a$ 9) 36 11) 71 13) 48 15) 0
17) -72 19) 406 21) 27 23) -179 25) -12 27) -126
29) -175 31) 0 33) 34 35) $-2b$ 37) $-2y$ 39) $-x + 4y$
41) -98 43) $\pm.003$ 45) $86 - (-11) = 97°$ 47) $41,000 profit

Exercise 3

1) -12 3) 40 5) -28 7) $-83{,}974$ 9) 120 11) 46,852
13) $-19{,}358{,}568$ 15) 2 17) $-.16$ 19) 4,557 21) 7,140 23) 11.91
25) 10.75 27) -438 29) -2.9166

Exercise 4

1) Composite 3) Prime 5) Prime 7) {3,7} 9) {2,3}
11) {2,3,7,11} 13) $3 \cdot 5 \cdot 13$ 15) $3 \cdot 5 \cdot 11$ 17) GCF 27, LCM 243
19) GCF 50, LCM 300 21) GCF 12, LCM 144 23) 52/7 25) 373/8
27) 12/5 29) 7/12 31) ⅓ 33) 23/30 35) 1/14 37) 1/6
39) 91/4 41) 9/10 43) 7/2 45) 3/77 47) 51/136 49) 176/135
51) 2 7/32 53) $a = 23/6, b = 25/6$ 55) $.57 57) 528 quarts
59) 13/40, 3 1/13 hr.

Exercise 5

1) 505.055 3) 450.7 5) .00475 7) 9⅜ 9) 450 7/10 11) 11.803
13) 14.087 15) 12499.34 17) 10369.755 19) 46.8 21) 689.976
23) 46.08632 25) 7949.438 27) 2 29) 12.1 31) 14 33) .0
35) 3 37) 3.276 39) .625 41) 4.865 43) 44.2 45) 41.25
47) 12½% 49) 66¢ 51) $4.07 53) $32.51

Exercise 6

1) 292723.2 3) 1077120 5) 445.5 7) 45.6 9) .5 11) 28.5
13) 4.5574 15) 90.85 17) 241.99 19) 35.274 21) 96.78 23) 19.59
25) 885.25 27) 729 29) 191.66 31) 2.244 33) 4.72 35) 98.63
37) 5.354 39) .258

Review Exercise Chapter 1

1) 77 3) 214.912 5) 59 7) 426,553 9) 2397 11) 280,192
13) 65 R 10 15) 488 R 136 17) 157/240 19) 8226.3894 21) 17.4068
23) 192 25) 1.7331 27) 34 29) −3.7965 31) 174,240 sq. ft.
33) 11.19 miles 35) 5.91 in.

Chapter 2

Exercise 7

1) $a^{m \cdot n}$ 3) a^m/b^m 5) a^{m-n} 7) 32 9) 625 11) 1 13) 216
15) 5.29 17) 25 19) 54 21) 144 23) 1 25) .001 27) .000001
29) 16/81 31) ⅓ 33) 256 35) 256/81 37) 2^4 39) $x \cdot y \cdot y \cdot y \cdot y$
41) $xy \cdot xy \cdot xy \cdot xy$ 43) $2x \cdot 2x$ 45) x^{14} 47) x^3y^2 49) x^5y^5
51) x^{10} 53) x^{12} 55) 1

Exercise 8

1) $1/x^2$ 3) 1/7 5) 1/5 7) x/y^2 9) $1/x^2y^2$ 11) x^2/y^2 13) 8/9
15) 9/4 17) $1/x^2$ 19) $-1/27x^6$ 21) y^4/x^3 23) $-1/8y^3$ 25) $36/x^2$
27) $5/y$ 29) $12/x^9$ 31) 2.641×10^3 33) 2.621×10^6
35) 3.927176×10^9 37) 2.97×10^{32} 39) 1.76×10^{11} 41) 2.4×10^{-4}
43) 2.97×10^{-7} 45) 4.56×10^{20} 47) 7.87815×10^{-9} approximately
49) $(1.25 \times 10^{-1})(5 \times 10^3) = 6.25 \times 10^2$ 51) 2×10^{-6}, 8×10^{-7}

Exercise 9

1) 23 3) 21 5) 4.2 7) 212 9) 29 11) 92 13) 3.2 15) 5.2
17) 16.47 19) 14.832 21) 13.416 23) 22.136 25) 4.123
27) 16.125 29) 20.248 31) 6.928 33) 12 35) Yes. The product of the square roots of two or more positive numbers is the square root of the product of the numbers. 37) 36 39) 144 41) 15,625 43) 11,664 45) 7.071

Exercise 10

1) 14.697 3) 15.969 5) 19.596 7) 17.205 9) 13.416
11) $\sqrt{45} = 6.708$ 13) $\sqrt{98} = 9.899$ 15) 49 17) 8 19) 3
21) 1/5 23) 4 25) −2.157 27) −.336

Exercise 11

1) $-9+0i$ 3) $1-i$ 5) $2+3i\sqrt{2}$ 7) $(1+i\sqrt{2})/2$ 9) $(2+8i)/3$ 11) $9+5i$ 13) $13+9i$ 15) $-3+4i$ 17) $8+5i$ 19) $10+3i\sqrt{2}$ 21) $5-7i\sqrt{3}/3$

Review Exercise Chapter 2

1) base, exponent 3) $1/A^3$ 5) multiplied 7) product (if the numbers are positive) 9) $5^{-4} = 1/5^4$ 11) x 13) $1/x^6y$ 15) 32 17) 2 19) 9 21) 20 23) 27.928 25) 69.282 27) 9 29) 64 31) $3+3i$ 33) $\sqrt{2}+2i$ 35) $2-3i$ 37) $1+i\sqrt{2}$

Chapter 3

Exercise 12

1) 514, 370 3) 131, 473 5) 1913, 366 7) 436, 191 9) 258, 172 11) 1229, 151 13) 3142, 312 15) 187, 350 17) 441, 922 19) 851, 618 21) The accuracy decreases 23) 1, or varies according to the rule 25) 2, or varies according to the rule 27) The sections are identical. 29) left and right index

Exercise 13

1) 32.5 3) 625,000 5) 105 7) 34.0 9) 23.0 11) 104.6 13) 227.2 15) .1616 17) 201 19) 10.97 21) 2140 23) 42.0 25) 47.6 27) 418 29) 17.85 31) 44.9 33) .00115 35) 15.32 37) 1.508 39) 35.8

Exercise 14

1) 15 3) 273 5) 184 7) 1050 9) 5140 11) 484 13) 1.62 15) 22.1 17) 35.2 19) 34.0 21) 1087 23) 106.3 25) 80.9 27) 6.86 29) 8200 31) 4.63 33) 133.9 35) 457 37) 3210 39) 277 41) 59.4 in. 43) $1062.00 45) 3840 47) $14.37 49) $353.00 51) 1.554

Exercise 15

1) 62.1 3) 3.66 5) 33.0 7) 7.04 9) .568 11) 257 13) 3.32 15) 4.04 17) 2,540 19) .1428 21) .833 23) .375 25) .594 27) .319 29) .562 31) .1415 33) .782 35) .698 37) .818 39) .941 41) 32.5 43) 1.507 45) 1.786 47) 1.83 49) .0034 51) 115.1 53) .250 55) 128.3 57) 16.57 59) 19,228.9

Exercise 16

1) 8.66 3) 2.28 5) .664 7) 6.64 9) 4.87 11) .935 13) 5.11 15) 1.648 17) .933 19) .354 21) 7.91 23) .1847 25) 2256 27) 361 29) .00471 31) 106.1 33) 502 35) 198,000 37) .00391 39) 181,500 41) 18.7

Exercise 17

1) 1.75 3) 7.68 5) 8.96 7) 30 9) 2.98 11) 3.54 13) .764 15) .44 17) 3.6 19) 3.51 21) 42.2 23) .521 25) .914 27) 4.44 29) 76.7 31) 83,400 33) 462 35) 46,500 37) .0000372 39) .00123 41) 17.62 43) -434 45) 3.32 47) 2.94 49) 3.56

Review Exercise Chapter 3

Most answers to slide rule accuracy except where inspection can determine the final digit
1) .125 3) .313 5) .297 7) 4030 9) .3760 11) 1.195 13) 1.734 15) .0234 17) 2.20 19) 7.86 21) 1.693 23) 6912 25) 29,570 lbs., 3563 gals.

Chapter 4

Exercise 18

1) $3 + a + d + f$ 3) $t + 9$ 5) $(x + y) + x \cdot y$ 7) $6(x - y)$ 9) $t - 5$ 11) $(2x)(7b)$ 13) $t - c$ 15) $2n + 7$, if n is the number 17) $\frac{2}{3}n + 2n$ 19) $5(n + 9)$ 21) monomial, polynomial 23) polynomial 25) monomial, polynomial 27) binomial, polynomial 29) polynomial 31) $2a$, $3a$ 33) $2ab$, $7ab$ 35) none 37) none 39) $(a + b)$, $3a + 3b$ 41) $24(x + y)$ 43) $n/2 + 5$ dollars 45a) $2x + 2(x + 5)$ 45)b $A = x(x + 5)$ 47) $150x$ 49) $40x + 8(\frac{3}{2}x)$ 51) $1.10x$ 53) $(s + 10)^2$

Exercise 19

1) 3 3) -11 5) -23 7) 6 9) 13 11) -5 13) $\frac{1}{2}$ 15) 129/4 17) 13 19) 24 21) 108 23) $-2b$ 25) $7ac$ 27) $10xy^2$ 29) $5x + 2y$ 31) $4a + 4b + 4c^2$ 33) $6a + 9b - c$ 35) $-2b$ 37) a 39) $3a^2 - 2b^2 + 2x$ 41) $-4a - b$ 43) $a^3 + 2a^2 - 2c$ 45) $c^2 - 4c^2 = -3c^2$ 47) $2x^2 - 11x + 7$ 49) $2x^2 + 9y$ 51) 1300 53) 1400 ft. 55a) 3 cu. in. 55b) 6.318 cu. in. 55c) 5 25/48 cu in.

Exercise 20

1) $x^2 + xy$ 3) $x^4 + x^2y^2$ 5) $5a^2 - 5ab^2$ 7) $2x^3 + 2x^2y$ 9) $3a^2b^2c - 6ab^4c - 3a^2b^4c^2$ 11) $-2a^3b^2c + 2ab^3c - 4ab^2c^3$ 13) $2x^2 - 3xy + y^2$ 15) $x^2 - 2xy - y^2$ 17) $ax + ay - bx - by$ 19) $a^2 - 2ab + b^2$ 21) $a^3 + b^3$ 23) $a^2 - b^2$ 25) $x^4 + 2x^3y + 2x^2 + x^2y^2 + 2xy + 1$ 27) $3a^3 - 3ab^2 - 2a^2b + 2b^3$ 29) $x^2 + 2ax + a^2$

31) $x^3 + ax^2 - a^2x - a^3$ 33) $x^3 + 3x^2 - 10x - 24$ 35) $a^2 - 2ab + 2ac + b^2 - 2bc + c^2$ 37) $25 - 9t^2$ 39) $(30 + 2)(30 - 2) = 896$ 41) $(100 + 2)\times(100 - 2) = 9996$ 43) $(20 + 5)^2 = 625$ 45) $(20 + 5)(20 - 5) = 375$

Exercise 21

1) $x + 2y + 3z$ 3) $4z^3 + 2z + 4$ 5) $2x^2 + 3 + y^2$ 7) $4a^2 + 3a - 4$ 9) $-3 - 6xy^3 + 9x^4y^7$ 11) $-6 + 9k^2 + 18k^4$ 13) $x + 1$ 15) $x + 2$ 17) $x - 7$ 19) $4x - 5$ 21) $x + 4 + 1/(x - 5)$ 23) $x + 3 + 4/(x - 1)$ 25) $y^2 + 2y - 8$ 27) $x^2 - 2x + 1$ 29) $r^2 - 2r + 1$ 31) $x^2 + 3x + 9$ 33) $x^2 - 2x + 1/(x - 1)$ 35) $2x^2 + 3x - 4$

Exercise 22

1) $x + 4 + 6/(x - 3)$ 3) $x - 2$ 5) $x + 4$ 7) $x^2 + 2x + 1$ 9) $2x^3 + 4x^2 + 8x + 13 + 32/(x - 2)$ 11) $x^3 - 4x^2 + 5x - 2$ 13) $x^5 + x^4 + x^3 + 4x^2 + 4x + 5 + 5/(x - 1)$ 15) $2x^3 + 10x^2 + 50x + 250 + 1257/(x - 5)$

Exercise 23

1) $x(a + b + c)$ 3) $x^2(3x^2 - 6 - 5x^7)$ 5) $ab(1 + ab + a^2b^2)$ 7) $2^2a(1 + 2a + 2^2a^2)$ 9) $4(x^2y^2z^2 + 4x - 8)$ 11) $7x^2(1 + 2y^2 + 14z)$ 13) $abc(a + b + c)$ 15) $b(9ac - 5a + 6c)$ 17) $(a - b)(5 + 7a)$ 19) $15m^2n^2y(3n - my)$ 21) $(a + 1)^2$ 23) $(x - 2) \cdot (x + 1)$ 25) $(x + 9)(x + 2)$ 27) $(x - 2)(x + 2)(x^2 + 4)$ 29) $(2x - 3a)(2x + 3a)$ 31) $a(3 - 5a)(3 + 5a)$ 33) $(x + 8)(x + 8)$ or $(x + 8)^2$ 35) $(2x + 3)(x - 4)$ 37) $(5m + n)^2$ 39) $z(3z - 1)(z - 1)$ 41) $-2(m + n)$ 43) $25(3x^2 - 5y^2)$

Exercise 24

1) $(a + b)(a - b)$ 3) $4(x + y)(x - y)$ 5) $4(2x^2 - m^2)(2x^2 + m^2)$ 7) $(m + 3)(m^2 - 3m + 9)$ 9) $(xy - r)(x^2y^2 + xyr + r^2)$ 11) $(a + b)(b + 1)$ 13) $7(x - 1)(x + 1)$ 15) $(x + y - z)(x + y + z)$ 17) $(5x - 4)(3x - 2y)$ 19) $(x + 1)(x - 1)(x^2 + x + 1)(x^2 - x + 1)$ 21) $3(2x - 3y)(4x^2 - 6xy + 9y^2)$ 23) Prime

Exercise 25

1) -1 3) i 5) $5 + 5i$ 7) $10i - 6$ 9) $1 + 4\sqrt{3}i$ 11) $-9 - 40i$ 13) 29 15) $25i$ 17) $-2 + 4i\sqrt{2}$ 19) $4/5 + (7/25)i$ 21) $-3 - i$ 23) $5 + 7i$

Review Exercise Chapter 4

1) $48x^4$ 3) $12mn$ 5) .0001 7) $a^2 + 2ab + b^2$ 9) $x^3 + 2x^2 + x + x^2y + 2xy + y$ 11) $x^2 + 3x + 5$ 13) $11x + 5y + 4x^4y^2$ 15) $13x^2(2 + 3xy + 4x^2)$ 17) $(x - 7)(x + 1)$ 19) $x(x - 4)(x + 4)$

Chapter 5

Exercise 26

1) 6 3) −1 5) 50 7) 152 9) 9 11) 10 13) 15 15) 17
17) .5 or ½ 19) −2 21) True for all H 23) −13 25) −4
27) −5

Exercise 27

1) 3 3) 7 5) 24 7) 13/5 9) −15 11) 9 13) 5
15) −6 17) .2 19) .54 21) 1/11 23) 104/3 25) $16V/9$

Exercise 28

1) 2 3) 1 5) −6 7) −1 9) 10/3 11) 9 13) 29
15) −4/3 17) 4 19) −⅝ 21) 20 23) 6 25) 7 27) 0
29) −7 31) −228 33) 7/11 35) 3 37) 1/10 39) 3
41) −13/11 43) −29 45) All real numbers

Exercise 29

1) 4/5 3) 5/3 5) 9/2 7) 99/82 9) 1406/105 11) 54/5
13) 86/9 15) 12000/31 17) 4 19) $x/3 = y/5$, $y/x = 5/3$, $5/y = 3/x$ and others 21) $5/x = 3/y$, $3/5 = y/x$, $x/y = 5/3$ and others 23) 7/3
25) 1/1 27) ±6 29) ±12 31) $\pm\sqrt{MR}$ 33) $\pm\sqrt{42}$
35) 1/96, 35 ft. 37) a) 2.571″ b) 4″ c) 12.19 cm 39) 330 RPM
41) 1800 43) 15

Exercise 30

1) $B = 180 - A - C$ 3) $r = C/2\pi$ 5) $t = D/r$ 7) $2A - y = x$
9) $r = cst/(st - ct - cs)$ 11) $Mt - Lt = g$, $(Mt - g)/t = L$ 13) $I/Pr = t$
15) $F/a = m$ 17) $E/e^2 = m$ 19) $2s/t^2 = a$ 21) $Fd^2/km = M$
23) $s(l - r) + rl = a$ 25) $wV/Fg = R$ 27) $DP(od) - 2 = N$

Exercise 31

1) {4,13,25} 3) {0,1,8} 5) All real numbers 7) 16 9) 4 11) 8
13) 27 15) 9 17) $(2t - 1)^2$ or $4t^2 - 4t + 1$ 19) $2t - 1$ 21) −5
23) {−2,−1,5} 25) {−12,9,49}

Exercise 32

1) I 3) IV 5) x axis 7) II 9) IV 11) y axis 13) Same straight line 15) Not on the same straight line

17)

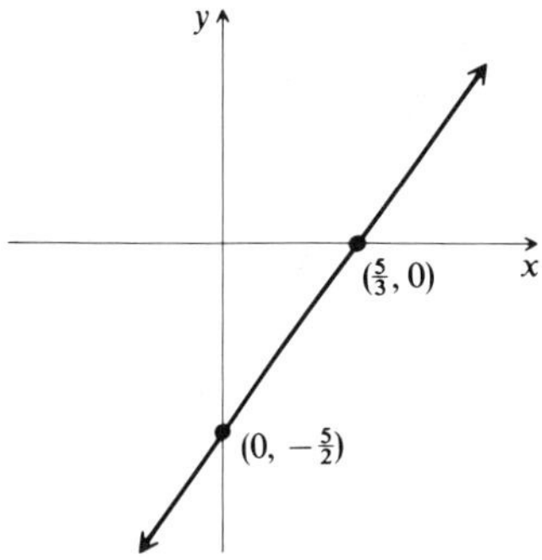

19)

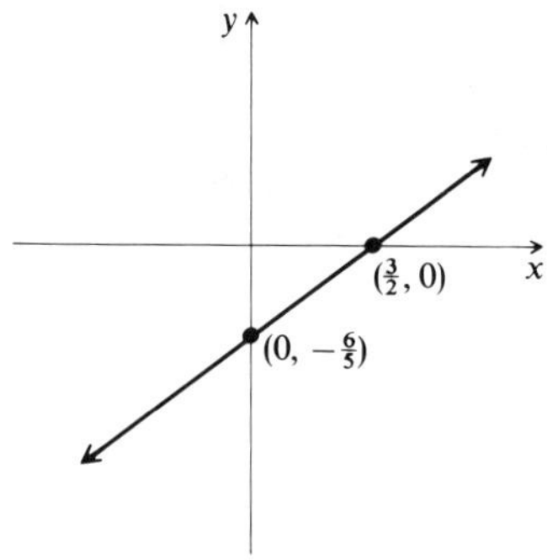

21)

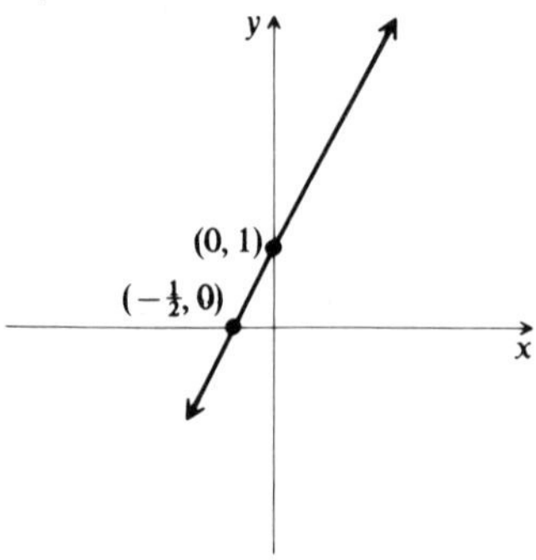

23)

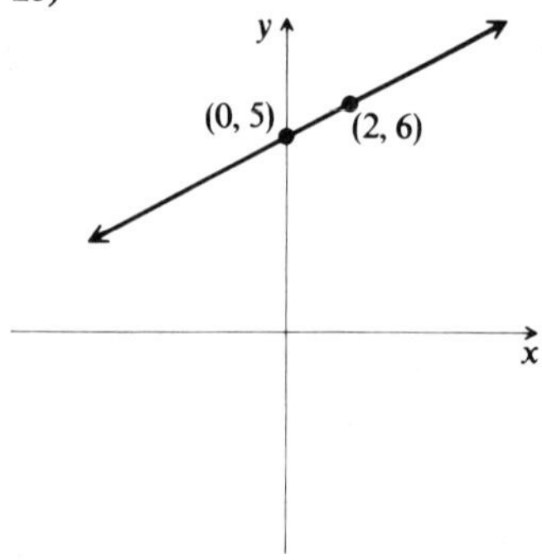

25)

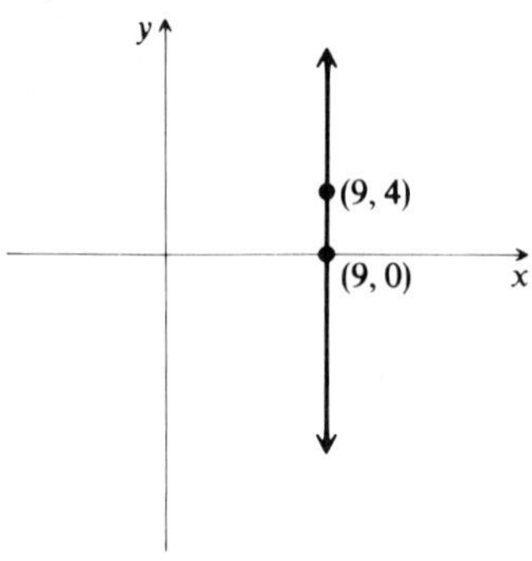

Exercise 33

1) $-2/5$ 3) $3/5$ 5) $-1/5$ 7) $-14/5$ 9) ⅔ 11) $y = 6x + 3$ 13) $y = -4x + 2$

15) Slope 3 int 9

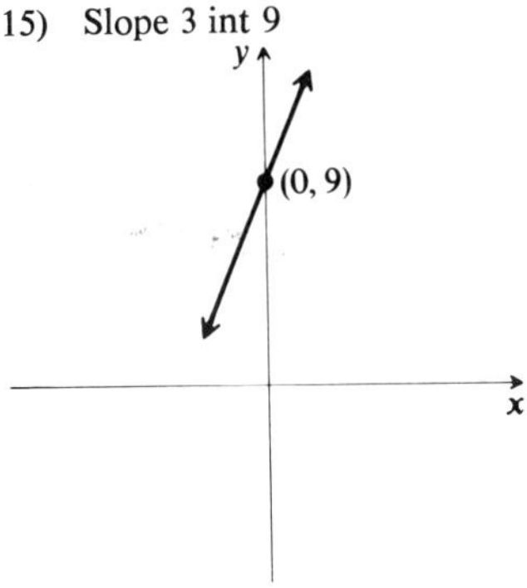

17) Slope 2 int -4

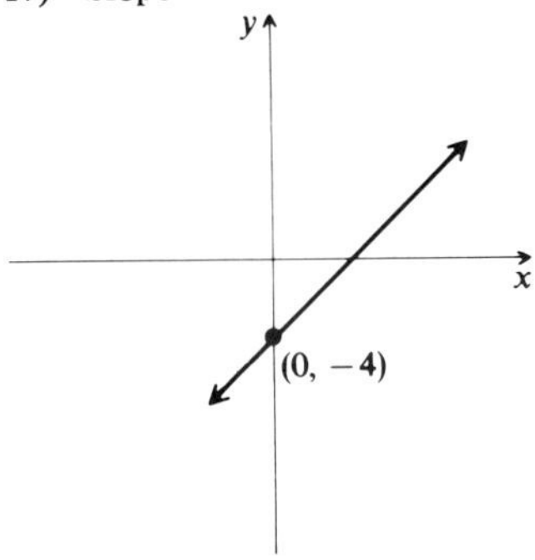

19) Slope 3 int 3/2

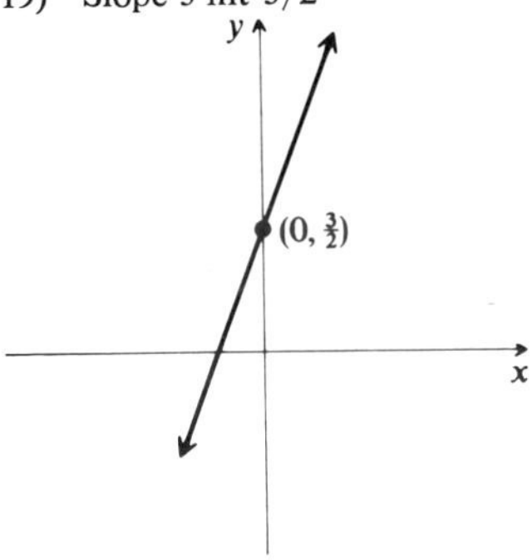

21) Slope 2/5 int $-11/5$

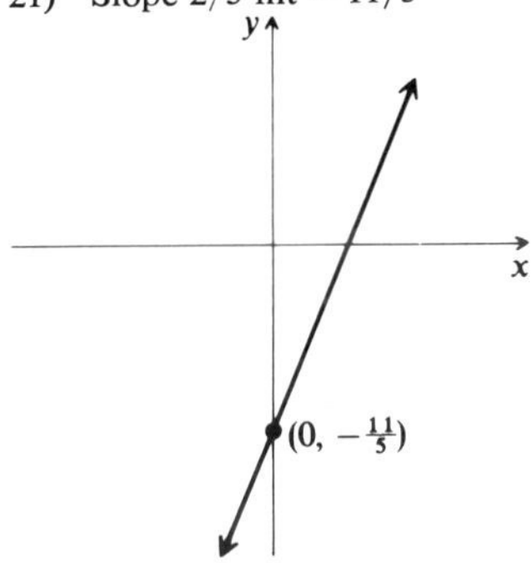

23) Slope $-3/2$ int 3

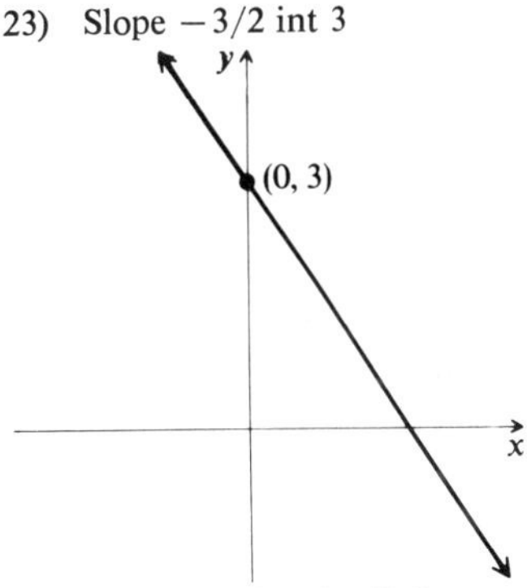

25) x int (8,0) y int (0,3)

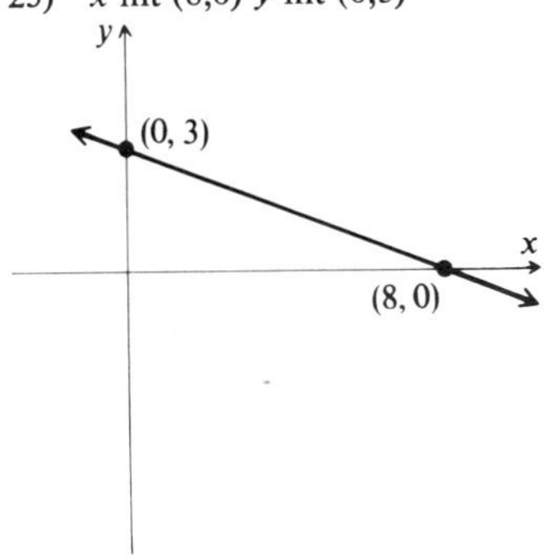

27) x int $(-2,0)$ y int (0,5)

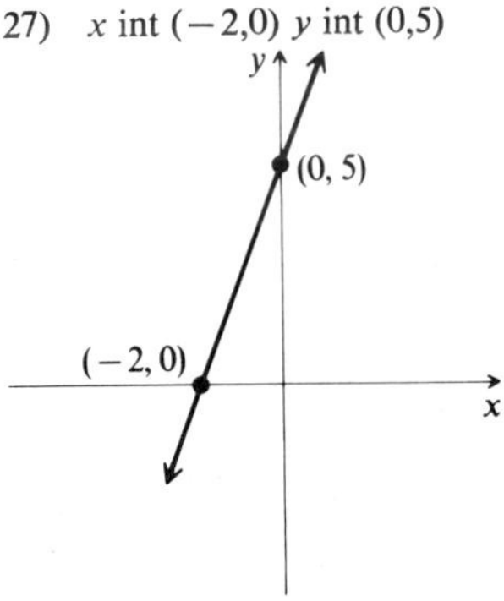

29) x int $(11/5,0)$ y int $(0,-11/9)$

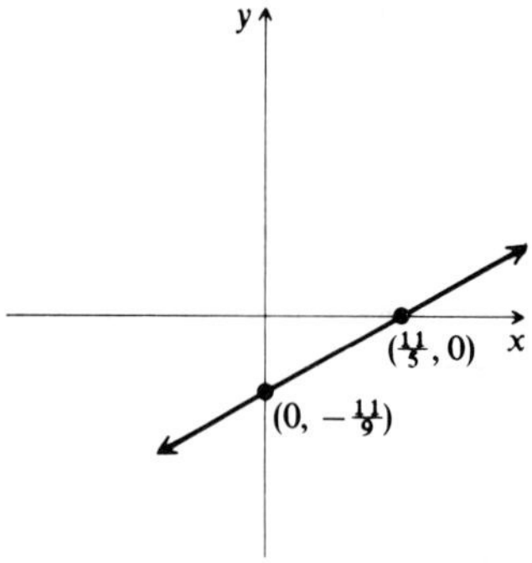

31)

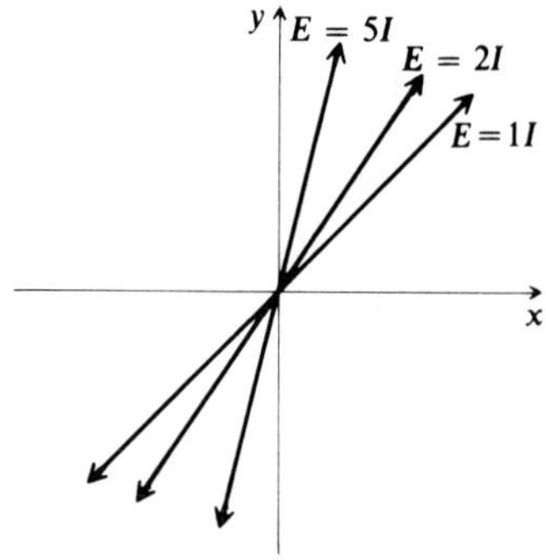

As R increases, the slope increases. Hence R acts as the slope.

Exercise 34

1) $3x + 8y - 41 = 0$ 3) $6x - 5y = 0$ 5) $y - 7 = 0$ 7) $x + y - 7 = 0$ 9) $y = 4x - 22$ 11) $y = \frac{2}{3}x - 6$ 13) $x = 5$ slope intercept form not applicable. 15) Graph passes through the points (0,0)(1,30.48)(2,60.96)(3,91.44)(4,121.92)(5,152.4)

Review Exercise Chapter 5

1) 24 3) 3/10 5) 4/7 7) 5/2 9) $520 - A - B$ 11) F/a 13) {11,12,13} 15) 48 17) 12 19) 3.1153 21) $-1,6,6$ 23) $\frac{1}{2},4,-2$ 25) $y = -\frac{1}{3}x + 1$

Chapter 6

Exercise 35

1) $y = 3^x$

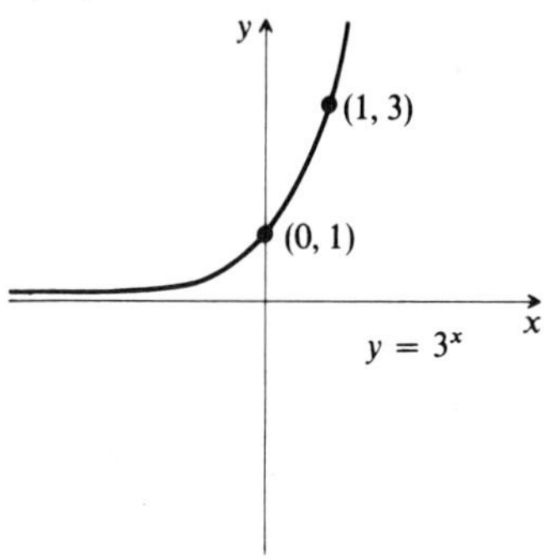

3) $y = (1/5)x$

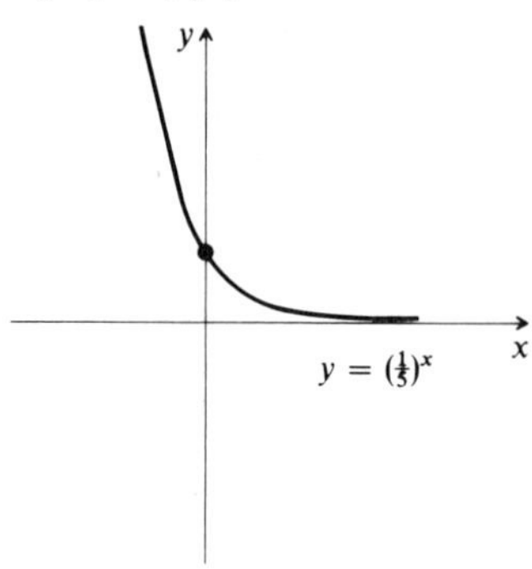

5) $y = \log_{10} x$

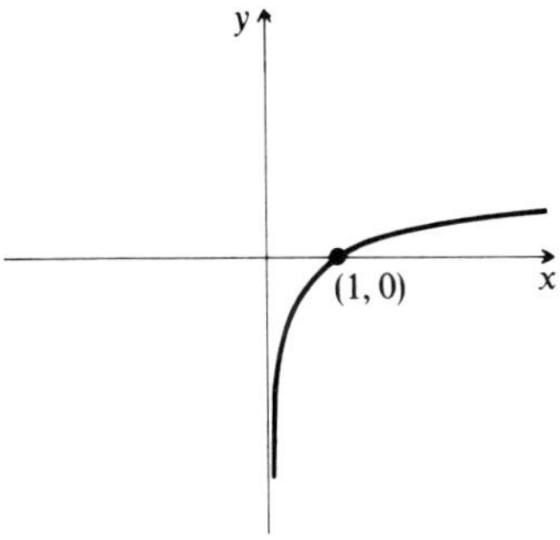

7) $\log_{25} 5 = \frac{1}{2}$ 9) $\log_3 9 = 2$ 11) $\log_2 64 = 6$ 13) $10^2 = 100$
15) $10^{-5} = .00001$ 17) $2^4 = 16$ 19) $5^2 = 25$ 21) 25 23) .0001
25) 2 27) $\frac{1}{3}$ 29) 9 31) $\log 5 + \frac{1}{2} \log 75 - \log 16$
33) $\log 57.5 + \log 8.35 - \frac{1}{3} \log 425$ 35) $\log(\sqrt{225}/\sqrt{27})$

Exercise 36

1) .6590 3) 2.6590 5) 6.8407 − 10 7) 5.2695 9) .4350 11) 65
13) 28,100 15) .214 17) .0449 19) .0007643 21) 39.83
23) 44,560 25) 19.89 27) 16,720 29) 55.43 31) .0156 33) 20.39

Review Exercise Chapter 6

1)

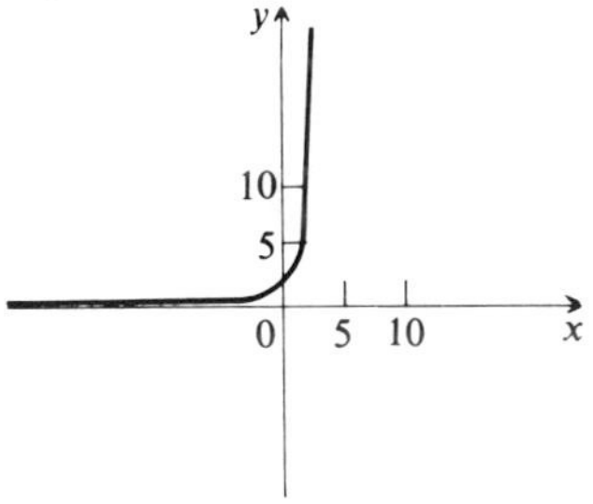

3)

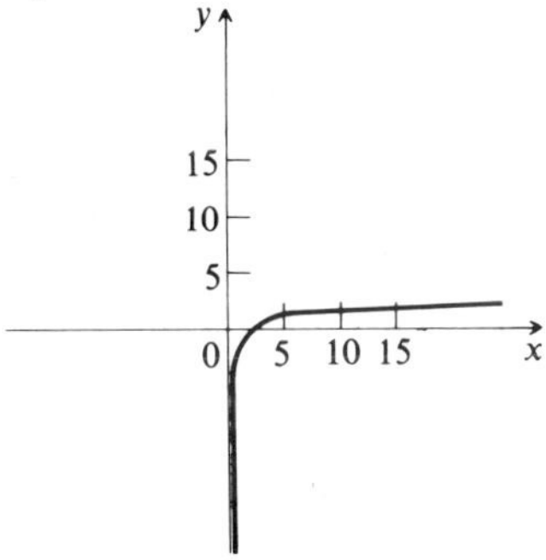

5) $\log_3 9 = 2$ 7) $\log_{10} 1/1000 = -3$ 9) $10^3 = 1000$ 11) 3 13) 64
15) 7.8768 − 10 17) 3.9352 19) 3.115 21) .007473 23) .3042
25) 20.86

Chapter 7

Exercise 37

1) 70° 3) No 5) 71° 7) right 9) isosceles 11) equilateral 13) scalene 15) right 17) right 19a) right 19b) obtuse 19c) acute 19d) right 19e) acute 19f) right 21) 13 23) $\sqrt{17.68} \approx 4.2$ 25) 389.5 27) 16.72 29) 7.42 31) 2.4

Exercise 38

1) congruent 3) neither 5) neither 7) 7½ 9) 7.25 11) 100 13) 14

Exercise 39

1) 100 3) 8 5) yes 7) the distance between the two parallel sides 9) bases 11) they are equal 13) not necessarily 15) 5 in. 17) 4.9 19) 10.82 in. 21) 5

Exercise 40

1) 270 sq. ft. 3) 39 sq. ft. 5) 1219.2 sq. cm. 7) 30 sq. ft. 9) 28 sq. ft. 11) the perpendicular distance from the base to the opposite angle (vertex) 13) It divides a parallelogram into two congruent triangles. 15) 28.44 17) 34.5 sq. units 19) 75 sq. in.

Exercise 41

1) 7 in., 153.86 sq. in., 43.96 in. 3) 12 rd., 113.04 sq. rd., 37.68 rd. 5) 7.5 cm., 176.625 sq cm., 47.10 in. 7) 94.2 cu. ft. 9) 3 ft. 11) 659.4 gal. 13) 36.53 cu. in. 15) $18\sqrt{2}$ 17) 33.33 cu. yd. 19) 6.88 cu. in. 21) 5265.08 cu. yd.

Exercise 42

1)

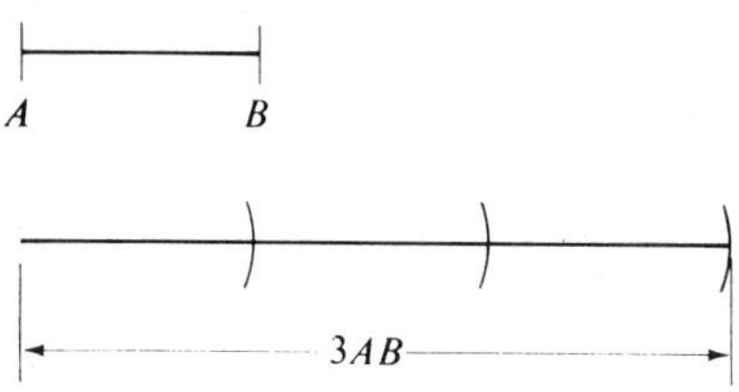

3)

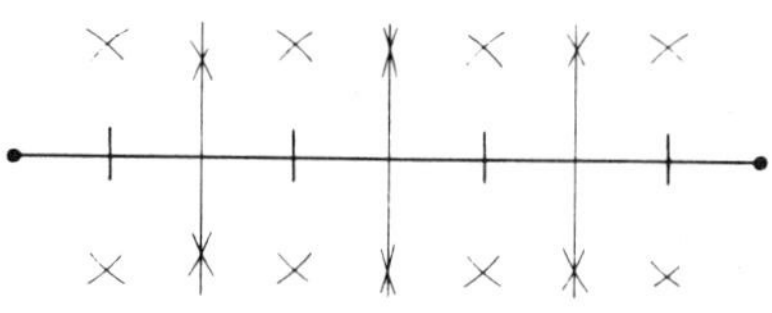

5)

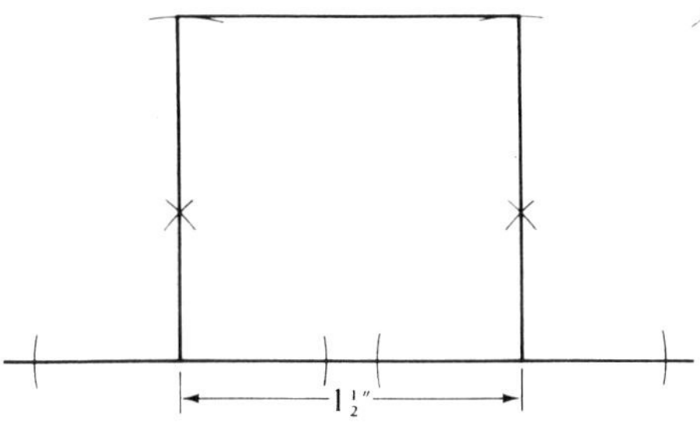

7)

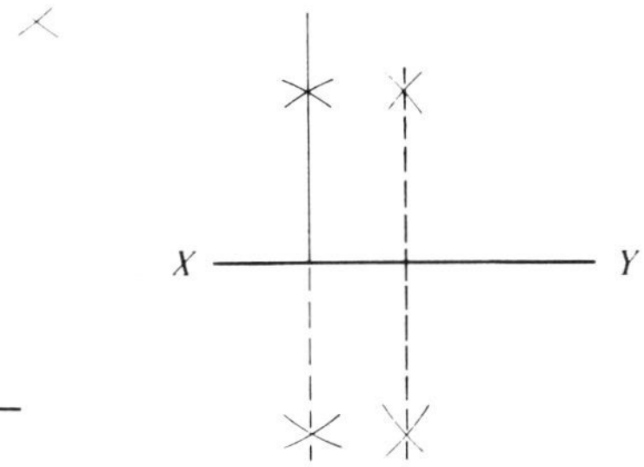

9)

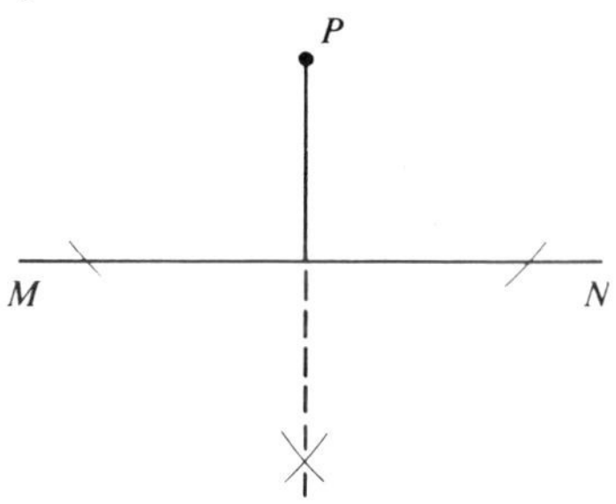

11)

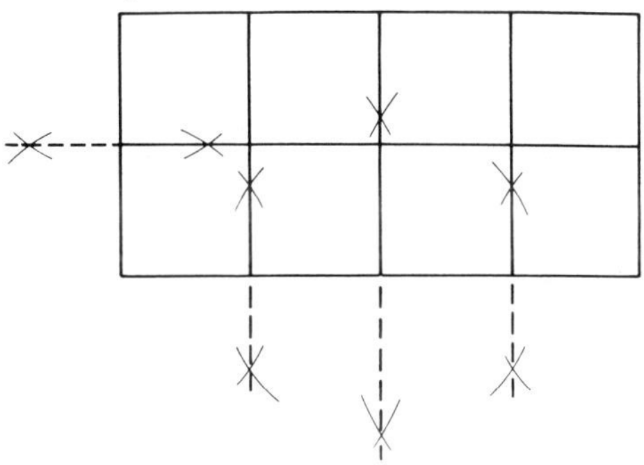

13) 25 mi., approximately 4.4 mi.

15) A square. $AP = BQ = \sqrt{2}$.

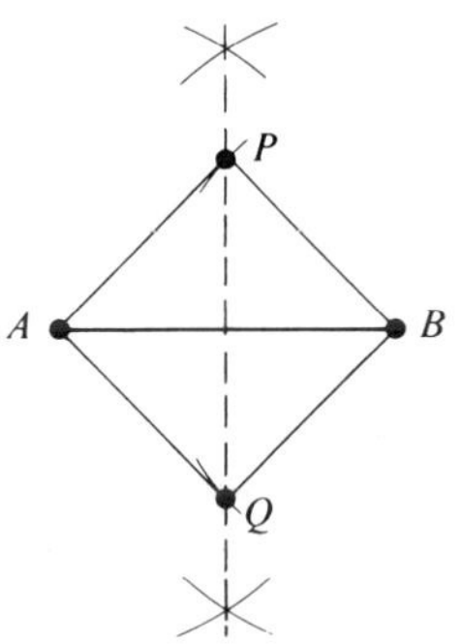

Exercise 43

Where constructions are carried out in the text, the answer will so indicate.

1)

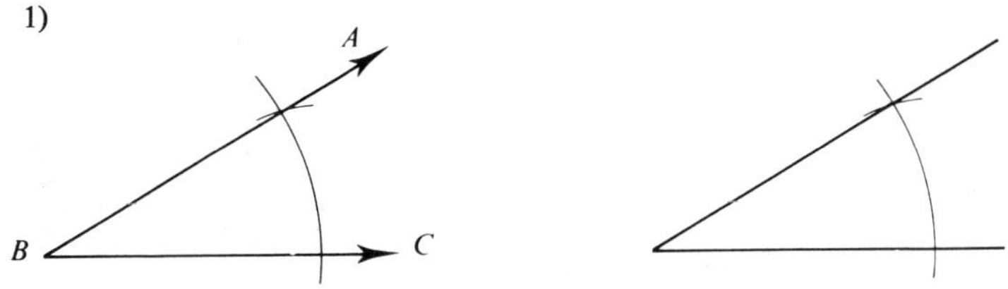

3)

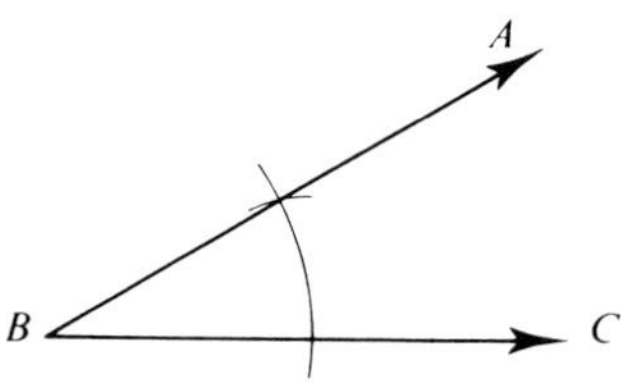

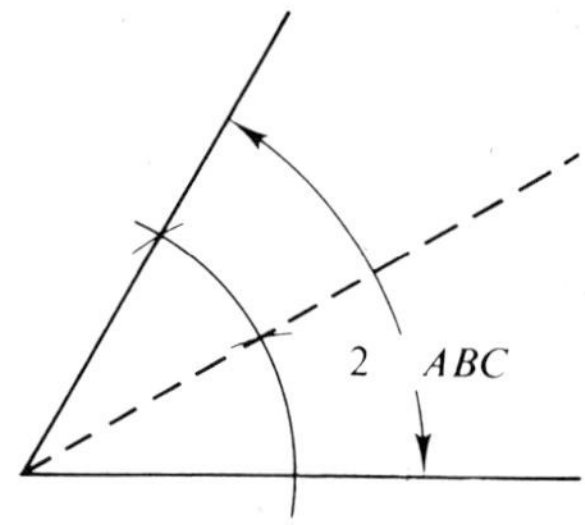

5) In text; construct two perpendicular lines. 7) In text; construct a triangle with three equal sides. Each angle is 60°. 9) Bisect the angle constructed in Problem 7. 11) In text. Construct a 60° angle. Erect a perpendicular at the vertex of this angle. Bisect the angle formed by the perpendicular and the terminal side of the 60° angle. This line completes the 75° angle. 13) Construct a 60° angle. Extend the side of the 60° angle. This extension and the other side of the 60° angle form a 120° angle. 15) Construct two perpendicular lines. Bisect one of the angles formed by these lines forming a 45° angle. The right angle added to one of these angles forms an angle of 135°. 17) Construct a 75° angle as in Problem 11. Construct a 22½° angle as in Problem 10. Cuplicate the 22½° angle on the interior of the 75° angle using one of the sides. The angle thus constructed contains 57½°.

19)

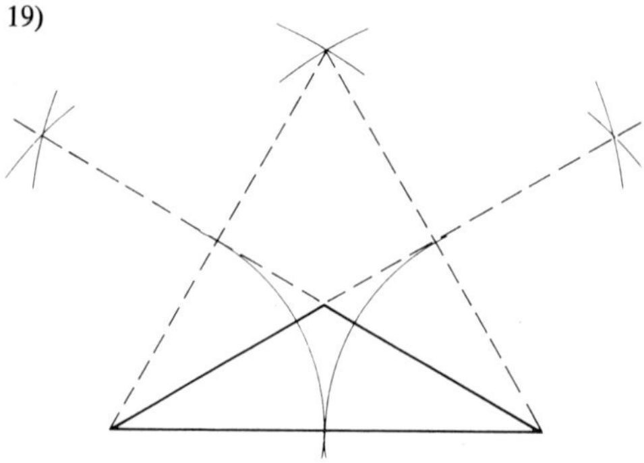

Exercise 44

1) Construct two perpendiculars to a line at two distinct points on the line. Using the distance between the intersections of the perpendiculars and the line as the length of the side of the square, lay off this distance on each perpendicular. Join the ends of these lines to complete the square. 3) Describe a circle using your compass. Use the radius of the circle as the length of the side of the hexagon. Starting at any point on the circle lay off six sides using the end point of each side as the beginning of the next side. Join these points to form the hexagon. 5) Construct a pentagon by the construction in the text. Bisect each side of the pentagon and extend the bisectors to intersect the circle. Join each of these points (intersections) to the adjacent vertices of the pentagon. 7) Bisect each side of the construction described in Problem 4. Extend the bisectors to intersect the circle. Join these points to the adjacent vertices. 9) Bisect each side of the hexagon constructed in Problem 3 and extend the bisectors to intersect the circle. Join each of these points to the adjacent vertices of the hexagon. 11) a star 13) the area of the circle 15) Construct an octagon as in Problem 4. Sketch only those sides that are illustrated in Problem 15. Using the side of the octagon as a radius, swing arcs from the adjacent vertices on opposite arms of the cross. Join the interesctions of these arcs to the vertices that served as centers for them.

Review Exercise Chapter 7

1) Definitions are in the text. 3) 13,188 5) $x = 25$ 7) 700 sq. units 9) 392.5 11) 2.84 cu. in. 13) 9.89 15) 6,104.2 cu. ft. 17) $h = 1.344$, $r = 1.344$ 19) 93.53

Chapter 8

Exercise 45

1) $x = 6, y = 2$ 3) $x = 1, y = 2$ 5) $x = 35/6, y = 15/2$

7)

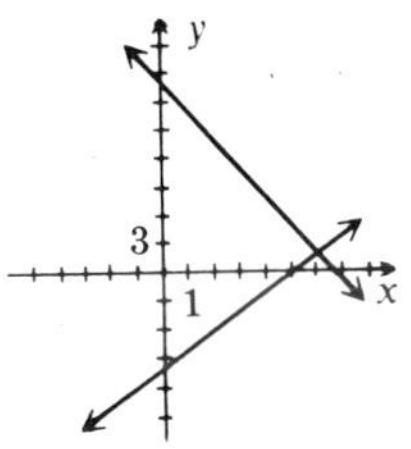

9)

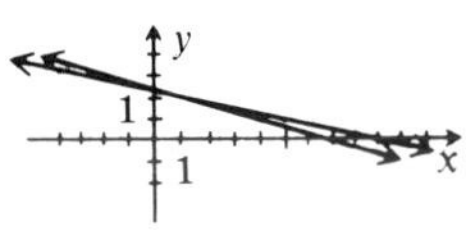

11)

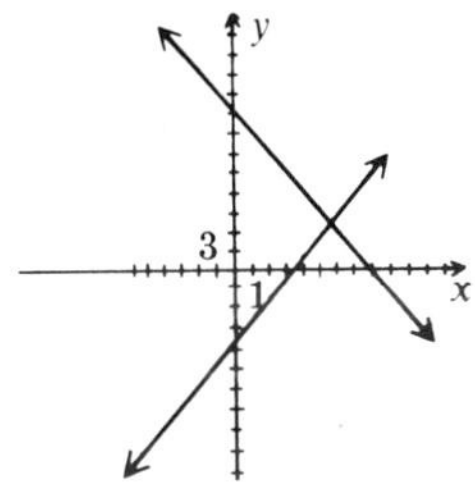

13) $x = 1, y = 6$ 15) $x = -8, y = 20$ 17) $x = 9, y = 2$ 19) $x = 26/7$, $y = 17/14$ 21) System is inconsistent. 23) $x = 6, y = -2$ 25) Smaller = 120, Larger = 200 27) 28°, 62°

Exercise 46

1) 20, 63 3) 20/24 5) 10 quarters, 20 nickels 7) 54 6/11 lbs cement, 65 5/11 lbs sand 9) 12 11) parts, \$5.00 labor, \$10.00 13) 6, 4 15) 7, 8 17) 250 cc 19) 8.1 lbs 21) 8, 12 23) 100 acres at each price 25) 30 mph 45 mph 27) 80 29) 5 5/33 hrs.

Exercise 47

1) (−0, 3/2, 9/2) 3) (7/6, 17/6, 2/3) 5) (3, 7/5, 12/5) 7) No solution 9) (1,1,1)

Exercise 48

1) $\begin{pmatrix} 3 & 6 \\ 3 & 9 \end{pmatrix}$ 3) $\begin{pmatrix} 3 & 9 \\ 6 & 21 \end{pmatrix}$ 5) $\begin{pmatrix} -4 \\ 7 \\ 3 \end{pmatrix}$ 7) (10) 9) $\begin{pmatrix} 3 & 5 & 6 \\ 4 & 8 & 10 \end{pmatrix}$

11) $\begin{pmatrix} 2 & 0 & 7 \\ 1 & 3 & 9 \\ 2 & 4 & 7 \end{pmatrix}$ 13) No solution. The elements in the matrix cannot be made the same.

15) $x = 0, y = -3$ 17) $x = -12, y = 1, z = 10$

Exercise 49

1) 3 3) 15 5) $ad - bc$ 7) 7 9) -11 11) 7 13) -11
15) 6 17) 53 19) -41

Exercise 50

1) $x = 6, y = 2$ 3) $x = 11/47, y = 7/47$ 5) $x = 13/6, y = 8$ 7) $x = 0, y = 3/2, z = 9/2$ 9) $a = 3/2, b = -½, c = 5/2$ 11) $x = -3/10, y = 7/5, z = 19/10, w = 2$ 13) $x = 2, y = 2, z = 2, w = 4$

Exercise 51

1) $x = 6, y = 2$ 3) $x = 1, y = 2$ 5) $x = 5/2, y = 7/2$ 7) $x = -8, y = 20$ 9) $x = ½, y = -½$ 11) $x = 0, y = 3/2, z = 9/2$ 13) $a = 1, b = 1, c = 1$ 15) $x = 2, y = 2, z = 4, w = 4$

Review Exercise Chapter 8

1) (2,0) 3) (12,2) 5) $(-5/16,23/16)$ 7) No solution, lines parallel
9) $(-5, -10/3)$ 11) $(1,2,-2)$ 13) -10 15) $(8,9/2)$ 17) (8,8)
19) No inverse exists.

Chapter 9

Exercise 52

1) $4, -1$ 3) $-7, 2$ 5) $-4, 3$ 7) $3, -3$ 9) 0, 5 11) -2
13) $-9, 7$ 15) $-7, -5$ 17) $-11, 8$ 19) $14, -7$ 21) $-5/3, 6$
23) $-⅔, 7/2$ 25) 4 seconds 27) 4 or $-10/3$

Exercise 53

1) ± 3 3) ± 4 5) $\pm\sqrt{11}$ 7) $\pm\sqrt{19}$ 9) $3, -5$ 11) $2, -10$
13) $-13 \pm 3\sqrt{2}$ 15) 11, 1 17) ± 3 19) $2, -8$ 21) $7/3, -1$
23) $-2, -6$ 25) $-2 \pm \sqrt{11}$

Exercise 54

1) $-2 \pm \sqrt{2}$ 3) $-9, 1$ 5) 4, 1 7) $(7 \pm \sqrt{73})/4$ 9) $2 \pm i$
11) $(-1 \pm \sqrt{5})/2$ 13) 3/2, 2 15) $-1 \pm 2i$ 17) $3 \pm \sqrt{11}$
19) $(-1 \pm 2i\sqrt{2})/3$ 21) $-6 \pm 3\sqrt{3}$ 23) $4 \pm 2\sqrt{6}$ 25) $(-4 \pm \sqrt{30})/2$
27) $(7 + \sqrt{329})/2, (-7 + \sqrt{329})/2$ 29) 12½ amps.

Exercise 55

1) Real and unequal 3) Complex conjugates 5) Real and unequal
7) Real and unequal 9) Real and unequal 11) Real and unequal

13) Real and unequal 15) $c = 13$ 17) $-3\sqrt{2}/4 < k < 3\sqrt{2}/4$
19) Sum = ¼, Product = 3/2

Exercise 56

1) $x^2 + y^2 = 9$
Circle with center at (0,0) radius 3

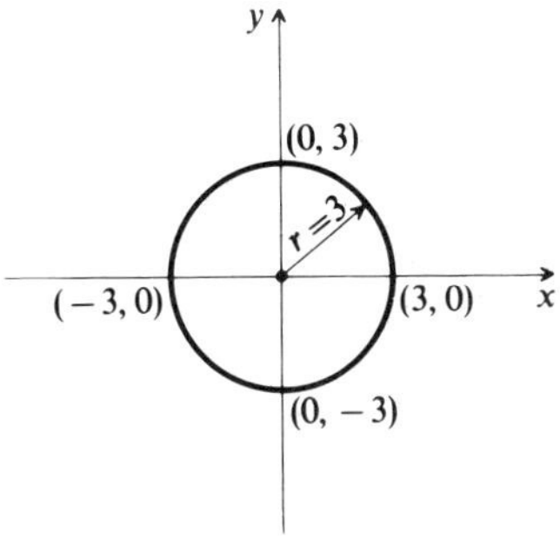

3) $(x - 2)^2 + (y - 4)^2 = 16$
Circle with center at (2,4) radius 4

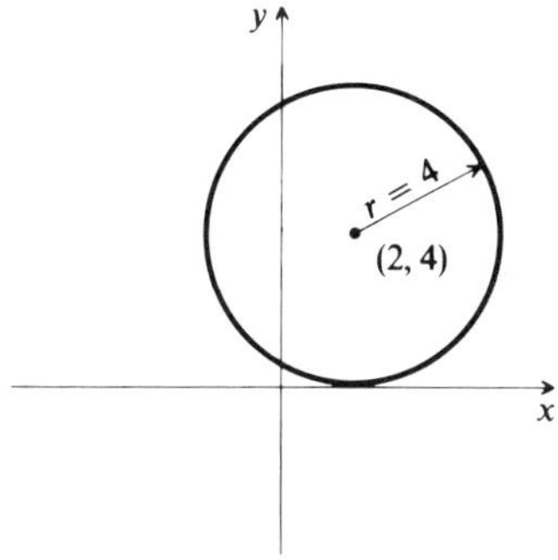

5) $(x + 2)^2 + (y - 4)^2 = 9$
Circle with center at (−2,4) radius 3

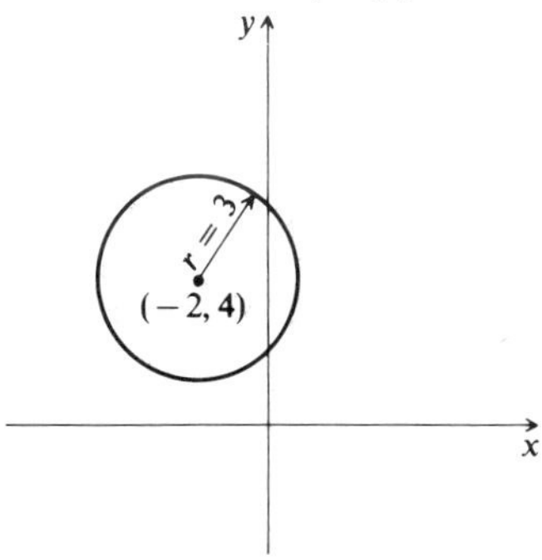

7) $\dfrac{(x - 2)^2}{25} + \dfrac{(y - 3)^2}{16} = 1$
Ellipse center at (2,3)
Ends of minor axis (2,7)(2,−1)
Ends of major axis (7,3)(−3,3)

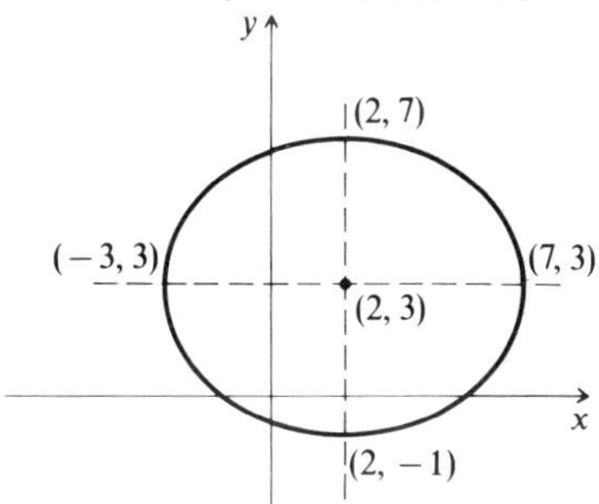

9) $x^2 + 6x + y^2 - 4y - 3 = 0$
or $(x + 3)^2 + (y - 2)^2 = 16$
Circle with center at (−3,2) radius 4

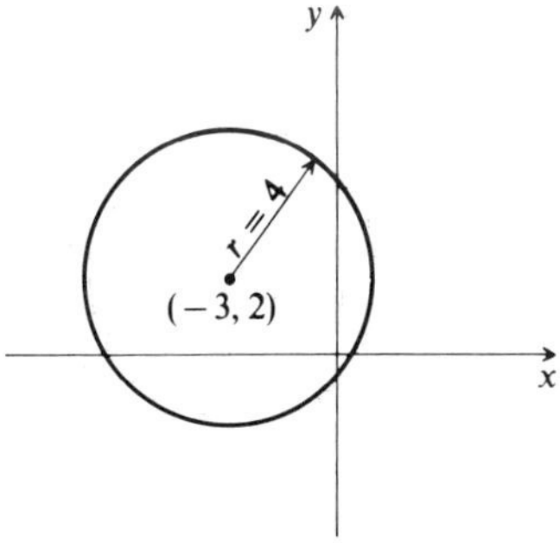

11) $\dfrac{(x - 4)^2}{4} + \dfrac{y^2}{1} = 1$
Ellipse with center at (4,0).
Ends of major axis (6,0), (2,0).
Ends of minor axis (4,1), (4,−1).

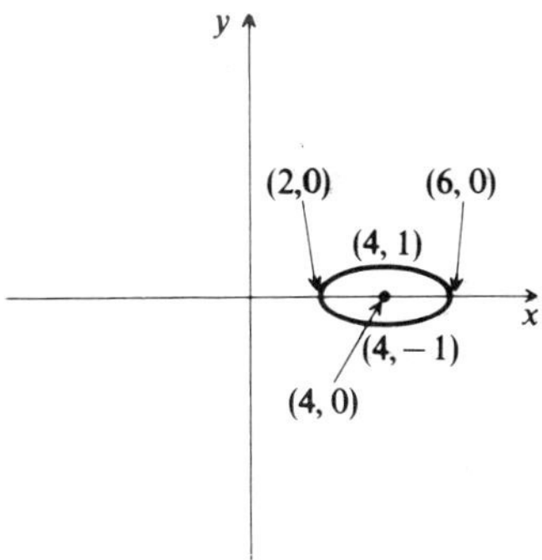

13) $\dfrac{x^2}{1/9} + \dfrac{y^2}{1/25} = 1$ Ellipse. Center at (0,0)
Ends of major axis (⅓,0), (−⅓,0).
Ends of minor axis (0,1/5), (0,−1/5).

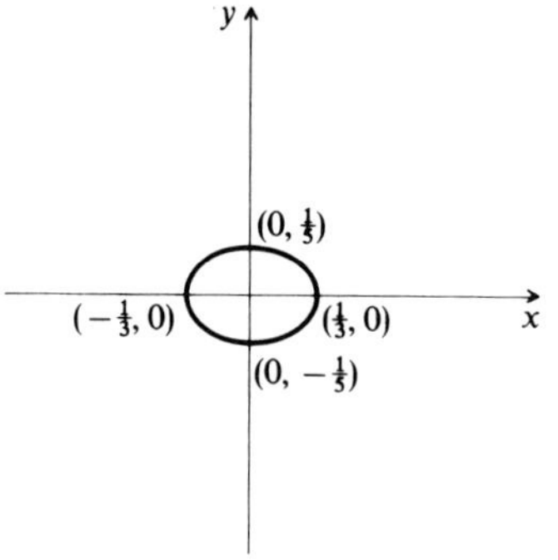

15) $(y - 2)^2 = (x - 3)^2 + 9$. Neither an ellipse nor circle since when rewritten the equation would be $(y - 2)^2 - (x - 3)^2 = 9$. Note the negative sign before $(x - 3)^2$.

Exercise 57

1) $y = x^2$ parabola

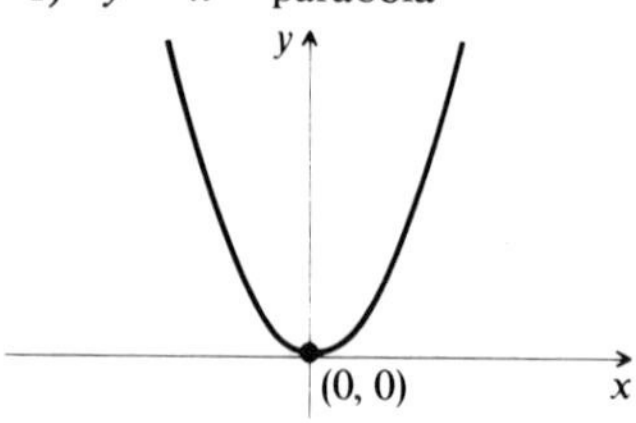

3) $y = x^2 + 3$ Parabola

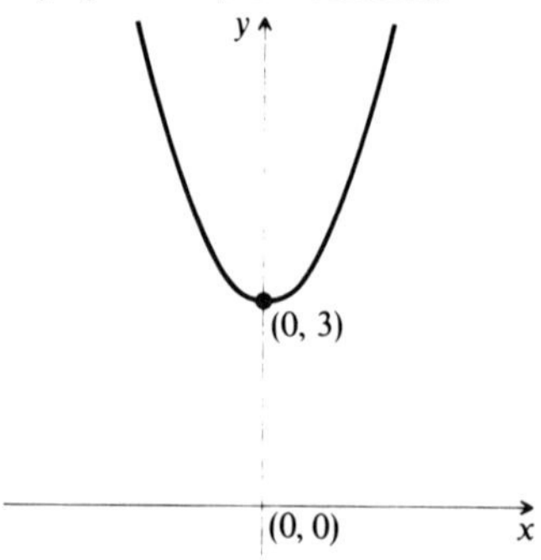

5) $y = -x^2 + 5$ Parabola

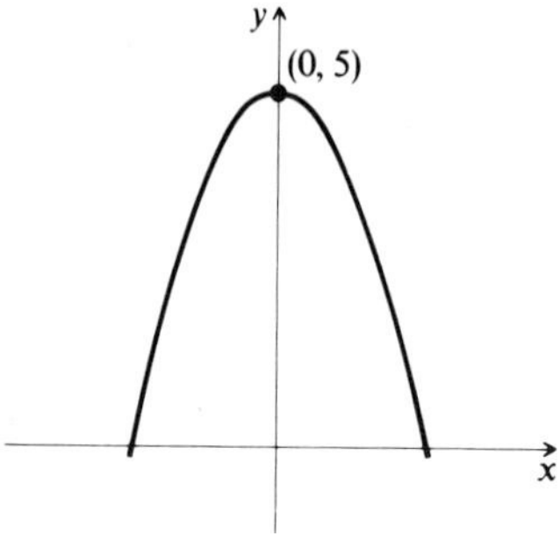

7) $y = x^2 - 3x + 2$ Parabola

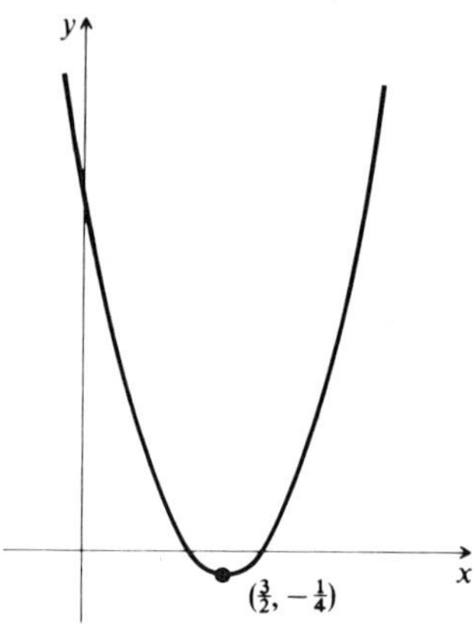

9) $y = \frac{1}{3}x^2 + 9$

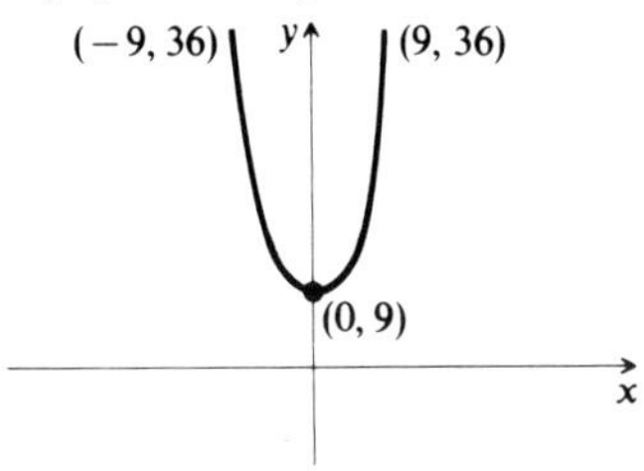

13) $\dfrac{(x-1)^2}{16} - \dfrac{(y+1)^2}{36} = 1x$ Hyperbola

Center at $(1, -1)$.

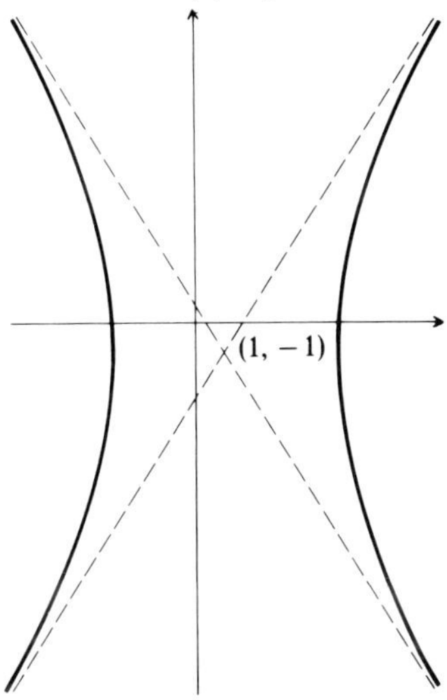

15) $\dfrac{x^2}{10} - \dfrac{y^2}{15} = 1x$ Hyperbola Center at (0,0)

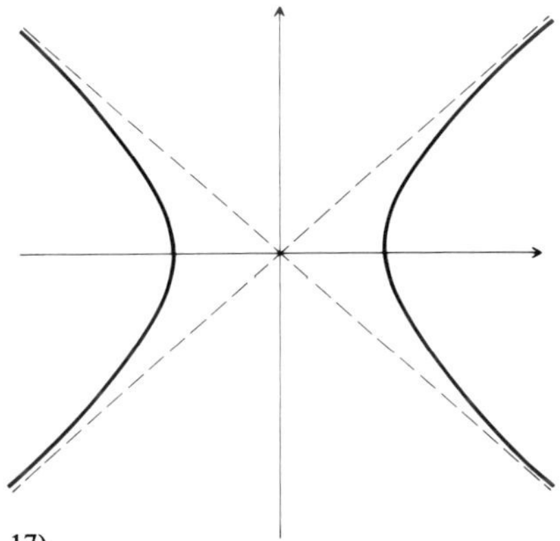

17)

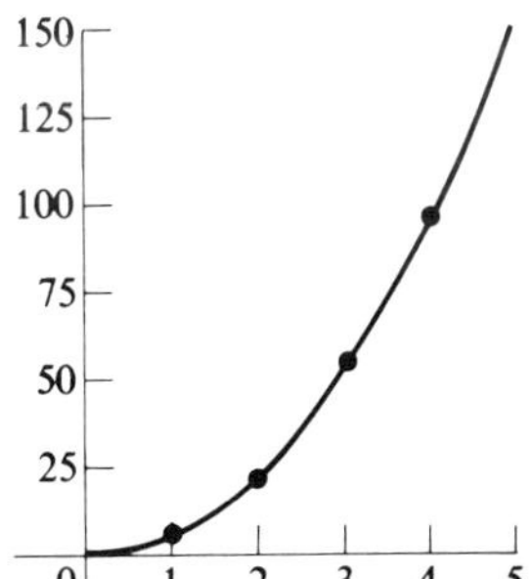

19)

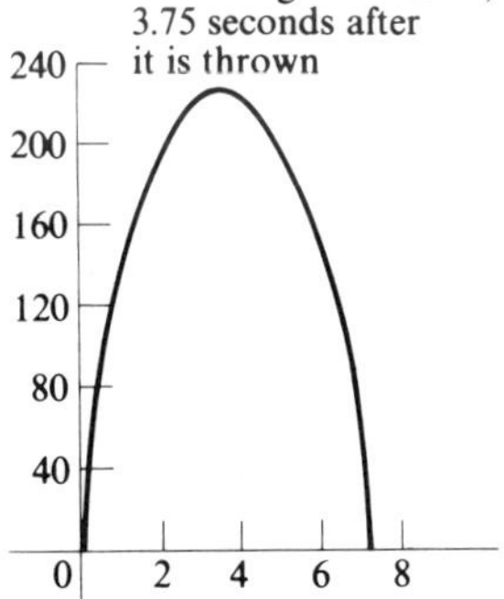

Exercise 58

1)

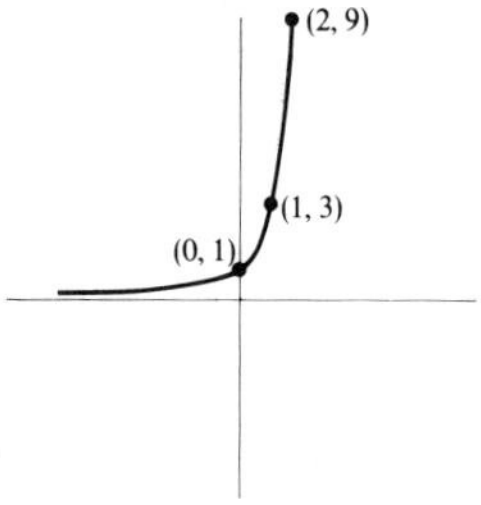

3)

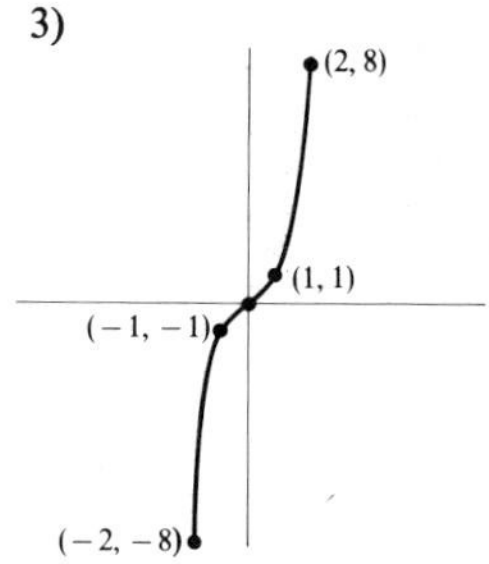

5)

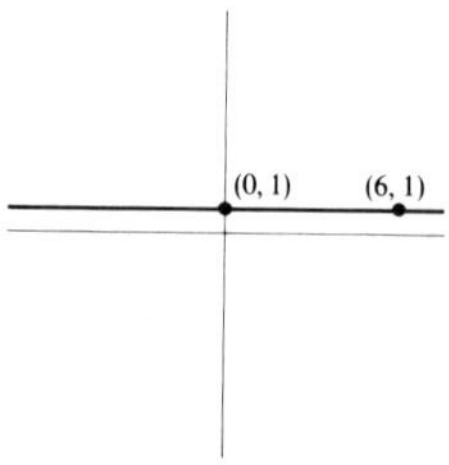

7)

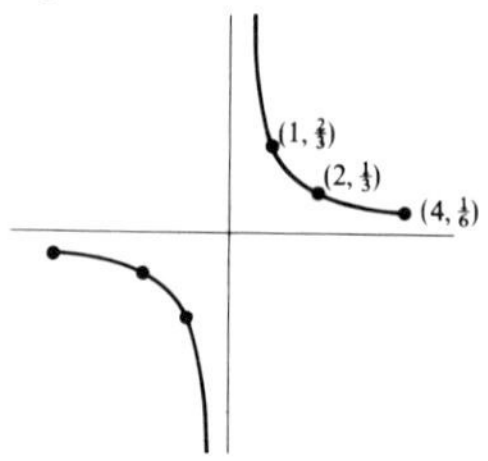

9)

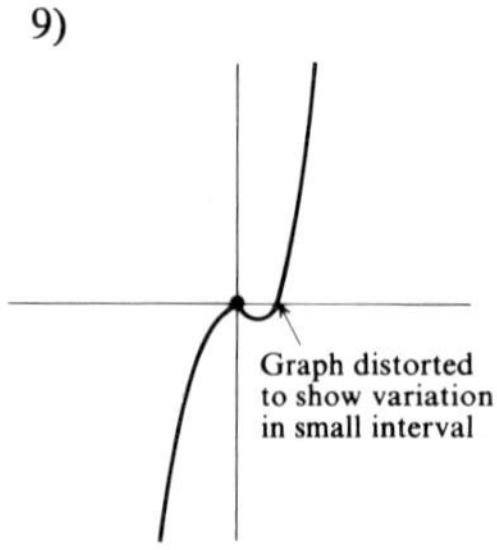

11)

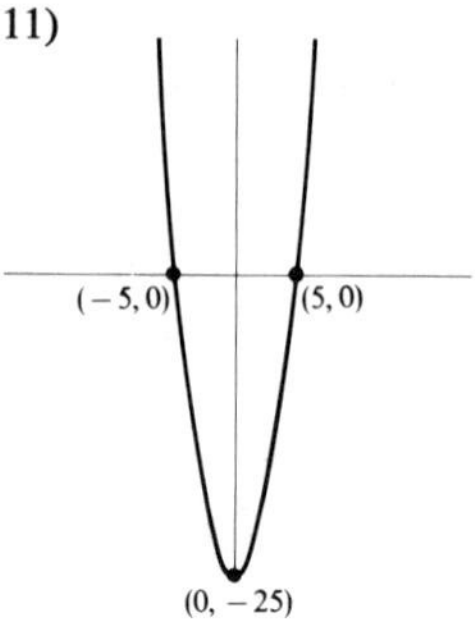

13)

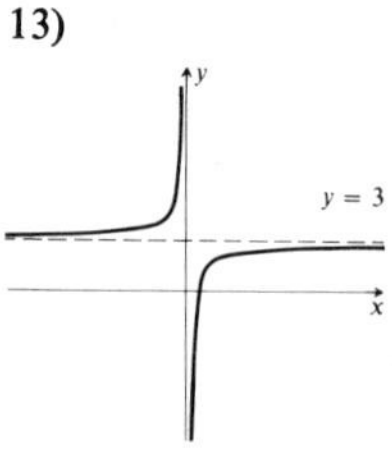

15)

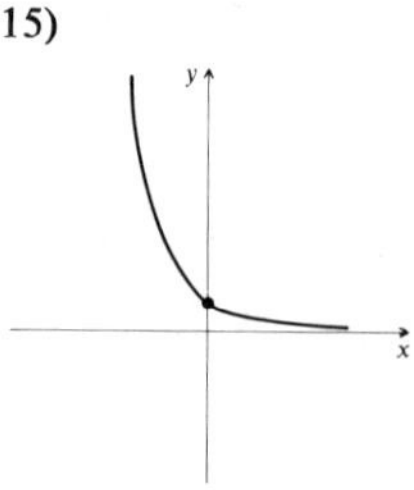

17)

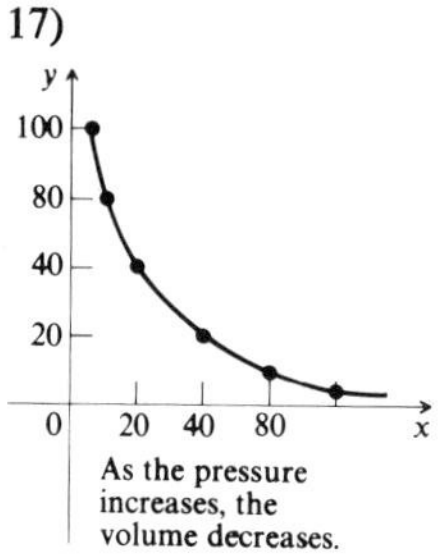

Review Exercise Chapter 9

1) $-2,-1$ 3) $4,-1$ 5) $3 \pm \sqrt{14}$ 7) $(5 \pm \sqrt{74})/3$ 9) $1,-\frac{1}{3}$
11) real and unequal, $-1/5$, -3 13) complex conjugates, $3/5$, $4/5$
15) Circle center at $(2,-6)$ radius 5

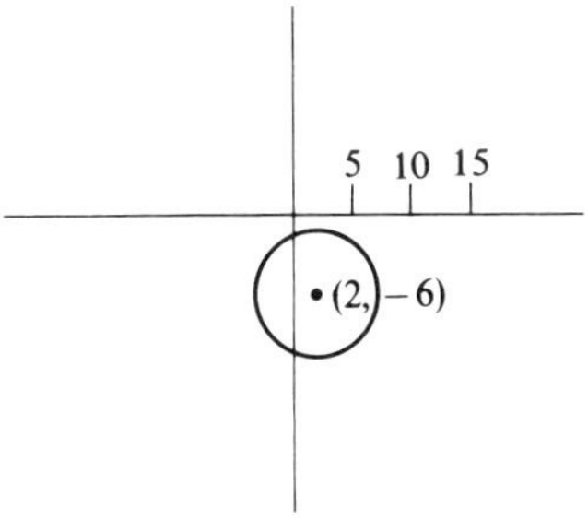

17) Parabola, vertex $(3,-4)$

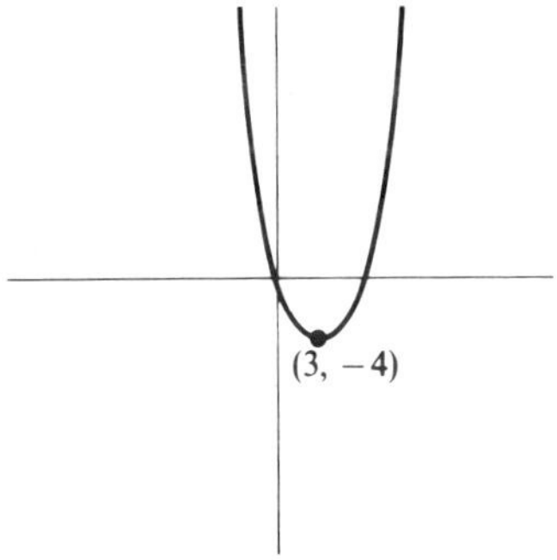

19)

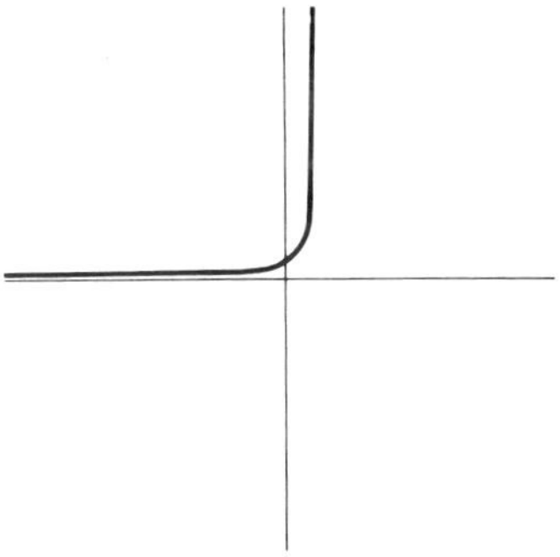

Chapter 10

Answers to Exercise 59

1) $\sqrt{2}/2$, $\sqrt{2}/2$ 3) $\sqrt{2}$, 1 5) 1, $\sqrt{2}$ 7) $\sqrt{3}/3$, $2\sqrt{3}/2$ 9) 2, $\sqrt{3}$
11) $3/5$, $3/5$ 13) $\sqrt{3}/2$, $\sqrt{3}$ 15) $40/41$, $40/41$ 17) $5/2$, $\sqrt{21}/2$
19) $2/3$, $2\sqrt{5}/5$ 21) $4/5$ 23) $\sqrt{3}/2$ 25) They are equal.

Exercise 60

Answers may vary depending on the tables used.
1) a) .2079 b) .4877 c) 1.0000 d) .7071 e) .8660 f) .8660 g) 1.0576
h) 1.3270 i) 2.1300 j) 1.0000 k) 1.0000 l) 5.6713 3) 31° 5) 60°
7) 30° 9) 3.4641 or $2\sqrt{3}$ 11) 12.618 13) 3.6528

Exercise 61

1) a) 4.067 b) 17.434 c) 16.639 d) 9.192 e) 11.547 f) 26.673 g) 16.08
3) 4.6353 5) 76.75° 7) .191 9) 6.2 11) 23.7° 13) 10.61
15) 5.8779 in. 17) 5458.56, R = 5458.56 G = 22.07

Exercise 62

1a I 1b) II 1c) I 1d) III 1e) IV 1f) on axis 1g) II 1h) I 1i) I 1j) II 1k) IV 1l) IV 1m) III 1n) IV 1o) I 3a) $\sqrt{13}$ 3b) $\sqrt{13}$ 3c) $2\sqrt{5}$ 3d) $2\sqrt{5}$ 3e) $2\sqrt{10}$ 3f) 3 3g) $3\sqrt{5}$ 3h) ⅝ 3i) 5 3j) $\sqrt{52}/3$ 3k) $2.5\sqrt{2}$ 3l) $3\sqrt{41}$ 3m) $\sqrt{193}$ 3n) $\sqrt{97}/2$ 3o) $11\sqrt{2}$

Exercise 63

1a) I 1b) II 1c) II 1d) I 1e) II 1f) I 3a) $\sin\theta = 5/\sqrt{26}$, $\cos\theta = 1/\sqrt{26}$, $\tan\theta = 5/1$, $\cot\theta = 1/5$, $\sec\theta = \sqrt{26}$, $\csc\theta = \sqrt{26}/5$ 3b) $\sin\theta = 1/\sqrt{2}$, $\cos\theta = 1/\sqrt{2}$, $\tan\theta = 1$, $\cot\theta = 1$, $\sec\theta = \sqrt{2}$, $\csc\theta = \sqrt{2}$ 3c) $\sin\theta = 3/\sqrt{10}$, $\cos\theta = -1/\sqrt{10}$, $\tan\theta = -3$, $\cot\theta = -\frac{1}{3}$, $\sec\theta = -\sqrt{10}$, $\csc\theta = \sqrt{10}/3$ 3d) $\sin\theta = 5/\sqrt{29}$, $\cos\theta = -2\sqrt{29}$, $\tan\theta = -5/2$, $\cot\theta = -2/5$, $\sec\theta = -\sqrt{29}/2$, $\csc\theta = \sqrt{29}/5$ 3e) $\sin\theta = \sqrt{3}/2$, $\cos\theta = -\frac{1}{2}$, $\tan\theta = -\sqrt{3}$, $\cot\theta = -1/\sqrt{3}$, $\sec\theta = -2$, $\csc\theta = 2/\sqrt{3}$ 3f) $\sin\theta = \sqrt{3}/\sqrt{7}$, $\cos\theta = 2/\sqrt{7}$, $\tan\theta = \sqrt{3}/2$, $\cot\theta = 2/\sqrt{3}$, $\sec\theta = \sqrt{7}/2$, $\csc\theta = \sqrt{7}/\sqrt{3}$ 5a) .8829, .4695, 1.8807 5b) .6018, .7986, .7536 5c) .8480, −.3299, −1.660 5d) .9848, −.1736, −5.6713 5e) .5000, .8660, .5774 5f) .3090, −.9511, −.3249 5g) .0698, −.9976, −.0699 5h) 0, 1, 0 5i) 1, 0, does not exist 7) No. $\sec\theta$ = hypotenuse/side adjacent to θ, and at 0° the adjacent side is 0. Division by zero is not defined. 9) 153.24 sq. in.

Exercise 64

1) $A = 85°$, $b = 8.83$, $a = 11.02$ 3) $C = 69°$, $a = 4.23$, $c = 11.03$ 5) $A = 99°$, $b = .56$, $c = .19$ 7) $C = 85°$, $b = 14.12$, $c = 14.56$ 9) $A = 43°$, $b = 51.5$, $c = 84.3$ 11) $B = 54°$, $a = .30$, $c = .52$ 13) $C = 78°$, $a = 6.95$, $b = 13.2$ 15) $C = 115°$, $b = 28.3$, $c = 36.3$ 17) 26.4 mi. from S tower, 19.6 mi. from N tower 19) 8.45 in. 12) .38

Exercise 65

1) a) 4/5, b) 12/13, c) .866 d) $\sqrt{40}/7$ 3) $a = 8.59$, $B = 50°$, $C = 104°$, 5) $c = 8.27$, $A = 25.6°$, $B = 49.4°$ 7) $h = 244.8$

Exercise 66

1) a) 40°, QIII b) 60°, QIII c) 39°, QIV d) 84°, QII e) 49°, QII f) 36°, QIII g) 0° h) 31°, QI i) 47°, QIII j) 30°, QII k) 10°, QIII l) 35°, QII m) 42°, QIV n) 47°, QIII 3) I and II 5) I and III 7) a) I b) II c) III d) I or III e) IV f) II g) I

Exercise 67

1) 5 3) $\csc x$ 5) $(1 + \cos^2 x)/\sin^2 x$ 7)–11) identities 13) 1 15) $2\sqrt{3}/3$ 17) $\sqrt{3}$ 19) $-\sqrt{3}/4$

Exercise 68

1) $(\sqrt{6}+\sqrt{2})/4$ 3) $(\sqrt{6}-\sqrt{2})/4$ 5) $-\sqrt{2}/2$ 7) 24/25
9) $3\sqrt{10}/10$ 11) $(2\sqrt{15}+\sqrt{5})/12$ 13) $(2\sqrt{15}-\sqrt{5})/12$ 15) $4\sqrt{5}/9$
17) $\sqrt{15}/8$ 19) $(\sqrt{15}+\sqrt{5})/(5\sqrt{3}-2)$

21) $$\tan(\theta+\phi)=\frac{\sin(\theta+\phi)}{\cos(\theta+\phi)}=\frac{\sin\theta\cos\phi+\cos\theta\sin\phi}{\cos\theta\cos\phi-\sin\theta\sin\phi}=\frac{\dfrac{\sin\theta\cos\phi}{\cos\theta\cos\theta}+\dfrac{\cos\theta\sin\phi}{\cos\theta\cos\theta}}{\dfrac{\cos\theta\cos\phi}{\cos\theta\cos\phi}-\dfrac{\sin\theta\sin\phi}{\cos\theta\cos\phi}}=\frac{\tan\theta+\tan\phi}{1-\tan\theta\tan\phi}$$

23) $$\tan 2=\frac{\sin 2\theta}{\cos 2\theta}=\frac{2\sin\theta\cos\theta}{\cos^2\theta-\sin^2\theta}=\frac{\dfrac{2\sin\theta\cos\theta}{\cos^2\theta}}{\dfrac{\cos^2\theta}{\cos^2\theta}-\dfrac{\sin^2\theta}{\cos^2\theta}}=\frac{2\tan\theta}{1-\tan^2\theta}$$

25) $$\tan(\theta-\phi)=\frac{\sin(\theta-\phi)}{\cos(\theta-\phi)}=\frac{\sin\theta\cos\phi-\cos\theta\sin\phi}{\cos\theta\cos\phi+\sin\theta\sin\phi}=\frac{\dfrac{\sin\theta\cos\phi}{\cos\theta\cos\phi}-\dfrac{\cos\theta\sin\phi}{\cos\theta\cos\phi}}{\dfrac{\cos\theta\cos\phi}{\cos\theta\cos\phi}+\dfrac{\sin\theta\sin\phi}{\cos\theta\cos\phi}}=\frac{\tan\theta-\tan\phi}{1+\tan\theta\tan\phi}$$

Exercise 69

1) .6428 3) .9668 5) .2924 7) .0297 9) .4348 11) 31.55°
13) 85.75° 15) 40.88° 17) 30° 19) 30° 21) 1.325 23) 2.856
25) 3.672 27) 1 29) 3.479 31) 32.72° 33) 12.5° 35) 22.83°
37) 32.92° 39) 87.4° 41) $B = 57°$ $a = .644$ $c = 2.69$ 43) $B = 36°$
$b = .494$ $a = .680$ 45) $A = 26.75°$ $B = 123.25°$ $b = 1003$ 47) $\theta = 32°$
$r = 14.2$ 49) $r = 34.5$

Review Exercise Chapter 10

1) $\cos x$ 3) 1 5) 1 7) $\sqrt{(1-\cos x)/2}$ 9)–11) identities
13) $\sqrt{3}$ 15) $\sqrt{2}/4-\sqrt{6}/4$ 17) $4/(-\sqrt{6}-\sqrt{2})$ 19) .904
21) .274

Chapter 11

Exercise 70

1) 90° 3) 120° 5) 450° 7) 157.5° 9) −10° 11) 660°
13) $\pi/6$ 15) $\pi/4$ 17) 4π 19) $-\pi/4$ 21) $5\pi/3$ 23) 309.4°
25) −40.1° 27) −263.6° 29) $\sqrt{2}/2$ 31)1 33) $-\sqrt{3}/2$
35) .676 37) 47° or 313°, .82 or 5.46 39) 246° or 294°, 4.29 or 5.13
41) 12°, .209 43) .0156 45) .1319

Exercise 71

1) 10.472 3) 40 ft. 5) 35.24 7) 2.749 9) 4.2 11) 1036.745 miles per hour 13) 10,560 degrees/sec 15) 33175 in./min. for each

17) alternator pulley 7650 R.P.M.
Waterpump pulley 7285 R.P.M.
Compressor pulley 5100 R.P.M.

Exercise 72

1)

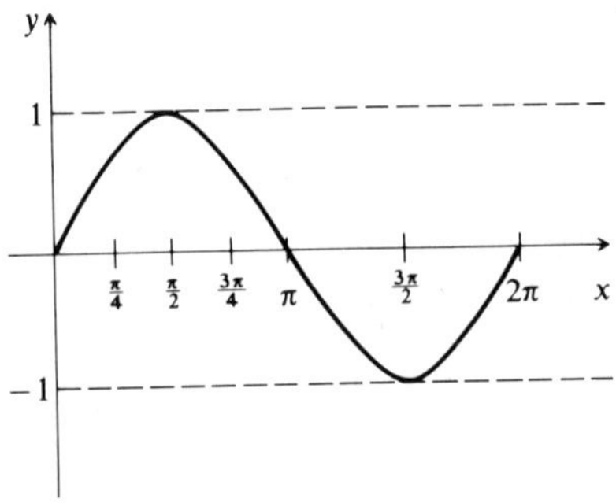

3)

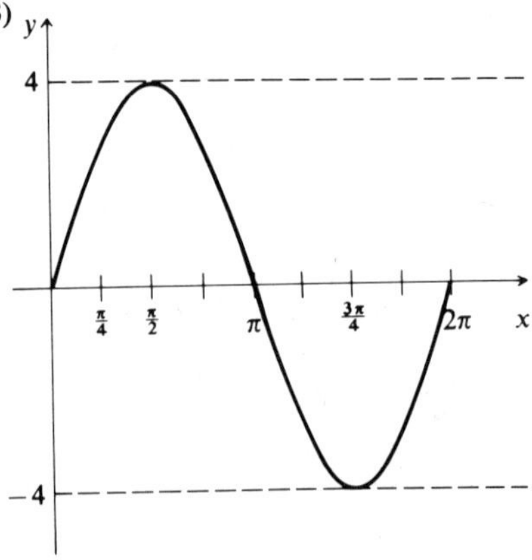

5)

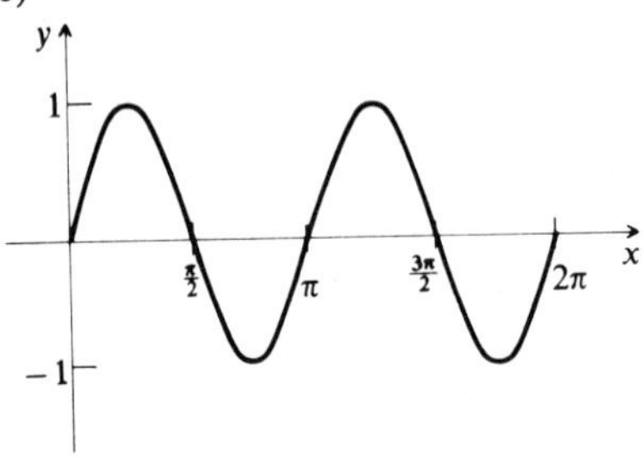

7)

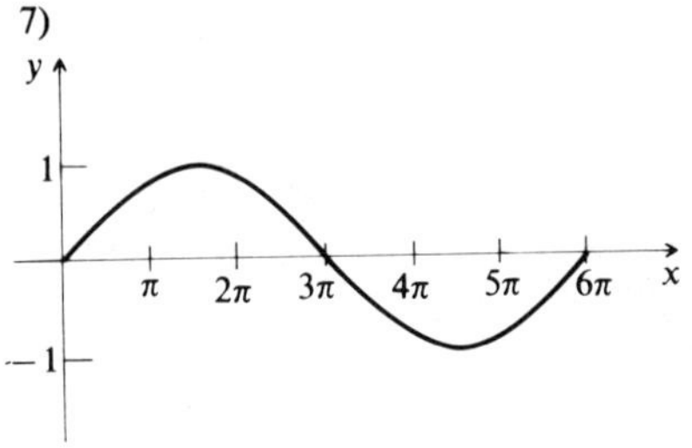

9)

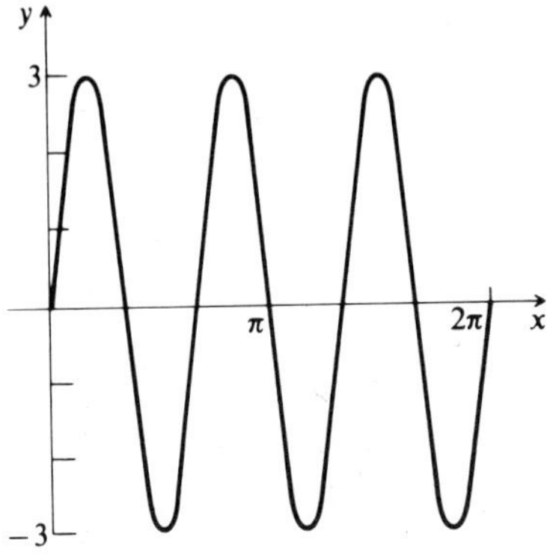

11)

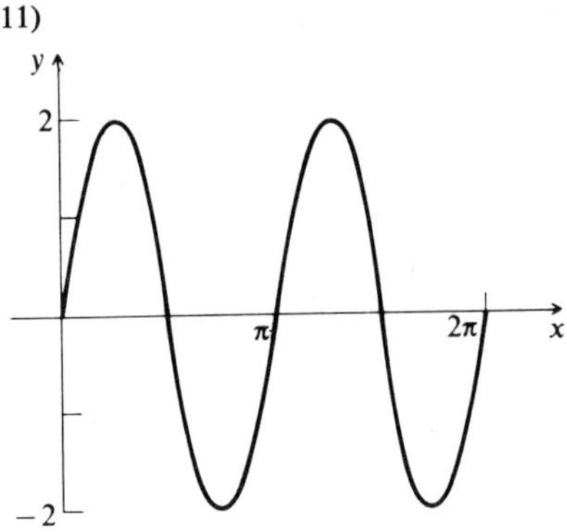

13)

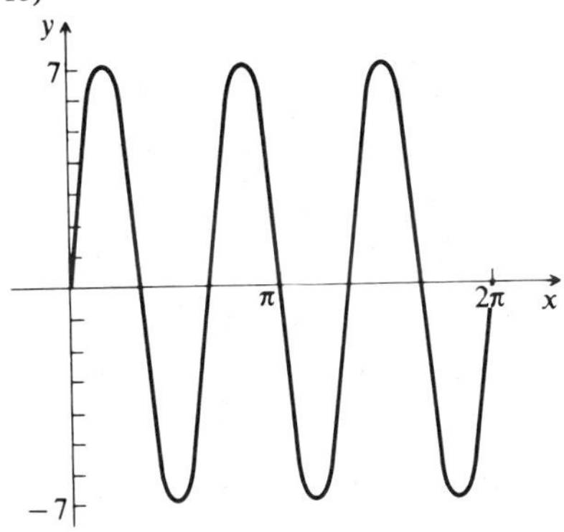

15)

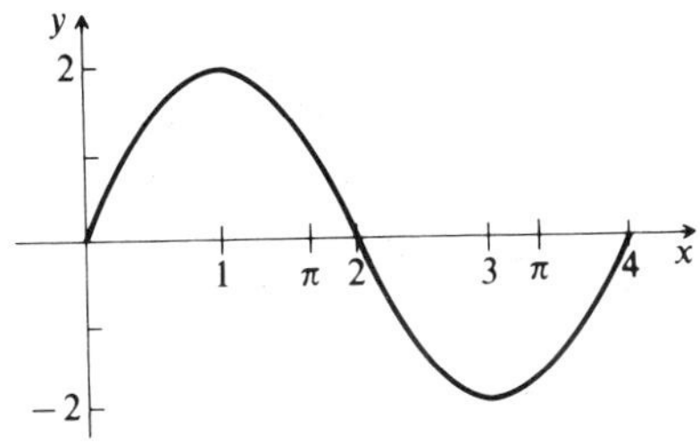

Exercise 73

1)

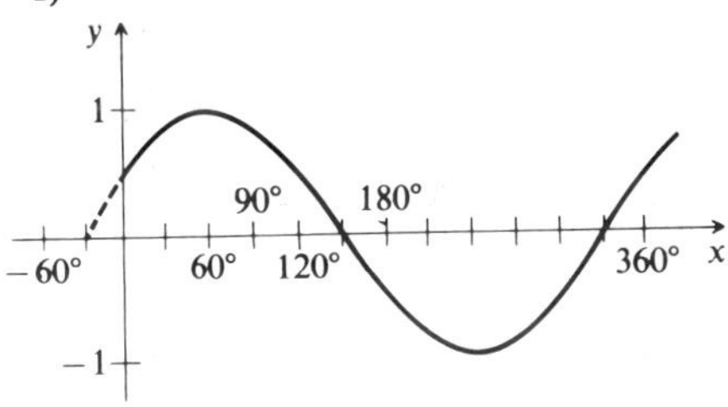

3)

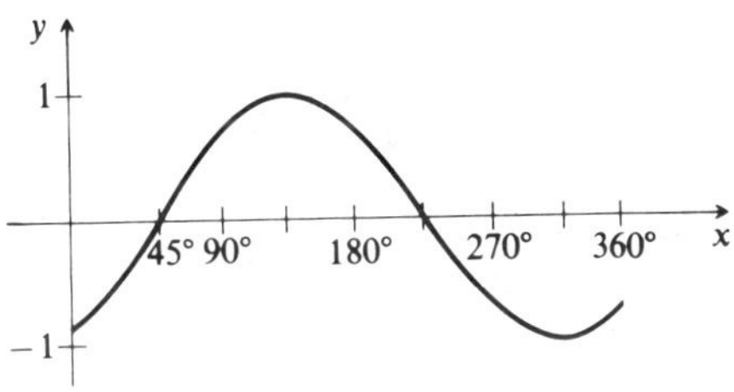

5)

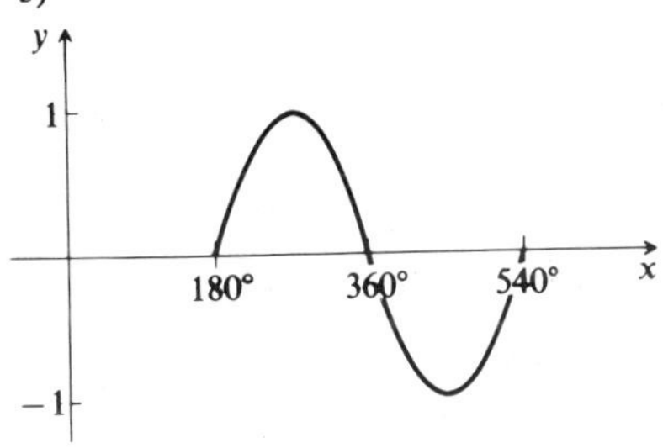

7)

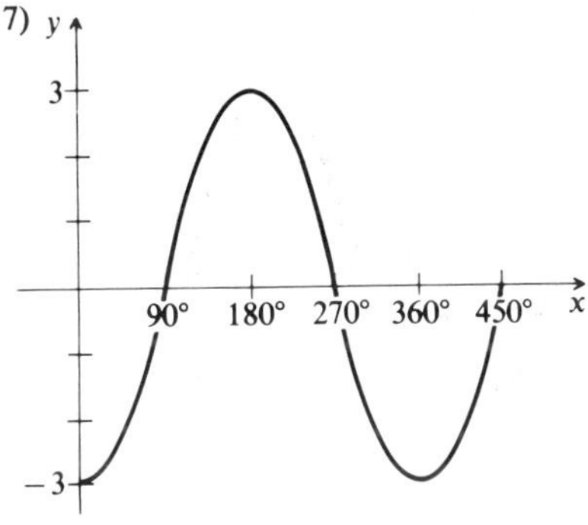

9)

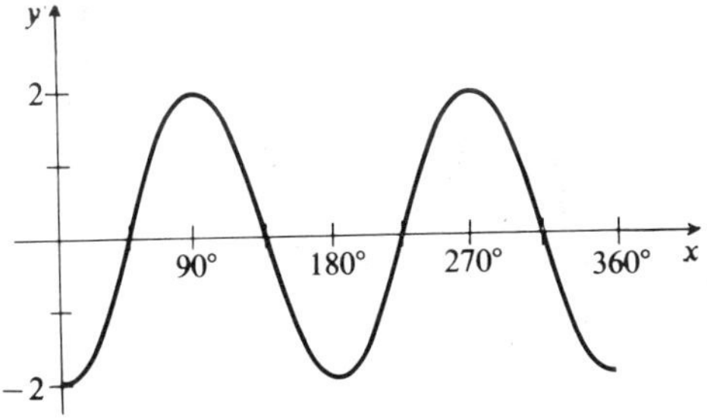

11)

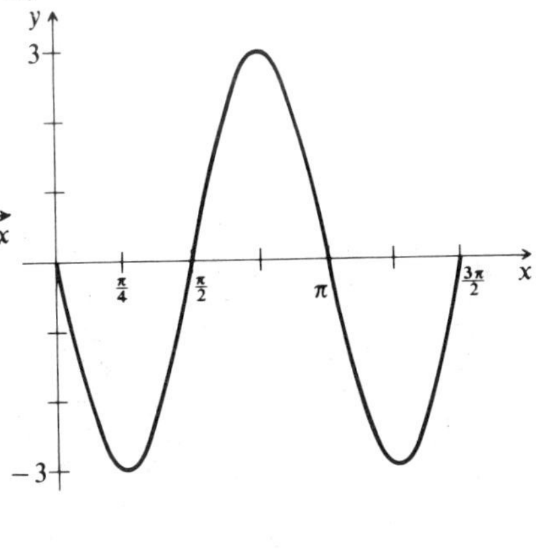

13) 0

Review Exercise Chapter 11

1) $17/36\pi$ 3) $-5/12\pi$ 5) $75°$ 7) $330°$ 9) 15 11) period $= \pi$, amp. $= 3$ 13) period $= \pi/2$, amp. $= \frac{1}{2}$, phase angle 3/2 radian 15) period 3π, amp. 3

Chapter 12

Exercise 74

1) 5 3) 4 5) 18.439 7) 190.3° 9) 211°
11) $\vec{u} + \vec{v} = (9,-7)$ $u - \vec{v} = (-3,3)$
13) $\vec{u} + \vec{v} = (-6,-11)$ $\vec{u} - \vec{v} = (-12,-1)$
15) $\vec{u} + \vec{v} = (3,-4)$ $\vec{u} - \vec{v} = (11,2)$ 17) 12.94 19) 207.704

Exercise 75

1)

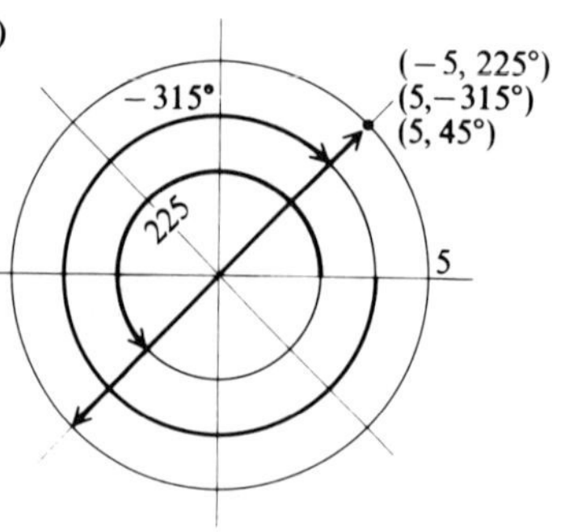

3)

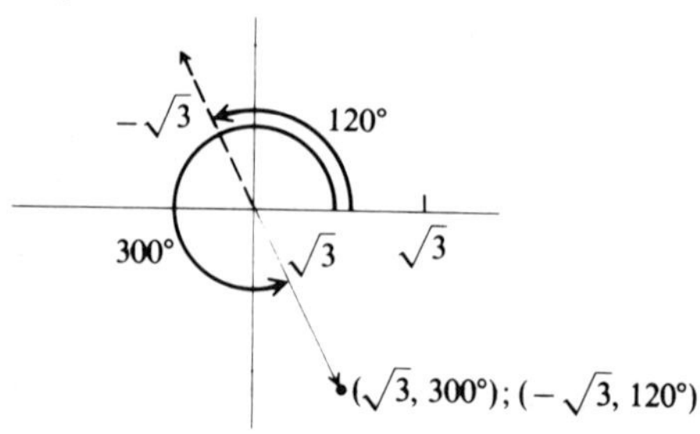

5)

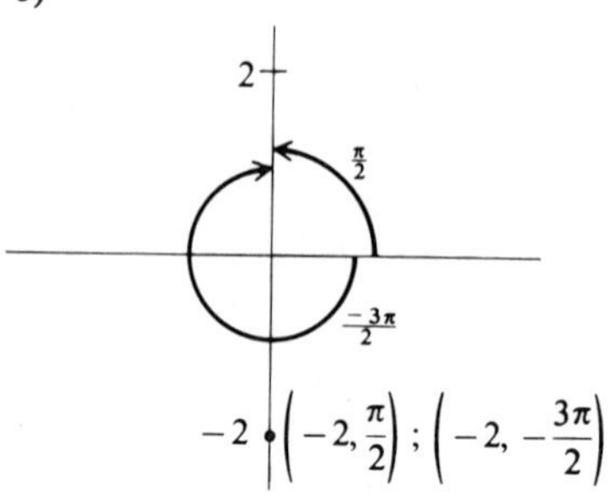

7)

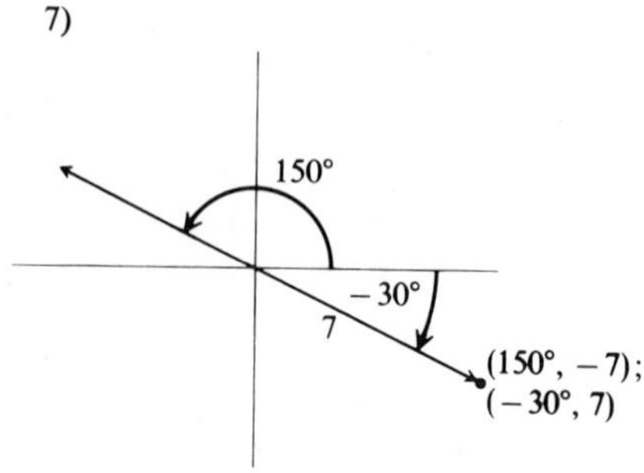

9)

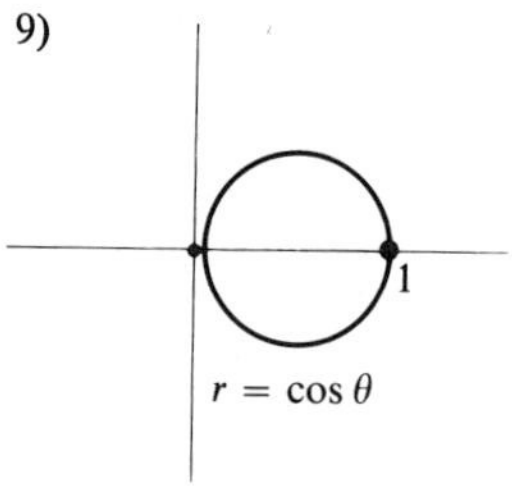

11)

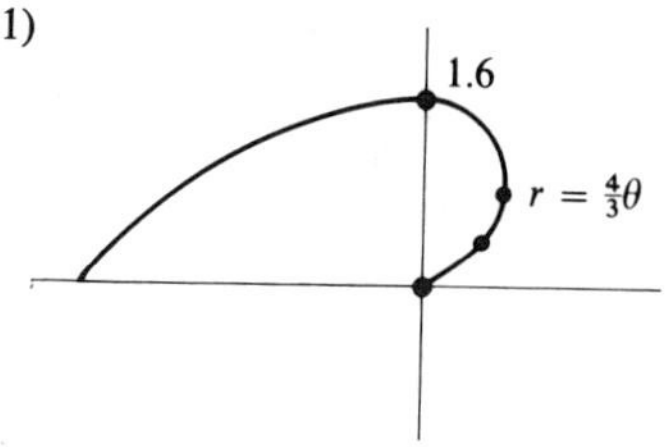

13)

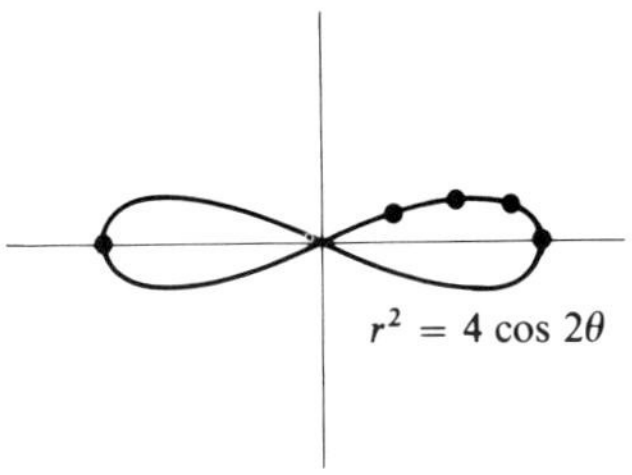

15)

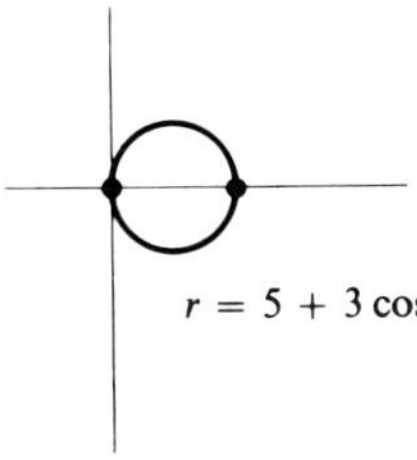

17)

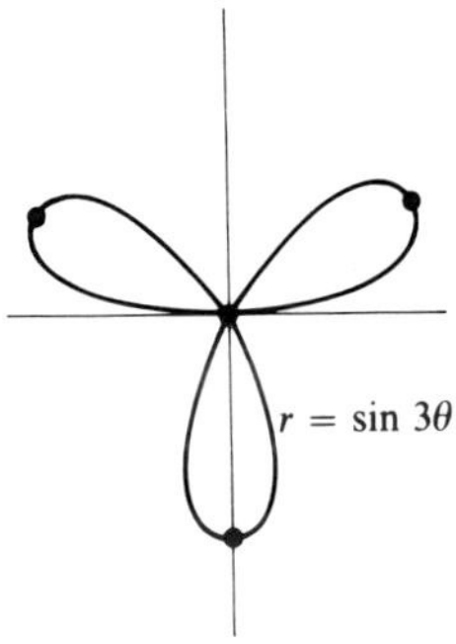

19) $r^2 = \sin 30$

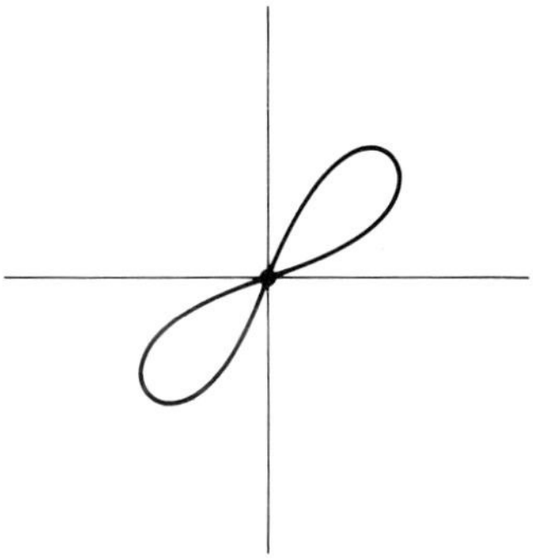

Exercise 76

1) $(3\sqrt{3}/2, 3/2)$ or $(2.598, 1.500)$ 3) $(-\sqrt{2}/2, \sqrt{2}/2)$ or $(-.707, .707)$
5) $(-2.457, 1.71)$ 7) $(.0268, -.2490)$ 9) $(4.8065, 1.3780)$
11) $(1.323, 49.1°)$ 13) $(5\sqrt{2}, 45°)$ 15) $(-7.211, 33.7°)$ 17) $(\sqrt{3}, 144.75°)$
19) $(15.3, 11.33°)$ 21) $(x^2 + y^2)^2 = 3x^2$ 23) $x^2 + y^2 = x + y$
25) $x^2 + y^2 = 25$ 27) $9y^2 = 7x^2$ 29) $2x = 0$ 31) $r = 5$
33) $r = 6/(4\cos\theta - 3\sin\theta)$ 35) $r = 6/\cos\theta$ 37) $r^2 = 36/(9\sin^2\theta + 4\cos^2\theta)$
39) $r = (\cos\theta - \sin\theta)/\sin^2\theta$

Exercise 77

1)

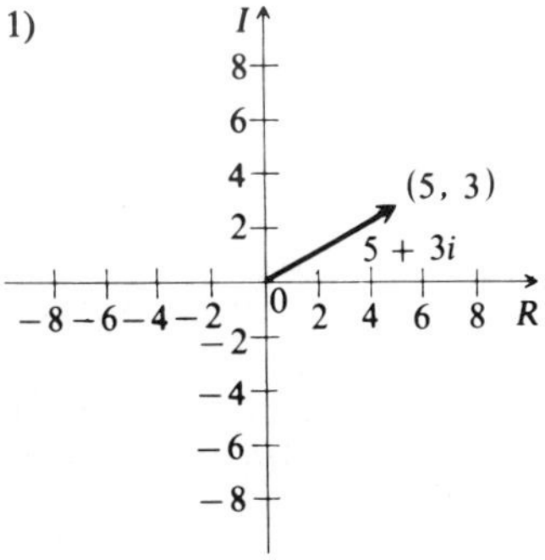

3)

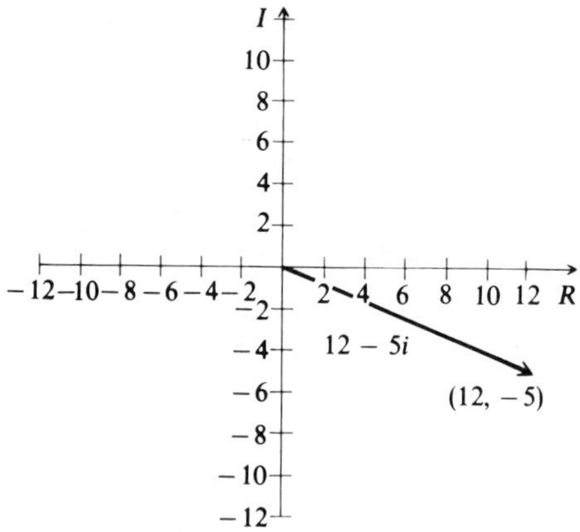

5)

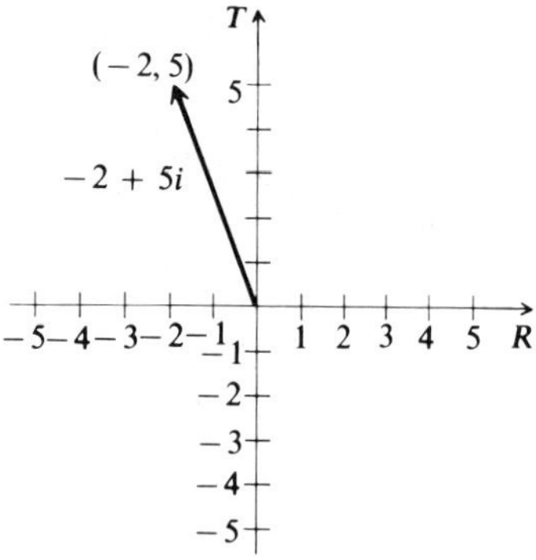

7)

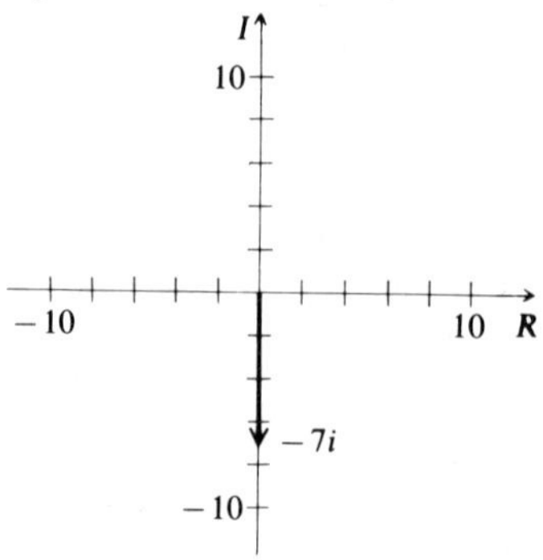

9)

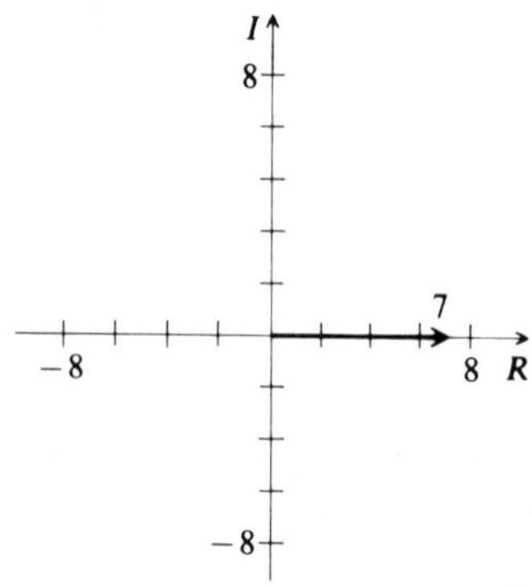

11) 2 cis 60° 13) $4\sqrt{2}$ cis 225° or 5.656 cis 215° 15) 2 cis 225°
17) 2 cis 270° 19) 5.831 cis (−31°) 21) $3/2 + 3i\sqrt{3}/2$ 23) $-1 - i$
25) $3\sqrt{3} + 3i$ 27) $-3\sqrt{2}/2 + 3i\sqrt{2}/2$ 29) $1.8542 + .7492i$

Exercise 78

Answers given in decimal form are approximate. 1) 6 2) $.4492 + 4.8784i$
3) $32 + 4i$ 4) $.7320 - 2.732i$ 5) $32 + 7i$ 6) 3 7) $5/2\,(1 - i\sqrt{3})$
8) $.1030 - .3183i$ 9) $.4482 - 1.629i$ 10) $.5 + .5i$ 11) $-7i$
12) $.9460 - .3242i$ 13) $-12{,}500(1 + i)$ 14) $32(1 + i\sqrt{3})$
15) $.2588 + .9659i$ 16) $6.0463 - 46.0065i$ 17) $81/2(-\sqrt{3} + i)$
18) $(8^{13}/2)(\sqrt{3} + i)$ 19) $-2, -1 \pm i\sqrt{3}$ 20) $\pm i$ 21) $.8090 + .5878i$, $.3090 + .9511i$, -1, $-.3090 - .9511i$, $.8090 - 5878i$ 22) $.9010 + .4339i$, $.2225 + .9749i$, $-.6236 + .7818i$, -2, $-.6236 - .7818i$, $.2225 - .9749i$, $.9019 - .4339i$
23) $2, 2i, -2, -2i$ 24) $(3/2)(1 + i\sqrt{3}), -3, (3/2)(1 - i\sqrt{3})$ 25) $1.6180 + 1.7150i$, $.6180 + 1.9022i$, -2, $.6180 - 1.9022$, $1.6180 - 1.7150i$

Review Exercise Chapter 12

1) mag $\sqrt{95}$, $\theta = 1.15$ 3) $(5/\sqrt{61})i + (6/\sqrt{61})j$ 5) sum $4i + 3j$; difference $8i + j$ 7) $(7, -345°)$; $(-7, 195°)$ 9) $(\sqrt{3}/2, 11\pi/6)$; $(-\sqrt{3}/2, 5\pi/6)$ 11) $2(x^2 + y^2)^{3/2} = \sqrt{x^2 + y^2} + x$ 13) $(\sqrt{2}, 450°)$ 15) $(1, 45°)$ 17) $r = 5/(\sin\theta - \cos\theta)$ 19) $(-5/2, -5\sqrt{3}/2)$ 21) $(4\sqrt{2}/12)(\sqrt{3} + 1)$, $(-4\sqrt{2}/12)(\sqrt{3} - 1))$ 23) $\sqrt{x^2 + y^2}$ = angle whose cos is $x/\sqrt{x^2 + y^2}$ 25) $\sqrt{7}(\text{cis}(-40.9°))$ 27) $\sqrt{5}(\text{cis}(-26.57°))$ 29) $-5\sqrt{2}/2 - (5/2)i$

Table I Approximate Square Roots and Cube Roots

n	$\sqrt{n}$	$\sqrt{10n}$	$\sqrt[3]{n}$	$\sqrt[3]{10n}$	$\sqrt[3]{100n}$
1	1.000	3.162	1.000	2.154	4.642
2	1.414	4.472	1.260	2.714	5.848
3	1.732	5.477	1.442	3.107	6.694
4	2.000	6.325	1.587	3.420	7.368
5	2.236	7.071	1.710	3.684	7.937
6	2.449	7.746	1.817	3.915	8.434
7	2.646	8.367	1.913	4.121	8.879
8	2.828	8.944	2.000	4.309	9.283
9	3.000	9.487	2.080	4.481	9.655
10	3.162	10.000	2.154	4.642	10.000
11	3.317	10.488	2.224	4.791	10.323
12	3.464	10.954	2.289	4.932	10.627
13	3.606	11.402	2.251	5.066	10.914
14	3.742	11.832	2.410	5.192	11.187
15	3.873	12.247	2.466	5.313	11.447
16	4.000	12.649	2.520	5.429	11.696
17	4.123	13.038	2.571	5.540	11.935
18	4.243	13.416	2.621	5.646	12.164
19	4.359	13.784	2.668	5.749	12.386
20	4.472	14.142	2.714	5.848	12.599
21	4.583	14.491	2.759	5.944	12.806
22	4.690	14.832	2.802	6.037	13.006
23	4.796	15.166	2.844	6.127	13.200
24	4.899	15.492	2.884	6.214	13.389
25	5.000	15.811	2.924	6.300	13.572
26	5.099	16.125	2.962	6.383	13.751
27	5.196	16.432	3.000	6.463	13.925
28	5.292	16.733	3.037	6.542	14.095
29	5.385	17.029	3.072	6.619	14.260
30	5.477	17.321	3.107	6.694	14.422
31	5.568	17.607	3.141	6.768	14.581
32	5.657	17.889	3.175	6.840	14.736
33	5.745	18.166	3.208	6.910	14.888
34	5.831	18.439	3.240	6.980	15.037
35	5.916	18.708	3.271	7.047	15.183
36	6.000	18.974	3.302	7.114	15.326
37	6.083	19.235	3.332	7.179	15.467
38	6.164	19.494	3.362	7.243	15.605
39	6.245	19.748	3.391	7.306	15.741
40	6.325	20.000	3.420	7.368	15.874
41	6.403	20.248	3.448	7.429	16.005
42	6.481	20.494	3.476	7.489	16.134
43	6.557	20.736	3.503	7.548	16.261
44	6.633	20.976	3.530	7.606	16.386
45	6.708	21.213	3.557	7.663	16.510
46	6.782	21.448	3.583	7.719	16.631
47	6.856	21.679	3.609	7.775	16.751
48	6.928	21.909	3.634	7.830	16.869
49	7.000	22.136	3.659	7.884	16.985
50	7.071	22.361	3.684	7.937	17.100

Table II Common Logarithms

n	0	1	2	3	4	5	6	7	8	9
1.0	.0000	.0043	.0086	.0128	.0170	.0212	.0253	.0294	.0334	.0374
1.1	.0414	.0453	.0492	.0531	.0569	.0607	.0645	.0682	.0719	.0755
1.2	.0792	.0828	.0864	.0899	.0934	.0969	.1004	.1038	.1072	.1106
1.3	.1139	.1173	.1206	.1239	.1271	.1303	.1335	.1367	.1399	.1430
1.4	.1461	.1492	.1523	.1553	.1584	.1614	.1644	.1673	.1703	.1732
1.5	.1761	.1790	.1818	.1847	.1875	.1903	.1931	.1959	.1987	.2014
1.6	.2041	.2068	.2095	.2122	.2148	.2175	.2201	.2227	.2253	.2279
1.7	.2304	.2330	.2355	.2380	.2405	.2430	.2455	.2480	.2504	.2529
1.8	.2553	.2577	.2601	.2625	.2648	.2672	.2695	.2718	.2742	.2765
1.9	.2788	.2810	.2833	.2856	.2878	.2900	.2923	.2945	.2967	.2989
2.0	.3010	.3032	.3054	.3075	.3096	.3118	.3139	.3160	.3181	.3201
2.1	.3222	.3243	.3263	.3284	.3304	.3324	.3345	.3365	.3385	.3404
2.2	.3424	.3444	.3464	.3483	.3502	.3522	.3541	.3560	.3579	.3598
2.3	.3617	.3636	.3655	.3674	.3692	.3711	.3729	.3747	.3766	.3784
2.4	.3802	.3820	.3838	.3856	.3874	.3892	.3909	.3927	.3945	.3962
2.5	.3979	.3997	.4014	.4031	.4048	.4065	.4082	.4099	.4116	.4133
2.6	.4150	.4166	.4183	.4200	.4216	.4232	.4249	.4265	.4281	.4298
2.7	.4314	.4330	.4346	.4362	.4378	.4393	.4409	.4425	.4440	.4456
2.8	.4472	.4487	.4502	.4518	.4533	.4548	.4564	.4579	.4594	.4609
2.9	.4624	.4639	.4654	.4669	.4683	.4698	.4713	.4728	.4742	.4757
3.0	.4771	.4786	.4800	.4814	.4829	.4843	.4857	.4871	.4886	.4900
3.1	.4914	.4928	.4942	.4955	.4969	.4983	.4997	.5011	.5024	.5038
3.2	.5051	.5065	.5079	.5092	.5105	.5119	.5132	.5145	.5159	.5172
3.3	.5185	.5198	.5211	.5224	.5237	.5250	.5263	.5276	.5289	.5302
3.4	.5315	.5328	.5340	.5353	.5366	.5378	.5391	.5403	.5416	.5428
3.5	.5441	.5453	.5465	.5478	.5490	.5502	.5514	.5527	.5539	.5551
3.6	.5563	.5575	.5587	.5599	.5611	.5623	.5635	.5647	.5658	.5670
3.7	.5682	.5694	.5705	.5717	.5729	.5740	.5752	.5763	.5775	.5786
3.8	.5798	.5809	.5821	.5832	.5843	.5855	.5866	.5877	.5888	.5899
3.9	.5911	.5922	.5933	.5944	.5955	.5966	.5977	.5988	.5999	.6010
4.0	.6021	.6031	.6042	.6053	.6064	.6075	.6085	.6096	.6107	.6117
4.1	.6128	.6138	.6149	.6160	.6170	.6180	.6191	.6201	.6212	.6222
4.2	.6232	.6243	.6253	.6263	.6274	.6284	.6294	.6304	.6314	.6325
4.3	.6335	.6345	.6355	.6365	.6375	.6385	.6395	.6405	.6415	.6425
4.4	.6435	.6444	.6454	.6464	.6474	.6484	.6493	.6503	.6513	.6522
4.5	.6532	.6542	.6551	.6561	.6571	.6580	.6590	.6599	.6609	.6618
4.6	.6628	.6637	.6646	.6656	.6665	.6675	.6684	.6693	.6702	.6712
4.7	.6721	.6730	.6739	.6749	.6758	.6767	.6776	.6785	.6794	.6803
4.8	.6812	.6821	.6830	.6839	.6848	.6857	.6866	.6875	.6884	.6893
4.9	.6902	.6911	.6920	.6928	.6937	.6946	.6955	.6964	.6972	.6981
5.0	.6990	.6998	.7007	.7016	.7024	.7033	.7042	.7050	.7059	.7067
5.1	.7076	.7084	.7093	.7101	.7110	.7118	.7126	.7135	.7143	.7152
5.2	.7160	.7168	.7177	.7185	.7193	.7202	.7210	.7218	.7226	.7235
5.3	.7243	.7251	.7259	.7267	.7275	.7284	.7292	.7300	.7308	.7316
5.4	.7324	.7332	.7340	.7348	.7356	.7364	.7372	.7380	.7388	.7396
n	0	1	2	3	4	5	6	7	8	9

Table II Common Logarithms

n	0	1	2	3	4	5	6	7	8	9
5.5	.7404	.7412	.7419	.7427	.7435	.7443	.7451	.7459	.7466	.7474
5.6	.7482	.7490	.7497	.7505	.7513	.7520	.7528	.7536	.7543	.7551
5.7	.7559	.7566	.7574	.7582	.7589	.7597	.7604	.7612	.7619	.7627
5.8	.7634	.7642	.7649	.7657	.7664	.7672	.7679	.7686	.7694	.7701
5.9	.7709	.7716	.7723	.7731	.7738	.7745	.7752	.7760	.7767	.7774
6.0	.7782	.7789	.7796	.7803	.7810	.7818	.7825	.7832	.7839	.7846
6.1	.7853	.7860	.7868	.7875	.7882	.7889	.7896	.7903	.7910	.7917
6.2	.7924	.7931	.7938	.7945	.7952	.7959	.7966	.7973	.7980	.7987
6.3	.7993	.8000	.8007	.8014	.8021	.8028	.8035	.8041	.8048	.8055
6.4	.8062	.8069	.8075	.8082	.8089	.8096	.8102	.8109	.8116	.8122
6.5	.8129	.8136	.8142	.8149	.8156	.8162	.8169	.8176	.8182	.8189
6.6	.8195	.8202	.8209	.8215	.8222	.8228	.8235	.8241	.8248	.8254
6.7	.8261	.8267	.8274	.8280	.8287	.8293	.8299	.8306	.8312	.8319
6.8	.8325	.8331	.8338	.8344	.8351	.8357	.8363	.8370	.8376	.8382
6.9	.8388	.8395	.8401	.8407	.8414	.8420	.8426	.8432	.8439	.8445
7.0	.8451	.8457	.8463	.8470	.8476	.8482	.8488	.8494	.8500	.8506
7.1	.8513	.8519	.8525	.8531	.8537	.8543	.8549	.8555	.8561	.8567
7.2	.8573	.8579	.8585	.8591	.8597	.8603	.8609	.8615	.8621	.8627
7.3	.8633	.8639	.8645	.8651	.8657	.8663	.8669	.8675	.8681	.8686
7.4	.8692	.8698	.8704	.8710	.8716	.8722	.8727	.8733	.8739	.8745
7.5	.8751	.8756	.8762	.8768	.8774	.8779	.8785	.8791	.8797	.8802
7.6	.8808	.8814	.8820	.8825	.8831	.8837	.8842	.8848	.8854	.8859
7.7	.8865	.8871	.8876	.8882	.8887	.8893	.8899	.8904	.8910	.8915
7.8	.8921	.8927	.8932	.8938	.8943	.8949	.8954	.8960	.8965	.8971
7.9	.8976	.8982	.8987	.8993	.8998	.9004	.9009	.9015	.9020	.9025
8.0	.9031	.9036	.9042	.9047	.9053	.9058	.9063	.9069	.9074	.9079
8.1	.9085	.9090	.9096	.9101	.9106	.9112	.9117	.9122	.9128	.9133
8.2	.9138	.9143	.9149	.9154	.9159	.9165	.9170	.9175	.9180	.9186
8.3	.9191	.9196	.9201	.9206	.9212	.9217	.9222	.9227	.9232	.9238
8.4	.9243	.9248	.9253	.9258	.9263	.9269	.9274	.9279	.9284	.9289
8.5	.9294	.9299	.9304	.9309	.9315	.9320	.9325	.9330	.9335	.9340
8.6	.9345	.9350	.9355	.9360	.9365	.9370	.9375	.9380	.9385	.9390
8.7	.9395	.9400	.9405	.9410	.9415	.9420	.9425	.9430	.9435	.9440
8.8	.9445	.9450	.9455	.9460	.9465	.9469	.9474	.9479	.9484	.9489
8.9	.9494	.9499	.9504	.9509	.9513	.9518	.9523	.9528	.9533	.9538
9.0	.9542	.9547	.9552	.9557	.9562	.9566	.9571	.9576	.9581	.9586
9.1	.9590	.9595	.9600	.9605	.9609	.9614	.9619	.9624	.9628	.9633
9.2	.9638	.9643	.9647	.9652	.9657	.9661	.9666	.9671	.9675	.9680
9.3	.9685	.9689	.9694	.9699	.9703	.9708	.9713	.9717	.9722	.9727
9.4	.9731	.9736	.9741	.9745	.9750	.9754	.9759	.9763	.9768	.9773
9.5	.9777	.9782	.9786	.9791	.9795	.9800	.9805	.9809	.9814	.9818
9.6	.9823	.9827	.9832	.9836	.9841	.9845	.9850	.9854	.9859	.9863
9.7	.9868	.9872	.9877	.9881	.9886	.9890	.9894	.9899	.9903	.9908
9.8	.9912	.9917	.9921	.9926	.9930	.9934	.9939	.9943	.9948	.9952
9.9	.9956	.9961	.9965	.9969	.9974	.9978	.9983	.9987	.9991	.9996
n	0	1	2	3	4	5	6	7	8	9

Table III Trigonometric Ratios

Angle	Tangent	Cosine	Sine	Angle	Tangent	Cosine	Sine
0°	.0000	1.0000	.0000	45°	1.0000	.7071	.7071
1°	.0175	.9998	.0175	46°	1.0355	.6947	.7193
2°	.0349	.9994	.0349	47°	1.0724	.6820	.7314
3°	.0524	.9986	.0523	48°	1.1106	.6691	.7431
4°	.0699	.9976	.0698	49°	1.1504	.6561	.7547
5°	.0875	.9962	.0872	50°	1.1918	.6428	.7660
6°	.1051	.9945	.1045	51°	1.2349	.6293	.7771
7°	.1228	.9925	.1219	52°	1.2799	.6157	.7880
8°	.1405	.9903	.1392	53°	1.3270	.6018	.7986
9°	.1584	.9877	.1564	54°	1.3764	.5878	.8090
10°	.1763	.9848	.1736	55°	1.4281	.5736	.8192
11°	.1944	.9816	.1908	56°	1.4826	.5592	.8290
12°	.2126	.9781	.2079	57°	1.5399	.5446	.8387
13°	.2309	.9744	.2250	58°	1.6003	.5299	.8480
14°	.2493	.9703	.2419	59°	1.6643	.5150	.8572
15°	.2679	.9659	.2588	60°	1.7321	.5000	.8660
16°	.2867	.9613	.2756	61°	1.8040	.4848	.8746
17°	.3057	.9563	.2924	62°	1.8807	.4695	.8829
18°	.3249	.9511	.3090	63°	1.9626	.4540	.8910
19°	.3443	.9455	.3256	64°	2.0503	.4384	.8988
20°	.3640	.9397	.3420	65°	2.1445	.4226	.9063
21°	.3839	.9336	.3584	66°	2.2460	.4067	.9135
22°	.4040	.9272	.3746	67°	2.3559	.3907	.9205
23°	.4245	.9205	.3907	68°	2.4751	.3746	.9272
24°	.4452	.9135	.4067	69°	2.6051	.3584	.9336
25°	.4663	.9063	.4226	70°	2.7475	.3420	.9397
26°	.4877	.8988	.4384	71°	2.9042	.3256	.9455
27°	.5095	.8910	.4540	72°	3.0777	.3090	.9511
28°	.5317	.8829	.4695	73°	3.2709	.2924	.9563
29°	.5543	.8746	.4848	74°	3.4874	.2756	.9613
30°	.5774	.8660	.5000	75°	3.7321	.2588	.9659
31°	.6009	.8572	.5150	76°	4.0108	.2419	.9703
32°	.6249	.8480	.5299	77°	4.3315	.2250	.9744
33°	.6494	.8387	.5446	78°	4.7046	.2079	.9781
34°	.6745	.8290	.5592	79°	5.1446	.1908	.9816
35°	.7002	.8192	.5736	80°	5.6713	.1736	.9848
36°	.7265	.8090	.5878	81°	6.3138	.1564	.9877
37°	.7536	.7986	.6018	82°	7.1154	.1392	.9903
38°	.7813	.7880	.6157	83°	8.1443	.1219	.9925
39°	.8098	.7771	.6293	84°	9.5144	.1045	.9945
40°	.8391	.7660	.6428	85°	11.4301	.0872	.9962
41°	.8693	.7547	.6561	86°	14.3007	.0698	.9976
42°	.9004	.7431	.6691	87°	19.0811	.0523	.9986
43°	.9325	.7314	.6820	88°	28.6363	.0349	.9994
44°	.9657	.7193	.6947	89°	57.2900	.0175	.9998
45°	1.0000	.7071	.7071	90°		.0000	1.0000

Index

Acute angle, 165
Additive inverse, 6
Adjacent angles, 165
Algebra, 86
Algebraic expressions, 91
Algebraic term, 88
Altitude
  of parallelogram, 174
  of trapezoid, 175
  of triangle, 176
Angle
  acute, 165
  adjacent, 165
  bisector of, 194
  central, 199
  construction of, 193
  obtuse, 165
  phase, 292
  reference, 295
  right, 165
  standard position of, 291, 348
  straight, 165
  symbol for, 164
  vertical, 165
Angular velocity, 339
Antilogarithm, 157
Argument, 373
Associative property
  for addition, 3
  for multiplication, 3
Axioms, 120
  of equality, 120
Axis of rotation, 256
Axis of symmetry, 255

Base
  of trapezoid, 178
  of triangle, 176
Binomial, 88
Board foot, 71

Center, 182
Central angle, 199
Characteristic, 156
Circle, 182, 255, 256
  chord of, 182
  circumference of, 182
  radius of, 182
  secant of, 183
  tangent to, 183
Circumference, 182
Coefficient
  literal, 88
  numerical, 88
Commutative property
  for addition, 3
  for multiplication, 3
Compass, 241
Complex numbers
  addition of, 114
  argument of, 373
  graphing, 371
  modulus of, 373
  multiplication of, 115
  powers of, 376
  product of, 375
  roots of, 376
  subtraction of, 115
  trigonometric form, 373
Composite, 15
Cone, 185
  axis of rotation, 266
  axis of symmetry, 255
  base of, 255
  lateral element of, 255
  nappe of, 255
Conic sections, 255
Congruent, 170
Conjugates, 251
Constants, 86
Constructions
  angle, 193
  angle bisector, 194
  perpendicular bisector, 189
  perpendicular to line, 190
  polygons, 199
Cosecant, 276
Cosine, 276
Cosine law, 306

Cotangent, 276
Cramer's rule, 231
Cube roots, 51
  on slide rule, 81
Cylinder, right circular, 238

Decagons, 173
Decimal fractions, 25
  addition of, 25
  conversion to percent, 30
  division of, 26
  multiplication of, 26
  subtraction of, 25
Decimal point, 25
de Moivres theorem, 376
Denominator, 17
  common, 19
  least common, 19
Determinant, 225
  cofactor of, 227
  minor of, 227
  order of, 226
  properties of, 228
Directed numbers, 6
Discriminant, 250
Distributive property, 4
Domain, 134

Element, 1
Ellipse, 255, 258
  definition of, 258
  major axis of, 259
  minor axis of, 259
Equality
  axioms of, 120
Equation, 119
  check of, 120
  conditional, 119
  exponential, 152
  identical, 119
  logarithmic, 153
  linear, 140
  quadratic, 240
Equation of line, 140
  general form, 149
  intercept form, 145
  one point slope form, 148
  slope intercept form, 144
  two point form, 145
Exponential equation, 152
Exponential numbers
  combined operations, 41
  dividing with, 39
  multiplying with, 39

Exponents, 37
  negative, 44
  zero as, 40
Extremes, 128

Factor, 14
  common, 15
  greatest common, 16
Factoring, 107
  by difference of cubes, 112
  by difference of squares, 112
  by grouping, 110
  solution of quadratics by, 240
Formulas, 132
Fraction, 17
  as mixed number, 20
  complex, 21
  decimal, 25
  decimal to arithmetic, 27
  division of, 21
  improper, 20
  lowest terms, 18
  proper, 20
  rule for multiplication, 20
  terms of, 17
Fulcrum, 216
Function, 134

Geometry, 163
Graphs, 136
Greatest common divisor, 16
Greatest common factor, 16

Hexagons, 173
Hyperbola, 255, 262
  definition of, 262
  conjugate axis of, 264
  transverse axis of, 264

Identity
  element for addition, 4
  element for multiplication, 4
  of equation, 119
Imaginary numbers, 55
  pure, 56
Impedance triangle, 291
Index, 51
Integers
  addition of, 6
  division of, 12
  subtraction of, 8
  multiplication of, 11

Inverse
  additive, 7
  multiplicative, 21
Irrational numbers, 51

Lateral element, 255
Least common denominator, 19
Least common multiple, 17
Left member, 119
Lever, 216
Lever arm, 216
Linear interpolation, 157
Line segment, 163
Linear velocity, 339
Literal numbers, 3, 86
Logarithm, 153
  characteristic of, 156
  common, 155
  mantissa of, 156

Mantissa, 156
Matrix, 222
  conformable, 224
  diagonal, 223
  determinant of, 225
  dimensions of, 222
  element of, 222
  identity, 223
  inverse of, 234
  multiplication of, 225
  nonsingular, 235
  singular, 235
  square, 222
  transpose of, 236
  unit, 223
Mean proportional, 128
Means, 128
Measurement
  conversion, 33
  English system, 33
  metric system, 33
Member, 1
Metric, 33
  conversion to English, 33
Mixed number, 20
  rule for conversion, 20
Modulus, 373
Monomial, 88
Multiple, 14
  least common, 17
Multiplicative inverse, 21

Nappe, 255
Natural numbers, 2
Negative number, 6
*n*th root, 51
Numbers
  complex, 55
  directed, 6
  imaginary, 55
  irrational, 51
  literal, 3
  mixed, 20
  natural, 2
  negative, 6
  rational, 14
  real, 51
  whole, 2
Numerator, 17

Obtuse angle, 165
Origin, 137

Parabola, 255, 266, 267
Parallel lines, 164
Parallelogram, 174
Percent, 29
  to decimal, rule for, 29
Perpendicular
  to line, 190
Perpendicular bisector, 189
  construction of 189
Pi (π), 183
Phase angle, 291, 348
Polar axis, 365
Polar coordinates, 362
  conversion to rectangular form, 368
Pole, 365
Polygons, 173
  construction of, 173
  diagonal of, 199
  equiangular, 173
  equilateral, 173
  regular, 173
Polynomial, 88
  multiplication by monomial, 96
  multiplication by polynomial, 96
Prime factored form, 15
Prime, 15
  relatively, 16, 21
Proportion, 128
  properties of, 128
Pyramid, 184
  vertex of, 184
Pythagorean identities, 316
Pythagorean theorem, 167, 282

Quadrants, 137, 182
Quadratic equation, 240
  product of roots, 252
  solution by completing square, 246
  solution by extraction of roots, 243
  solution by factoring, 241
  solution by formula, 247
  sum of roots, 252
Quadratic formula, 247
  discriminant of, 250
Quadrilateral, 173

Radian measure, 332
  conversion to degree, 334
Radical, 48
Radicand, 48
Radius, 182
Range, 134
Ratio, 128
Rational numbers, 14
Ray, 164
Real numbers, 51
Reciprocal, 21
Rectangle, 174
Rectangular coordinate system, 137
Reference angle, 295
Resultant, 356
Right angles, 165
Right circular cylinder, 238
Right member, 119
Rhombus, 174

S Scale, 327
Scalar, 353
Scientific notation, 45
Secant
  of angle, 276
  of circle, 229
Sector, 336
Semicircle, 182
Set, 1
  empty, 2
  finite, 1
  infinite, 2
  well-defined, 1
Similar, 171
Sine, 276
Sine law, 299
Sine wave, 343
  phase angle of, 348
Slide rule
  combined operations with, 73
  cube root, 81
  cubing a number, 83
  division with, 72
  finding square root, 78
  multiplication with, 68
  parts of, 60
  primary divisions, 61
  reciprocal functions on, 328
  secondary divisions, 61
  squaring a number, 80
Slope, 142
Simultaneous equations, 206
Solutions of equations, 119
  by factoring, 240
Solution set, 119
Square, 174
Square root, 48
  on slide rule, 78
  use of tables, 49
ST scale, 327
Standard reference triangles, 279
Stated problems, 214
Straight angle, 165
Straight edge, 241
Synthetic division, 104
  analysis of, 106
Systems of equations
  consistent, 206
  dependent, 206
  inconsistent, 206
  independent, 206
  linear, 205
  simultaneous, 206

Tangent
  of an angle, 276
  to a circle, 229
Terms
  algebraic, 88
  similar, 88
Trapezoid, 174
  altitude of, 175
  isosceles, 175
  median of, 174
Triangle, 166
  acute, 166
  congruent, 170
  equiangular, 166
  equilateral, 166
  isosceles, 166
  obtuse, 166
  right, 166
  scalene, 166
  similar, 171
  vertex angle of, 166
Trigonometry, 275

Trigonometric identities, 314
  double angle, 322
  for negative angles, 318
  half angle, 323
  sum and difference, 319
Trigonometric ratios, 275
Trigonometric tables, Appendix
  use of, 283
Trinomial, 88

Unit vector, 359

Variables, 86, 205
Vector, 353
  difference of, 358
  sum of, 353
  unit, 359

Velocity
  angular, 339
  linear, 339
Vertex
  of an angle, 164
  of pyramid, 184
  of triangle, 166
Vertical angles, 165
Volume, 230

Whole numbers, 2

x axis, 137

y axis, 137

Zero product property, 4

1 2 3 4 5 6 7 8 9 10 11 12 13 14 15 16 17 18 19 20 21 22 23 24 25 82 81 80 79 78 77 76 75 74

Greek Alphabet

Α	α	alpha
Β	β	beta
Γ	γ	gamma
Δ	δ	delta
Ε	ϵ	epsilon
Ζ	ζ	zeta
Η	η	eta
Θ	θ	theta
Ι	ι	iota
Κ	κ	kappa
Λ	λ	lambda
Μ	μ	mu
Ν	ν	nu
Ξ	ξ	xi
Ο	o	omicron
Π	π	pi
Ρ	ρ	rho
Σ	σ	sigma
Τ	τ	tau
Υ	υ	upsilon
Φ	ϕ	phi
Χ	χ	chi
Ψ	ψ	psi
Ω	ω	omega

Common Decimal Equivalents

1/32	= .03125	1/2	= .50000
1/16	= .06250	17/32	= .53125
3/32	= .09375	9/16	= .56250
1/8	= .12500	19/32	= .59375
5/32	= .15625	5/8	= .62500
3/16	= .18750	21/32	= .65625
7/32	= .21875	11/16	= .68750
1/4	= .25000	23/32	= .71875
9/32	= .28125	3/4	= .75000
5/16	= .31250	25/32	= .78125
11/32	= .34375	13/16	= .81250
3/8	= .37500	27/32	= .84375
13/32	= .40625	7/8	= .87500
7/16	= .43750	29/32	= .90625
15/32	= .46875	15/16	= .93750
		31/32	= .96875

Table of Conversions

Linear Measure—English System

	1 inch (in.)	= 2.54 cm. (Metric)
12 in.	= **1 foot (ft.)**	= 30.48 cm.
3 ft.	= **1 yard (yd.)**	= .9144 m.
5½ yd.	= **1 rod (rd.)**	= 5.0292 m.
5280 ft.	= **1 mile (mi.)**	= 1.6093 km.

Linear Measure—Metric System

	1 millimeter (mm.)	= .0394 in. (English)
10 mm.	= **1 centimeter (cm.)**	= .3937 in.
10 cm.	= **1 decimeter (dm.)**	= 3.937 in.
10 dm.	= **1 meter (m.)**	= 39.37 in.
10 m.	= **1 dekameter (dkm.)**	= 393.7 in.
10 dkm.	= **1 hectometer (hm.)**	= 328.083 ft.
10 hm.	= **1 kilometer (km.)**	= .6214 mi.

Liquid Measure—English System

	1 gill (gi.)	= .1183 l. (Metric)
4 gi.	= **1 pint (pt.)**	= .4732 l.
2 pt.	= **1 quart (qt.)**	= .9463 l.
4 qt.	= **1 gallon (gal.)**	= 3.7853 l.
42 gal.	= **1 barrel (bbl.)**	= 158.98 l.

Capacity Measure—Common Household

3 teaspoons	= **1 tablespoon (tbsp.)**
16 tablespoons	= **1 cup (c.)**
2 cups	= **1 pint (pt.)**
2 pints	= **1 quart (qt.)**
4 quarts	= **1 gallon (gal.)**

Capacity Measure—Metric System

	1 milliliter (ml.)	= .0610 cu. in. (English)
10 ml.	= **1 centiliter (cl.)**	= .6102 cu. in.
10 cl.	= **1 deciliter (dl.)**	= 6.102 cu. in.
10 dl.	= **1 liter (l.)**	= 61.02 cu. in.
		= 1.057 qt. liquid
10 l.	= **1 dekaliter (dkl.)**	= 610.2 cu. in.
10 dkl.	= **1 hectoliter (hl.)**	= 6102.5 cu. in.
10 hl.	= **1 kiloliter (kl.)**	= 264.178 gal. liquid

Measure of Weight—Metric System

	1 milligram (mg.)	= .0154 gr. (English)
10 mg.	= **1 centigram (cg.)**	= .1543 gr.
10 cg.	= **1 decigram (dg.)**	= 1.543 gr.
10 dg.	= **1 gram (gm.)**	= 15.43 gr.
10 gm.	= **1 decagram (dkg.)**	= .3527 oz.
10 dkg.	= **1 hectogram (hg.)**	= 3.527 oz.
10 hg.	= **1 kilogram (kg.)**	= 2.2046 lb.

Measure of Weight—Avoirdupois

	1 grain (gr.)	= .0648 gm. (Metric)
27.34 gr.	= **1 dram (dr.)**	= 1.77 gm.
16 dr.	= **1 ounce (oz.)**	= 28.35 gm.
16 oz.	= **1 pound (lb.)**	= 453.59 gm.
2000 lb.	= **1 short ton (s.t.)**	